APPLIED MATHEMATICAL MODELING

EETU N. VIRTANEN
EDITOR

Nova Science Publishers, Inc.
New York

LIBRARY OF CONGRESS CATALOGING-IN-PUBLICATION DATA

Applied mathematical modeling / Eetu N. Virtanen (editor).
p. cm.
Includes index.
ISBN-13: 978-1-60021-975-7 (hardcover)
ISBN-10: 1-60021-975-6 (hardcover)
1. Engineering--Mathematical models. 2. Environmental engineering--Mathematical models. I. Virtanen, Eetu N.
TA342.A65 2007
620.001'5118--dc22 2007034878

Published by Nova Science Publishers, Inc. ✦ New York

APPLIED MATHEMATICAL MODELING

CONTENTS

PREFACE

This new book focuses on important research related to the mathematical modelling of engineering and environmental processes, manufacturing, and industrial systems. It includes heat transfer, fluid mechanics, CFD, and transport phenomena; solid mechanics and mechanics of metals; electromagnets and MHD; reliability modelling and system optimization; finite volume, finite element, and boundary element procedures; decision sciences in an industrial and manufacturing context; civil engineering systems and structures; mineral and energy resources; relevant software engineering issues associated with CAD and CAE; and materials and metallurgical engineering.

Short Communication - The mean and turbulent flow properties of a turbulent heated round jet discharged in a cold co-flowing ambient of the same fluid have been obtained computationally using the standard k-ε model and Reynolds-stress transport (RST) model. The two turbulence models have been chosen to cover a wide spectrum of the modeling options that are available. The present flow configuration is an important test case for turbulence models, because it is more complex compared to that of a standard turbulent round jet in a quiescent ambient of the same fluid with both at the same temperature. The present predictions are compared with the experimental data reported in literature. The results suggest that self-preservation does not hold in this flow. Overall the k-ε model provides better agreement with the experimental data compared to that by the RST model, which overpredicts. The present chapter clearly shows that the RST model, though more elaborate, does not have an advantage over the standard k-ε model in the prediction of the present flow configuration.

Chapter 1 - Mathematical modeling is a powerful tool for solving optimization problems in chemical engineering. In this work several models are proposed aimed at helping to make decisions about different aspects of the processes lifecycle, from the synthesis and design steps up to the operation and scheduling. Using an example of the Sugar Cane industry, several models are formulated and solved in order to assess the trade-offs involved in optimization decisions. Thus, the power and versatility of mathematical modeling in the area of chemical processes optimization is analyzed and evaluated.

Chapter 2 - In the context of computational fluid dynamics (CFD), modeling low-pressure subcooled boiling flow is of particular significance. A review is provided in this article of the various numerical modeling approaches that have been adopted to handle subcooled boiling flow. The main focus in the analysis of such a challenging problem can be broadly classified according into two important categories: (i) Heat transfer and wall heat flux partitioning during subcooled boiling flow at the heated wall and (ii) Two-phase flow and bubble behaviors in the bulk subcooled flow away from the heated wall. For the first category, details of both empirical and mechanistic models that have been proposed in the literature are given. The enhancement in heat transfer during forced convective boiling attributed by the presence of both sliding and stationary bubbles, force balance model for bubble departure and bubble lift-off as well as the evaluation of bubble frequency based on fundamental theory depict the many improvements that have been introduced to the current mechanistic model of heat transfer and wall heat flux partitioning. For the second category, details of applications of various empirical relationships and mechanistic model such as population balance model to determine the local bubble diameter in the bulk subcooled liquid that have been employed in the literature are also given. A comparison of the predictions with experimental data is demonstrated. For the local case, the model considering population balance and improved wall heat partition shows good greement with the experimentally measured radial distributions of the Sauter mean bubble diameter, void fraction, interfacial area concentration and liquid velocity profiles. Significant weakness prevails however over the vapor velocity distribution. For the axial case, good agreement is also achieved for the axial distributions of the Sauter mean bubble diameter, void fraction and interfacial area concentration profiles. The present model correctly represents the plateau at the initial boiling stages at upstream, typically found in low-pressure subcooled boiling flows, followed by the significant rise of the void fraction at downstream.

Chapter 3 - In this chapter, a second-order scheme of precise time-step integration (PTI) method is introduced for dynamic analysis with respect to long-term integration and transient responses while spatial discretization is realized with the differential quadrature method. Rather than transforming into first-order equations, a recursive scheme is presented in detail for direct solution of the homogeneous part of second-order differential and algebraic equations. The sine and cosine matrices involved in the scheme are calculated using the so-called 2^N algorithm, and the corresponding particular solution is also presented where the excitation vector is approximated by the truncated Taylor series. The performance and numerical behaviors of the second-order scheme of the PTI method are tested by a series of numerical examples in comparison with the first-order scheme or with the traditional time-marching Newmark-β method as the reference. The issue of spurious high-frequency responses resulting from spatial discretization for shock-excited structural dynamic analysis is also studied in the framework of the second-order PTI method. The effects of spatial discretization, numerical damping and time step on solution accuracy are explored by analyzing longitudinal vibrations of a shock-excited rod with rectangular, half-triangular and Heaviside step impact.

The chapter is organized as follows: In Section 1, previous research into time-marching algorithms is briefly reviewed. Following a brief introductory overview of the first-order scheme, the second-order PTI method is introduced in detail in Section 2 together with an analysis of the accuracy that can be achieved, the treatment of particular solutions, and solution behavior with the second-order scheme. Numerical examples for smooth responses of various wave equations and wave propagations are presented using the second-order PTI method in Section 3. Numerical examples of impact responses are presented and studied extensively in Section 4 with respect to the effect of spatial discretization, modal types that the discretization can afford, and numerical damping, etc., with some concluding remarks presented in Section 5.

Numerical tests show that for the analysis of smooth and shock-excited responses, both efficiency and accuracy can be improved considerably with the second-order PTI method. For the analysis of shock-excited responses, spatial discretization can provide a reasonable number of model types for any given error tolerances. An appropriate spatial discretization scheme for a given structure is usually required in order to obtain an accurate and meaningful numerical solution, especially for describing the fine detail of traction responses with sharp changes. Under the framework of the PTI method, the numerical damping that is often required in traditional integration schemes is found to be unnecessary, and there is no restriction on the size of time steps, because the PTI method can usually produce results with machine-like precision and is an unconditionally stable explicit method.

Chapter 4 - This chapter is designed to present a relatively simple, but realistic exposition of some of the ideas in the author's previous works. All the authors calculations are explicit and analytic. By considering an elastic beam of narrow rectangular cross section, the two beam theories, the refined theory and the decomposition theorem, are exact in the sense that one-dimensional (1D) beamlike equations are obtained from the three-dimensional (3D) equations of linear elasticity without any hypotheses or approximations whatsoever. Thus, the exact theories are not asymptotic and no appeal is made to small slenderness ratios. To set up the general framework of the beam theories, the appropriate stress and mixed boundary conditions accurate to all order for the beam of general edge geometry and loadings are formulated explicitly.

Chapter 5 - The single server queues with repeated attempts have been extended to include several types of extensions and generalizations, to which attention has been paid by several researchers. The pioneering studies of retrial queues present the concept of retrial time as an alternative to classical models of telephone systems. The interest in such types of models have been further enhanced in recent years because of their theoretical structures as well as their applications in many practical situations such as telephone switching systems, telecommunication networks and computer compacting to gain service from a central processing unit. However, the authors' intention in this paper is to look into a mathematical model, that includes many interesting features such as second optional service, Bernoulli schedule vacation and repeated attempts. An extensive stationary analysis of the model has been carried out including discrete time process, continuous time process and calculation of some important performance measures.

Chapter 6 - Particle science and technology is a rapidly developing interdisciplinary research area with its core being the understanding of the relationships between micro- and macro-scopic properties of particulate/granular matter – a state of matter that is widely encountered but poorly understood. The macroscopic behaviour of particulate matter is controlled by the interactions between individual particles as well as interactions with surrounding fluids. Understanding the microscopic mechanisms in terms of these interaction forces is therefore key to leading to truly interdisciplinary research into particulate matter and producing results that can be generally used. This aim can be effectively achieved via particle scale research based on detailed microdynamic information such as the forces acting on and trajectories of individual particles in a considered system. In recent years, such research has been rapidly developed worldwide, mainly as a result of the rapid development of discrete particle simulation technique and computer technology. This article introduces such simulation techniques and their application to the study of some typical particulate systems.

Chapter 7 - Compact heat exchangers are used in a wide variety of applications. Typical among these are automobile radiators, air-conditioning evaporators and condensers, electronic cooling devices, recuperators and regenerators, and cryogenic exchangers. The need for lightweight, space saving, and economical heat exchangers has driven the development of compact surfaces. A mathematical model to predict the overall performance of such heat exchanges has been presented here. The computational domain considered was a three-dimensional rectangular duct with an array of circular pin fins in the staggered arrangement and mounted on the heated channel bed maintained at a constant temperature. The values of different parameters chosen were based on the practical working conditions. The governing Reynolds-averaged Navier-Stokes and thermal energy equations for the complex turbulent convective heat transfer were numerically solved using the control volume approach and by applying appropriate thermal boundary conditions within the domain. The numerical simulations were performed using two different turbulent models, namely, the standard k-ε model and the RNG k-ε model along with the wall functions to treat the turbulence in the vicinity of the solid walls. The details of the governing equations, numerical procedure, mesh employed, and grid independence study have been presented in this chapter. The predictions have been compared with the experimental data reported in the literature and a good agreement between the two has been observed. The temperature and flow fields have been investigated for different inlet velocities. The present investigation shows that choosing an appropriate turbulence model and near-wall treatment has a significant influence on thermal and fluid dynamics behaviour of the heat exchanger. The results also enable us to present the effects of the various operating parameters on the performance of the heat exchanger.

Chapter 8 - Fluid turbulence is one of the biggest challenges to scientists and engineers, who need to understand turbulent flows in order to control them and design for their adverse effects or utilize them for their best effects. In all cases, prediction is an important element in all engineering design calculations. The performance of six different turbulence models, namely, the standard k-ε model, realizable k-ε model, renormalization group k-ε model, standard k-ω model, shear stress transport k-ω model, and Reynolds stress transport model, is assessed in this chapter for the simulation of a turbulent single-phase flow generated by a grid disc impeller in a stirred vessel. The measurements of three components of the mean velocity and turbulent kinetic energy at different locations within the stirred vessel reported in the literature are used to validate the numerical predictions. The numerical simulations were carried out with a commercial flow solver FLUENT 6.1.22 using the multiple reference frames (MRF) approach. A fairly good agreement was observed between the predicted and experimental velocity components along axial, radial and tangential directions. The predictions of the standard k-ε model were better than the other turbulence models considered in the present work. The performances of other turbulence models were found to be approximately similar.

Chapter 9 - The Non-linear AutoRegressive Moving Average with eXogenous input (NARMAX) model set has been extensively studied in theory and gradually adopted in applications. There are two main steams of sub-model sets, polynomial and rational. The rational model represents an extension of the polynomial model set, and is defined as a ratio of two polynomial expressions. The introduction of the denominator polynomial terms makes the rational NARMAX model to be non-linear in both the parameters and the regression terms. The justification for using the rational model is that it provides a very concise and parsimonious representation for complex non-linear systems and has excellent interpolatory and extrapolatory properties. However model identification and controller design are much more challenging compared to the polynomial models. This has been a new and fascinating research trend in the area of mathematical modelling and applications, but still within a limited research community. This chapter brings several representative algorithms together, developed by the author and his colleagues, to form an easily referenced archive for promotion of awareness, tutorial, applications, and even further research expansion.

In: Applied Mathematical Modeling
Editor: Eetu N. Virtanen, pp. 1-3
ISBN: 978-1-60021-975-7

Expert Commentary

MODELING AND SIMULATION OF FLUID TURBULENCE: INTRODUCTION, STATE-OF-THE-ART AND FUTURE TRENDS

Anupam Dewan*
Department of Mechanical Engineering, Indian Institute of Technology Guwahati
Guwahati – 781039, India

Most fluid flows encountered in engineering applications are turbulent. Turbulent flows are characterized by unsteadiness, vorticity, three-dimensionality, dissipation, wide spectrum of scales, and large mixing rate. Fluid turbulence remains one of the biggest challenges to scientists and engineers. With the emergence of computers and different computational techniques, the computational fluid dynamics (CFD) has become an important tool for investigating turbulent flows. In CFD, the governing equations are numerically solved over a computational domain along with suitable boundary conditions to obtain the flow and other fields.

The variables in the governing Navier-Stokes equation can be decomposed into mean and fluctuating components (termed as the Reynolds decomposition). By this decomposition and performing time averaging the Navier-Stokes equations reduce to an equivalent set of equations, termed as the Reynolds-Averaged Navier-Stokes (RANS) equations. The RANS equations represent transport equations for the mean flow variables only. The Reynolds-averaging, however, produces additional unknown terms termed as the Reynolds stresses, that need to be modeled. This is termed as the closure problem. There are two different approaches to compute the Reynolds stresses in the RANS equations: use of (a) the eddy-viscosity based models and (b) the Reynolds stress transport (RST) models. In the former, the Boussinesq eddy-viscosity hypothesis is employed to estimate the Reynolds stresses, according to which the Reynolds (turbulent) stresses are assumed to be proportional to the mean strain rate, the constant of proportionality being the isotropic eddy viscosity. The

* Professor. Email: adewan@iitg.ernet.in, Phone: 0091-361-2582656, FAX: 0091-0361-2582699 & 2690762 http://www.iitg.ernet.in/adewan

second approach is also termed as the second order closure in which the transport equations for individual Reynolds stress components are numerically solved.

The use of the isotropic eddy viscosity in the first approach is usually inconsistent with the highly anisotropic nature of turbulent flows. The eddy viscosity based turbulence models are categorized by the number of transport equations employed in addition to the time-averaged momentum equations and these include the zero equation models, one equation models, and two equation models. Of these, the two equation models, the simplest complete turbulence models, are based mostly on the transport equation for the turbulent kinetic energy (k) and a suitable second transport equation. The standard k-ε model is the most commonly used and one of the simplest two-equation models. A large number of other two equation models exist in the literature and are continuously being developed. However, most of these models fail in the situations that are significantly different from the ones used to obtain the values of the closure coefficients in the new transport equation used in the turbulence model. Some other popular two equation models include the renormalization group (RNG) k-ε model, realizable k-ε model, k-ω model, and the shear stress transport k-ω model. Overall the k-ε model seems to be the most versatile model.

The use of the RST models results in seven additional transport equations (six for the components of Reynolds stresses and one for the dissipation of Reynolds stresses) in three-dimensional flows and five in two-dimensional flows. The exact transport equation for the Reynolds stresses accounts for several physical effects such as the streamline curvature, swirl, rotation, and rapid changes in the strain rate in a more rigorous manner than the one-equation and two-equation models. Therefore, it is expected that the use of the transport equations for Reynolds stresses will have a good potential to provide accurate predictions for complex flows. However, the accuracy of the RST model predictions depends on the closure assumptions employed to model various terms in the exact transport equation and RST models might not always yield results that are superior to the other simpler models in all classes of flows. This can be proved by considering a plethora of papers appearing in the literature to compare the performance of different turbulence models under a wide range of complex flow situations, including stirred vessel used in chemical industries, heat exchangers, turbulent free shear and wall bounded flows, atmospheric flows, aerodynamics, and combustion, etc. Therefore it may not be always wise to use the computationally intensive RST models when relatively simpler two equation models can provide predictions of at least similar or sometimes better accuracy than those by RST models.

The other and the most accurate approach to turbulence simulations is to numerically solve the Navier-Stokes equations for the instantaneous and three-dimensional flow without averaging or approximation and this is termed as the direct numerical simulation (DNS). But the limitations of the computational resources (storage capacity and speed) restrict the use of DNS in many practical flow problems. For DNS the grid to be used must be fine enough to resolve both spatially and temporally the smallest eddies whose sizes are of the order of the minuscule Kolmogorov scale. In a large-eddy simulation (LES), the large-scale three-dimensional, instantaneous motions are resolved and the small scale motions, termed as the subgrid scales, are modeled. This approach is based on the idea that the large-scale motions are dependent on the boundary conditions and must be computed, whereas the small scale motions are relatively universal everywhere and can be easily modeled. In LES the governing equations are filtered in space. The numbers of grid points needed in LES is somewhat less than that in DNS. The accuracy of LES depends upon the modeling of the subgrid scales.

Several models for the subgrid scales exist in the literature, including the simplest Smagorinsky model, dynamic model, similarity model, and mixed model. If a LES resolves nearly 80% of the entire energy spectrum, it can be termed as fairly accurate. It may be noted that performing DNS and LES in complex situations such as in the vicinity of solid walls may pose difficulties because the size of the smallest scales both spatially and temporally may be quite small than that far away and needless to say that such scales need to be resolved. LES is comparatively quite expensive than the use of RANS modeling.

The generation of accurate initial conditions at all points within the computational domain and boundary conditions at all instants for all flow variables for three-dimensional flow field corresponding to a particular flow configuration is a challenge in DNS and LES. This is due to the fact that in engineering applications one is usually interested in time-averaged flow field and therefore this is what that can be easily specified in terms of initial and boundary conditions and not the three-dimensional, instantaneous flow field. The problem is further compounded by the fact that the flow field 'remembers' the initial conditions for several eddy turnover times and therefore the prescription of inaccurate conditions can result in unphysical computed flow field. The use of DNS and LES also produces huge data which contains all computed flow properties at all grid points within the computational domain and at all instants both of which can be several millions in number. The data needs to be stored to obtain the time-averaged flow field. No such problems exist while employing the RANS equations to compute the mean flow field. In conclusion we can say that no better RANS based turbulence model than the existing ones is likely to be developed in the foreseeable future. Further RANS based models are likely to be used for engineering flow computations. However, LES has a good potential to be used for the investigation of practical fluid flow problems.

In: Applied Mathematical Modeling
Editor: Eetu N. Virtanen, pp. 5-22

ISBN: 978-1-60021-975-7

Short Communication

INVESTIGATION OF HEATED ROUND JET IN A COLD CO-FLOWING AMBIENT USING TWO TURBULENCE MODELS

Anupam Dewan*, Anas Viquar, Gaurav Jindal and Niranjan Sahoo
Department of Mechanical Engineering
Indian Institute of Technology Guwahati, Guwahati – 781039, India

ABSTRACT

The mean and turbulent flow properties of a turbulent heated round jet discharged in a cold co-flowing ambient of the same fluid have been obtained computationally using the standard *k-ε* model and Reynolds-stress transport (RST) model. The two turbulence models have been chosen to cover a wide spectrum of the modeling options that are available. The present flow configuration is an important test case for turbulence models, because it is more complex compared to that of a standard turbulent round jet in a quiescent ambient of the same fluid with both at the same temperature. The present predictions are compared with the experimental data reported in literature. The results suggest that self-preservation does not hold in this flow. Overall the k-ε model provides better agreement with the experimental data compared to that by the RST model, which overpredicts. The present chapter clearly shows that the RST model, though more elaborate, does not have an advantage over the standard *k-ε* model in the prediction of the present flow configuration.

Keywords: RANS equations; Convective heat transfer; Turbulence model; CFD

* Corresponding author. Professor. Email: adewan@iitg.ernet.in, FAX: 0361-2582699, Ph: 0361-2582656, http://www.iitg.ernet.in/adewan

NOMENCLATURE

U	Mean velocity of jet in the axial (streamwise) direction
V	Mean velocity of jet in the radial direction
r	Radial coordinate
U_j	Velocity of jet at the discharge
U_1	Co-flow velocity
U_o	Velocity of jet at the axis
U_{co}	Centerline excess velocity of jet
$\overline{u'v'}$	Reynolds shear stress
k	Turbulent kinetic energy
ε	Turbulent dissipation rate
$\overline{u'^2}$	Reynolds axial normal stress
$\overline{v'^2}$	Reynolds radial normal stress
b	Width of the jet
$r_{1/2}$	Half width of the jet
u_i	Instantaneous velocity component
$\overline{u_i}$	Mean velocity component
$\overline{u'_i}$	Fluctuating velocity component
μ_t	Eddy viscosity
δ_{ij}	Kronecker delta
υ_t	Kinematic eddy viscosity
μ	Molecular viscosity
ρ	Mass density
σ_k	Turbulent Prandtl number for k
σ_ε	Turbulent Prandtl number for ε
C_{ij}	Convection term
P_k	Rate of production of k
$D_{T,ij}$	Turbulent diffusive transport term
G_{ij}	Buoyancy production
ϕ_{ij}	Pressure-strain term
ε_{ij}	Dissipation tensor

1. INTRODUCTION

Fluid turbulence is one of the biggest challenges to scientists and engineers, who need to understand turbulent flows in order to control them and to design for their adverse effects or utilize them for their best effects (Hinze [1]). In all cases, prediction is an important element in all engineering design calculations. The flow field of turbulent jet in a co-flow is encountered in several engineering and environmental applications (Dewan et al. [2], Bradbury and Riley [3]). It also provides simplification to some complex problems such as the design of combustors when the recirculation of hot combustion products is replaced with a co-flow and both the jet and co-flow fluids are assumed to be the same. Further this is an important test case for turbulence modeling as it is more complex compared to that of a standard turbulent round jet in a quiescent ambient of the same fluid with both at the same temperature (Cho and Chung [4]). The objective of the present chapter is to report on the computational investigation of the characteristics of a heated turbulent round jet discharged in a cold co-flowing ambient of the same fluid. It is well known that the mean and turbulent flow profiles of a jet in a quiescent ambient are self-preserving (see for example, Cho and Chung [4] Panchapakesan and Lumley [5] and Webster et al. [6]). However, the self-preservation does not hold for a jet in a co-flow and the flow far downstream depends strongly on the complete flow history, unlike that for self-similar flows [7, 8]. Due to the presence of the co-flowing ambient, the flow field of a round jet experiences more advection and diffusion than that in a still ambient. From the experimental work, Antonio and Bilger [7] observed that although the mean velocity profiles show remarkable similarity at small values of x/D, the turbulence field does not attain a universal self-preserving state. Further, the maximum excess velocity in the co-flowing round jet is almost inversely proportional to the distance from the jet exit and not to the jet width (Maczynski [9]). No study of heated turbulent round jet in a co-flow has been reported in the literature. We intend to fill this gap through the present chapter.

In the present chapter the flow field of a slightly heated turbulent round jet discharged in a cold co-flowing ambient is investigated numerically using the standard k-ε and RST models (Pathak et al. [10]). The predicted mean and turbulent flow fields are compared with the experimental data of Antonio and Bilger [7], which corresponds to an unheated turbulent round jet discharged in a co-flow making momentum as the only driving force with negligible buoyancy effects. It is first established that there is negligible change in the flow properties on slightly heating the jet as compared to the unheated jet and therefore the flow remains only momentum driven with negligible buoyancy effects. Thereafter, the predictions of the heated jet are compared with the experimental data of Antonio and Bilger [7]. The predictions of some flow properties which were absent in the experimental observations [7] are also reported in the present chapter.

2. PROBLEM DESCRIPTION

A heated turbulent round jet with the discharge velocity Uj issues out from a cylindrical pipe of the diameter D in a cold co-flowing ambient which has a velocity of U_1 (Figure 1). The jet and the co-flowing fluid are taken to be the same (air) in the present case. The computed flow field comprises of a non-turbulent co-flow away from the axis and a turbulent region close to the axis. The discharged jet was at a slightly higher temperature than that of the co-flow with a temperature difference of 5°C. The value of the exit buoyancy Richardson number, $R_{bj} = (\Delta\rho_j gD)/(\rho_a U_1)$, due to the heating was quite low thus ensuring a negligible buoyancy effect with heat simply playing the role of a passive scalar. In this condition, a passive scalar such as the temperature is transported only by the forced convective flow. It is logical to use the momentum ratio, given by $(\rho_j U_j^2)/(\rho_a U_1^2)$, in the formulation of the investigation of heated jet in co-flow. But, since both the jets and the co-flow are of the same fluid (air), the change in density for a small temperature difference was negligible. Thus, the momentum ratio is equivalent to the velocity ratio.

3. Governing Equations

The Reynolds-averaged Navier-Stokes (RANS) and energy equations for the steady flow form the governing equations. As the flow was assumed to be axisymmetric, the corresponding governing equations in the cylindrical co-ordinates may be expressed as

Continuity equation:

$$\frac{\partial U}{\partial x}+\frac{1}{r}\frac{\partial}{\partial r}(rV)=0 \tag{1}$$

Momentum equation in the axial direction:

$$U\frac{\partial U}{\partial x}+V\frac{\partial U}{\partial r}=-\frac{1}{\rho}\frac{\partial \bar{p}}{\partial x}+\frac{\partial}{\partial x}\left[\left(-\overline{u'^2}\right)\right]+\frac{1}{r}\frac{\partial}{\partial r}\left[r\left(-\overline{u'v'}\right)\right] \tag{2}$$

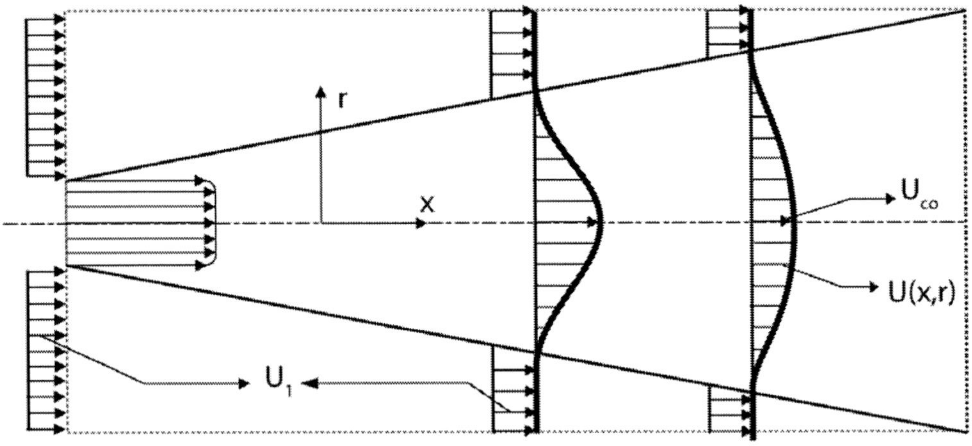

Figure 1. Flow configuration.

Momentum equation in the radial direction:

$$U\frac{\partial V}{\partial x}+V\frac{\partial V}{\partial r}=-\frac{1}{\rho}\frac{\partial \overline{p}}{\partial r}+\frac{\partial}{\partial x}\left[\left(-\overline{u'v'}\right)\right]+\frac{1}{r}\frac{\partial}{\partial r}\left[r\left(-\overline{v'^2}\right)\right] \tag{3}$$

Energy equation:

$$u\frac{\partial T}{\partial x}+v\frac{\partial T}{\partial r}=\frac{\partial}{\partial x}(-\overline{u't'})+\frac{1}{r}\frac{\partial}{\partial r}\left[r(-\overline{v't'})\right] \tag{4}$$

The standard *k-ε* model and Reynolds stress transport (RST) model were used for the closure of the Reynolds stress tensor in the momentum equation and turbulent heat flux in the energy equation. These were chosen to cover a wide spectrum of the modeling options that are available for the prediction. This is so because, on one hand the *k-ε* model is the simplest complete model known to handle many engineering flows with reasonable accuracy and on the other hand RST model is the most elaborate turbulence model containing several additional terms to take care of the anisotropy, streamline curvature, compressibility and rotation effects [11]. However, the computational cost and time involved in the use of RST model is quite high compared to that of the standard *k-ε* model.

Standard k-ε Model

The standard *k-ε* model is a two equation model (Launder and Spalding, 12) and here the turbulent stresses are modeled as

$$-\overline{u'v'}=\nu_t\left(\frac{\partial U}{\partial r}+\frac{\partial V}{\partial x}\right),\ -\overline{u'^2}=2\nu_t\frac{\partial U}{\partial x}-\frac{2}{3}k,\ -\overline{v'^2}=2\nu_t\frac{\partial V}{\partial r}-\frac{2}{3}k,$$
$$-\overline{u't'}=-\frac{\nu_t}{\mathrm{Pr}_t}\frac{\partial T}{dx},\ -\overline{v't'}=-\frac{\nu_t}{\mathrm{Pr}_t}\frac{\partial T}{dr} \tag{5}$$

where $\nu_t = C_\mu \frac{k^2}{\varepsilon}$ (6)

Here ν_t is the eddy viscosity, C_μ a k-ε model constant, *k* the turbulent kinetic energy and ε the dissipation rate of turbulent kinetic energy. The following modelled transport equations for *k* and ε were used

$$U\frac{\partial k}{\partial x}+V\frac{\partial k}{\partial r}=\frac{\partial}{\partial x}\left[\left(\frac{\nu_t}{\sigma_k}\right)\frac{\partial k}{\partial x}\right]+\frac{1}{r}\frac{\partial}{\partial r}\left[r\left(\frac{\nu_t}{\sigma_k}\right)\frac{\partial k}{\partial r}\right]+P_k-\varepsilon \tag{7}$$

$$U\frac{\partial \varepsilon}{\partial x}+V\frac{\partial \varepsilon}{\partial r}=\frac{\partial}{\partial x}\left[\left(\frac{\nu_t}{\sigma_\varepsilon}\right)\frac{\partial \varepsilon}{\partial x}\right]+\frac{1}{r}\frac{\partial}{\partial r}\left[r\left(\frac{\nu_t}{\sigma_\varepsilon}\right)\frac{\partial \varepsilon}{\partial r}\right]+C_1\frac{\varepsilon}{k}P_k-C_2\frac{\varepsilon^2}{k} \tag{8}$$

where P_k is the rate of production of k given by

$$P_k=\left[\left(\frac{\partial U}{\partial r}+\frac{\partial V}{\partial x}\right)^2+2\left(\frac{\partial U}{\partial x}\right)^2+2\left(\frac{\partial V}{\partial r}\right)^2+2\left(\frac{V}{r}\right)^2\right] \tag{9}$$

The standard values of the model constants used in the above equations are $C_\mu = 0.09$, $C_1 = 1.44$, $C_2 = 1.92$, $\sigma_k = 1.0$, $\sigma_\varepsilon = 1.3$ [13].

Reynolds Stress Transport (RST) Model

The Reynolds stress transport model [10, 11] involves the computation of the individual Reynolds stresses ($\rho\overline{u_i'u_j'}$) using differential transport equations. For the present flow configuration five additional transport equations were numerically solved to obtain the flow field using a RST model. The general transport equation for Reynolds shear stress $\rho\overline{u_i'u_j'}$ is as follows

$$\frac{\partial}{\partial x_k}\left(u_k\overline{u_i'u_j'}\right)=-\frac{\partial}{\partial x_k}\left[\left(\overline{u_i'u_j'u_k'}+\overline{\frac{p'}{\rho}\left(\delta_{kj}u_i'+\delta_{ik}u_j'\right)}\right)\right]+\frac{\partial}{\partial x_k}\left[\nu\frac{\partial}{\partial x_k}\left(\overline{u_i'u_j'}\right)\right]$$
$$+G_{ij}+\phi_{ij}-2\nu\overline{\frac{\partial u_i'}{\partial x_k}\frac{\partial u_j'}{\partial x_k}} \tag{10}$$

where G_{ij} represents the stress production and Φ_{ij} the pressure-strain given as

$$G_{ij}=-\left(\overline{u_i'u_k'}\frac{\partial u_j}{\partial x_k}+\overline{u_j'u_k'}\frac{\partial u_i}{\partial x_k}\right) \tag{11}$$

$$\phi_{ij}=\overline{\frac{p'}{\rho}\left(\frac{\partial u_i'}{\partial x_j}+\frac{\partial u_j'}{\partial x_i}\right)} \tag{12}$$

The pressure strain term Φ_{ij} in the above equation was modelled according to the proposals by Launder and Gibson [14], Launder [15], Launder et al. [16] and Fu et al. [17]. The turbulent heat flux term was modelled as

$$\overline{u_j t} = -C_t \left(\overline{u_j u_k} \frac{\partial T}{\partial x_k} \right) \text{ where } C_t = 0.3 \text{ is a model constant [18].} \qquad (13)$$

4. Computational Procedure

A commercial package Gambit 2.1.6 was used to generate the geometry and the mesh for the computational domain. The simulations were carried out using the commercial flow solver FLUENT 6.1.22 which is based on the finite volume method. The pressure equation was discretized using the standard scheme which interpolates the values at the faces using the momentum equation coefficients. The momentum, Reynolds stress transport, turbulence kinetic energy, turbulence dissipation and energy equations were discretized using the fist order upwind scheme. The pressure-velocity coupling was solved using the SIMPLE algorithm.

The streamwise length of the computational domain was taken as 300D and width in the radial direction as 30D where D = 0.00528 m according to [7]). Here D denotes the diameter of the slot through which the jet effuses out (Figure 2). The ratio of the velocity of the jet to the velocity of co-flow was taken to be 4.5 and both the fluids were assumed to be air according to the experimental conditions [7]. For the co-flow and the jet discharge, the velocity inlet types of boundary conditions were used. At the jet discharge *ab* (Figure 2) the flow was assumed to be fully turbulent flow with a top hat velocity profile of 137 ms^{-1} [7]. The turbulence was specified in terms of the turbulence intensity and hydraulic diameter and it was given a value of 4.14% based on the relation: $u'/u_{avg} = 0.16\ Re_D^{-1/7}$ where u_{avg} denotes the average jet velocity, Re_D the Reynolds number (= U_jD/υ) and ν the kinematic viscosity. The co-flow was assumed to be non-turbulent with a constant velocity of 30.5 ms^{-1} [7]. The temperatures of the jet discharge and that of co-flow were set to be 303 K and 298 K, respectively. For the freestream boundary *cd*, the u-velocity was specified = 30.5 ms^{-1} [7] and the v-velocity was assumed to be zero. The boundary *de* was assigned the outflow boundary condition where the normal gradient of all the parameters were assumed to be zero. The implicit and segregated solver was used for the solution of the system of governing equations. Mean axial velocity (*U*), mean radial velocity (*V*), turbulence kinetic energy (*k*) and turbulence dissipation rate (ε) were the variables for which the convergence criterion was applied for the solution using the *k*-ε model. For the RST model, normal stresses ($\overline{u'^2}$, $\overline{v'^2}$) and Reynolds shear stress ($\overline{u'v'}$) were also checked for the convergence. For all cases the convergence criteria of 1×10^{-5} was applied which is a more stringent convergence criterion compared to that recommended by the FLUENT 6.1.22 (1×10^{-3}) for these kinds of turbulent flows.. The computations were performed on a Pentium 4 machine with 512 MB RAM, 2.8 GHz processor speed and it took approximately 5 hours and 7 hours of CPU time for the k-ε model and the RST model, respectively, to achieve the converged solution for the grid with 60000 mesh elements.

4.1. Grid Independence Study

A non-uniform grid meshed with rectangular elements was used for the simulation and the flow field was obtained for three different grid resolutions with 30000, 60000 and 120000 number of mesh elements. The mesh was fine near the jet discharge and in the vicinity of the centerline where the flow is complex and it became coarser as one moves away from this region (Figure 3).

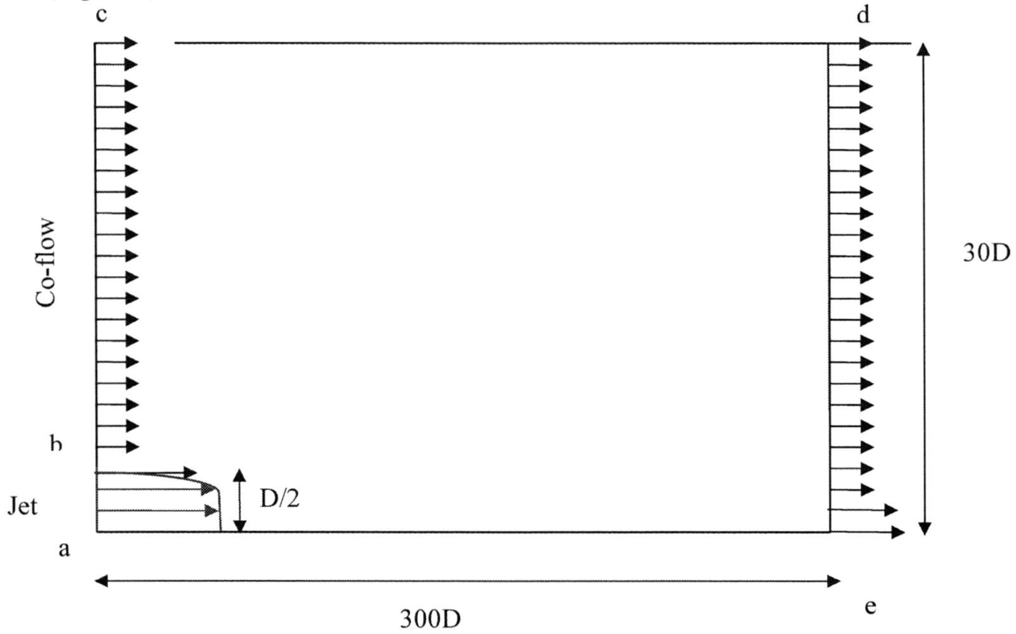

Figure 2. Computational domain used in the present investigation (not to scale).

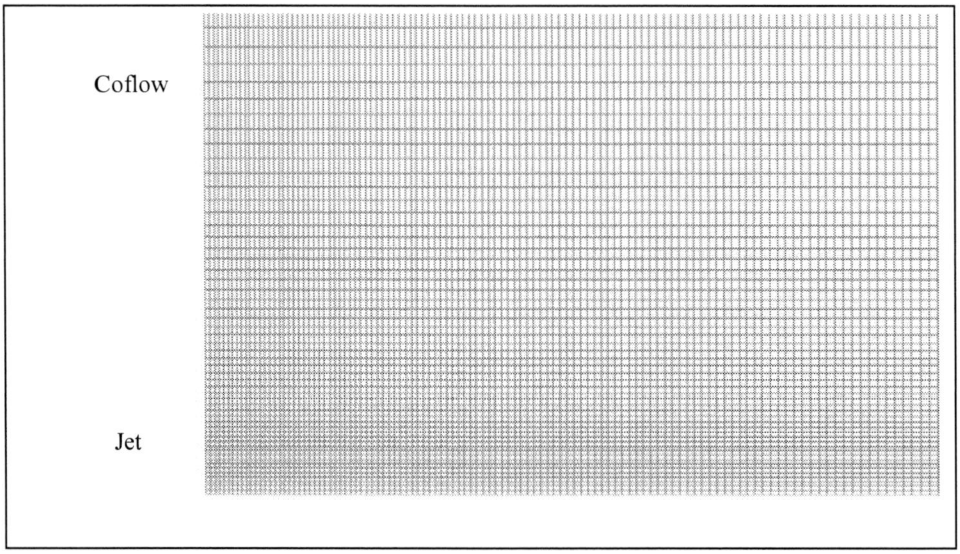

Figure 3. Non-uniform grid with 60000 mesh points.

For the grid independence study the *k*-ε model was employed and the Reynolds shear stress profiles at x/D = 38 were considered because this variable encapsulates the important information in turbulent flows. It was found that the results were grid independent after a size of 60000 grid points (Figure 4a) and therefore the results presented in the next section are using this mesh. Further to affirm the accuracy of the mesh chosen, mean velocity profiles at the same downstream location (x/D = 38) were also checked for the grid independence (Figure 4b).

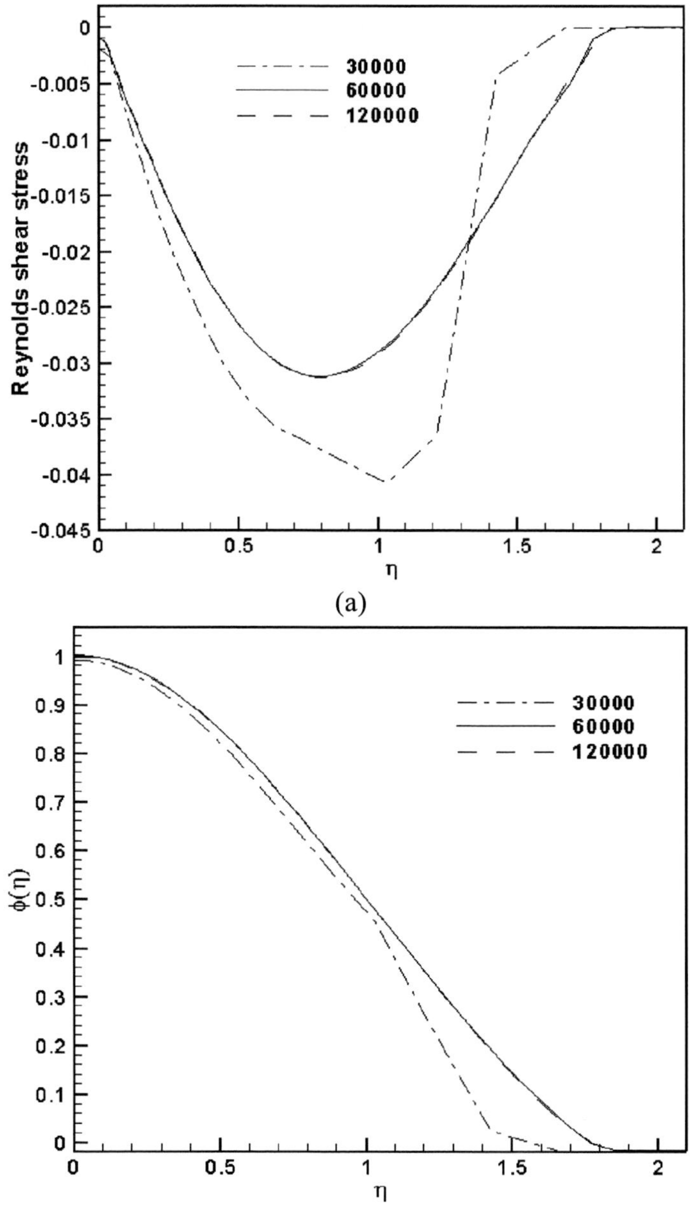

(b)

Figure 4. Grid independence study using standard *k*-ε model: a comparison of (a) Non-dimensional Reynolds shear stress and (b) Non-dimensional mean velocity profiles at x/D = 38. Here $\varphi(\eta) = (U-U_1)/(Uo-U_1)$ and $\eta = r/r_{1/2}$.

5. Results and Discussion

The profiles of the mean velocity, mean temperature, Reynolds shear stress and turbulence intensity were computed using both the *k*-ε and RST models. The predicted results were compared with the experimental data of Antonia and Bilger [7] on the unheated turbulent jet discharged in a co-flow. The centerline velocity excess U_{co} (defined as the difference of the jet centerline velocity U_o and co-flow velocity U_1) and the velocity half-width $r_{1/2}$ (defined as the radial distance from the centerline where the jet velocity excess over the co-flow is half of the centerline excess) were taken as the characteristic velocity and length scales, respectively, and were used to normalize the results.

5.1. Heated and Unheated Turbulent Jets

Figure 5 shows a comparison of the mean velocity profiles for the heated and unheated turbulent jets in co-flow along with the experimental data at various downstream locations. The velocity profiles do not get much affected by slightly heating the jet and the difference in the profiles between unheated and heated jets is negligible (Figure 5). It is due to the fact that as the jet is heated slightly, the variation of the density and hence the buoyancy effect is negligible keeping the flow only the momentum driven. Having established this fact, we are now able to compare the predicted profiles for the heated turbulent jet in co-flow with the experimental data of Antonio and Bilger [7].

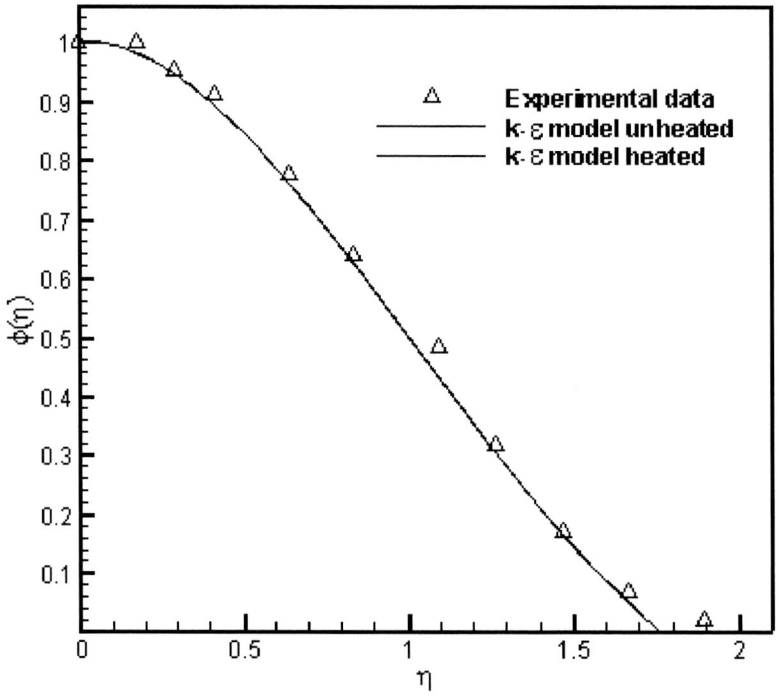

(a)

Figure 5. Continued on next page.

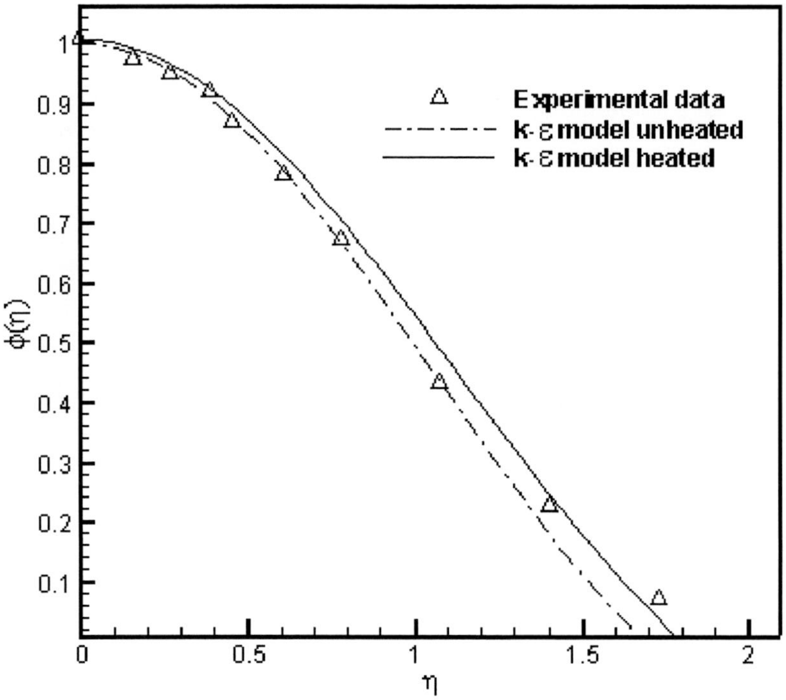

(b)

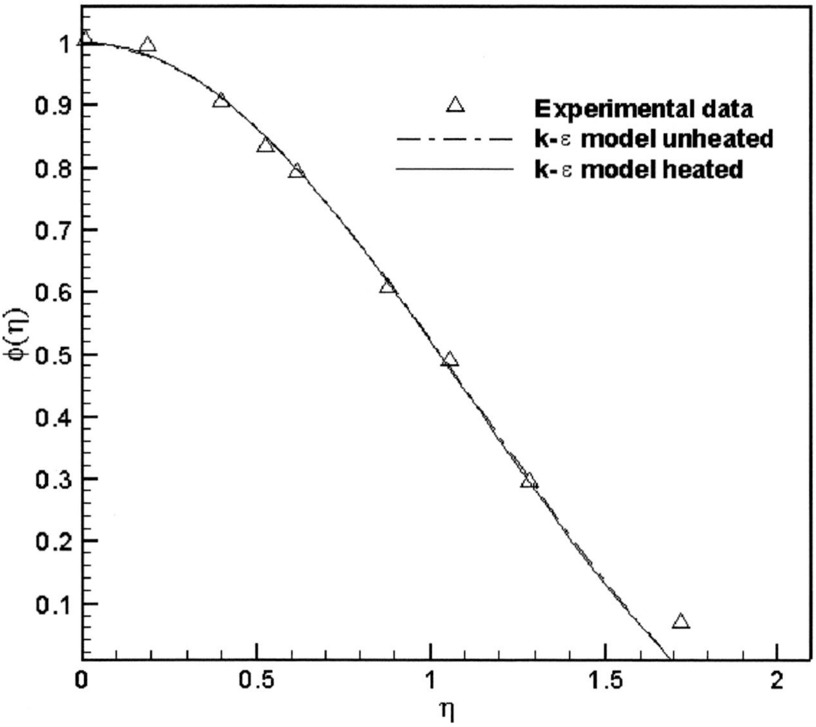

(c)

Figure 5. Continued on next page.

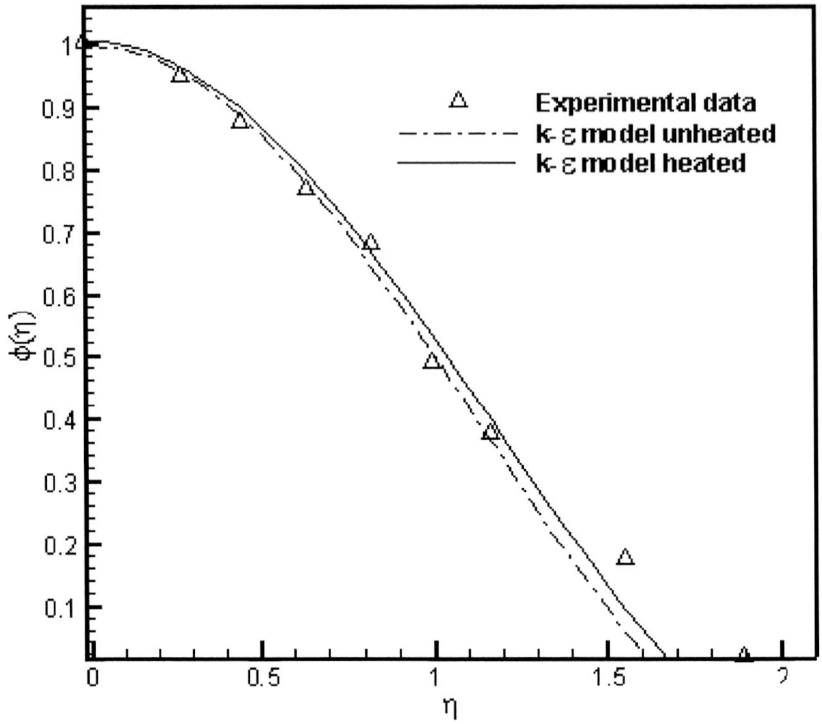

(d)

Figure 5. Non-dimensional mean velocity distributions for heated and unheated turbulent jet in a co-flow at (a) x/D = 38, (b) x/D = 76, (c) x/D = 152, and (d) x/D = 266. Here $\varphi(\eta) = (U-U_1)/(U_o-U_1)$ and $\eta = r/r_{1/2}$. Experimental data corresponds to that of [7].

5.2. Mean Flow and Temperature Fields

The predictions of the mean velocity profiles by both the *k-ε* and RST models are consistent with the experimental data (Figure 6). The predictions match well with the experimental data at the low x/D value (Figure 6a) but deviate somewhat at the high x/D value (Figure 6b). However, the *k-ε* model provides better predictions than those by RST model at the far downstream location (Figure 6b).

The predicted mean temperature profiles at two downstream locations are presented in Figure 7. The predictions of *k-ε* model and RST model match well at low x/D value (x/D = 38) but deviate somewhat at high x/D value (x/D = 266). Due to the unavailability of an experimental data, the predictions are not compared with measurements. The peak of the mean temperature is at the axis which indicates the high temperature of the jet stream. Further downstream, the magnitude of the highest temperature goes on decreasing with the profile becoming more and more flat indicating the entrainment of the co-flow.

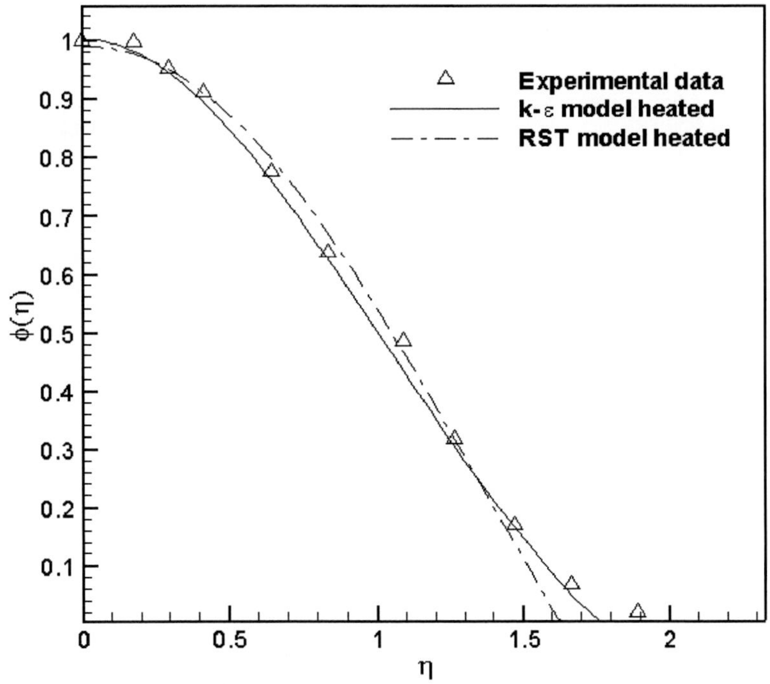

(a)

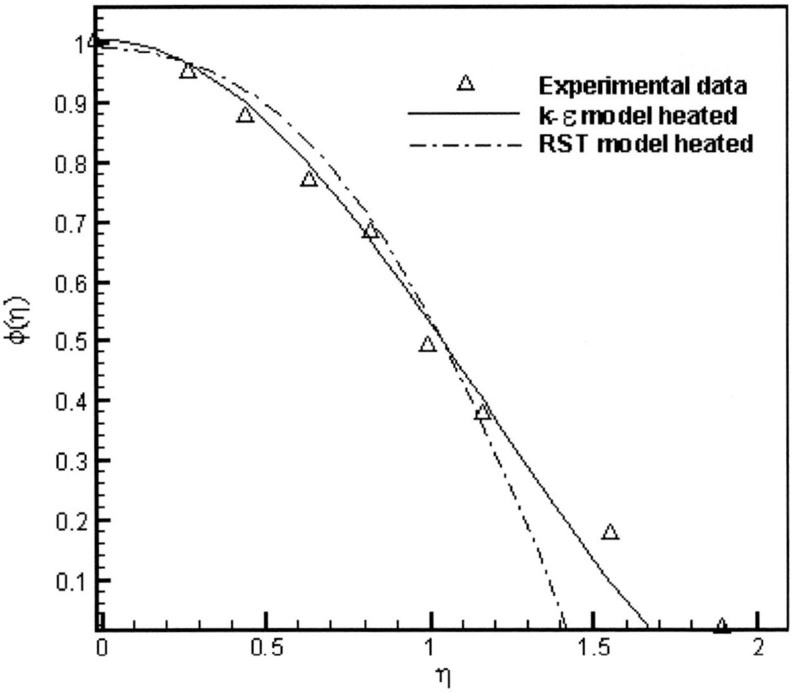

(b)

Figure 6. Non-dimensional mean velocity distributions at (a) x/D = 38 (b) x/D = 266. Here $\varphi(\eta) = (U-U_1)/(Uo-U_1)$ and $\eta = r/r_{1/2}$. Experimental data corresponds to that of [7].

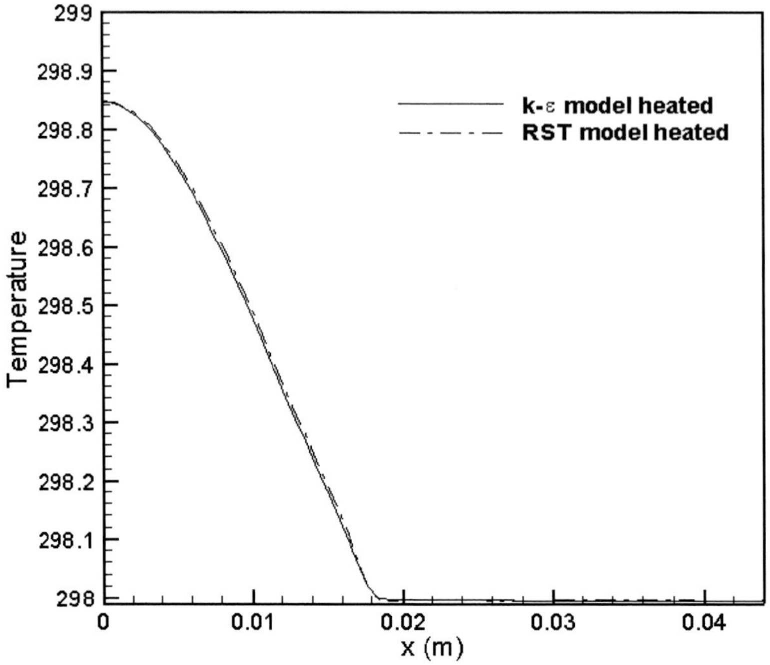

(a)

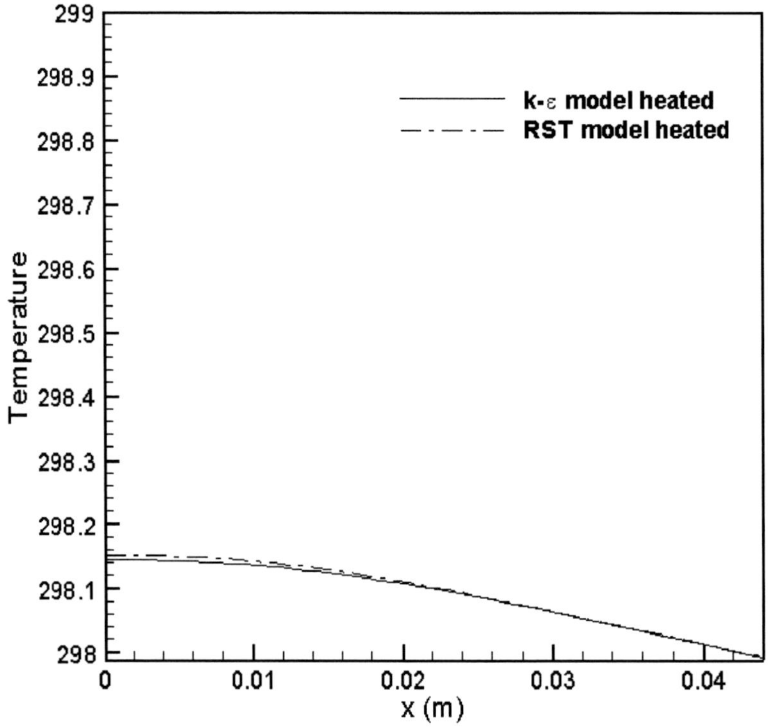

(b)

Figure 7. Predicted mean temperature profiles (in K) at (a) x/D = 38 (b) x/D = 266.

5.3. Turbulent Quantities

The experimental data [7] of Reynolds shear stress at different downstream locations shows that the self-preservation does not exist for this flow configuration (Figure 8). The *k-ε* model provides better agreement with the data of Antonia and Bilger [7] than that by RST model, which predicts profiles whose numerical values are higher in magnitude than the experimental ones (Figure 8).

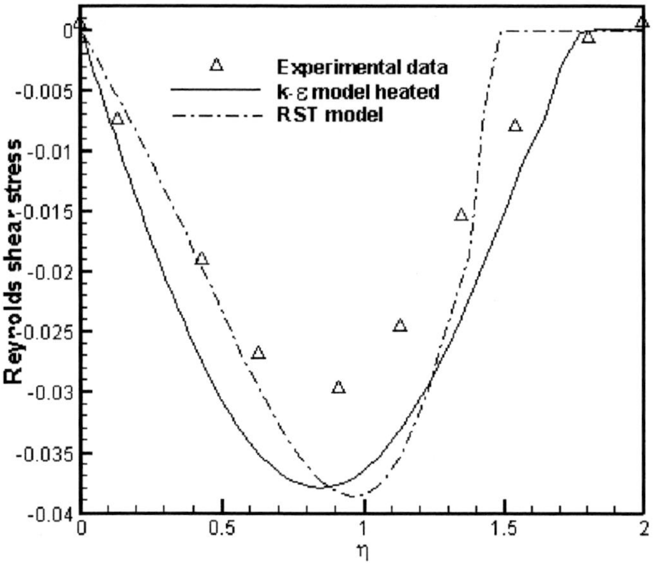

(a)

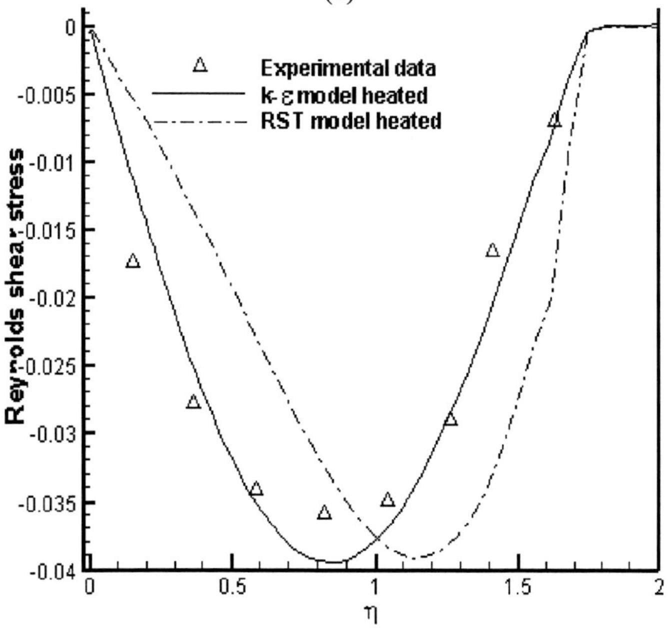

(b)

Figure 8. Non-dimensional Reynolds shear stress profiles at (a) x/D = 152 and (d) x/D = 186. Here $\varphi(\eta) = (U-U_1)/(Uo-U_1)$ and $\eta = r/r_{1/2}$. Experimental data corresponds to that of [7].

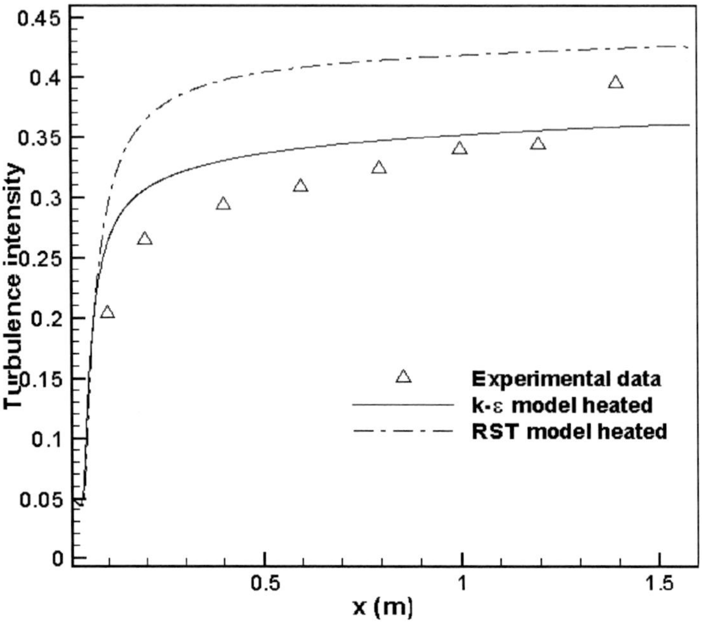

(a)

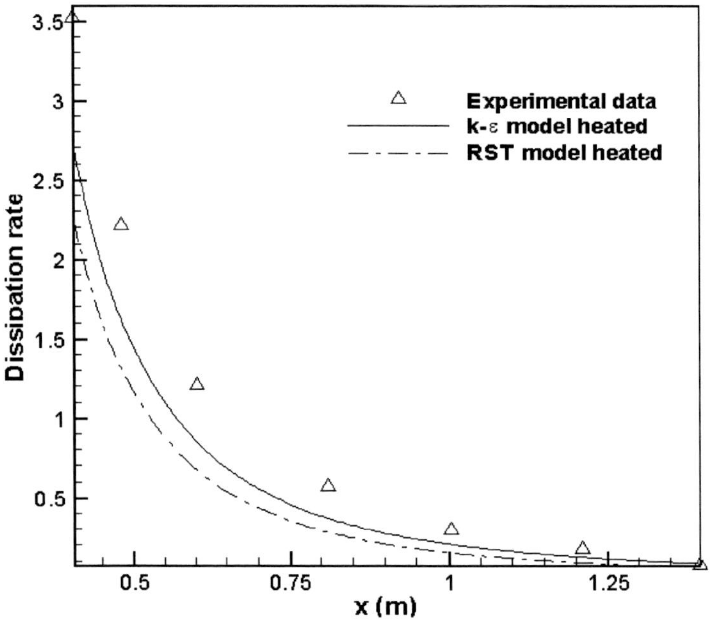

(b)

Figure 9. Variation of (a) non-dimensional turbulence intensity ($\left(\overline{u'^2}\right)^{1/2} / U_{co}$) and (b) dissipation rate ($\varepsilon \times 10^{-3}$) along the axis of symmetry. Experimental data corresponds to that of [7].

The value of the root mean square longitudinal turbulence intensity $\left(\overline{u'^2}\right)^{1/2}$ presented in Figure 9a was obtained from the expression $\overline{u'^2}$ = 2k/3 with the assumption of isotropy for both the models. The ratio $\left(\overline{u'^2}\right)^{1/2} / U_{co}$ rises fairly sharply as the end of the potential core is reached but increases only slightly for x/D > 190 (which corresponds to x > 1 m) attaining approximately a constant value further downstream. Both the models predict a non-linear variation of the intensity in accordance with the data of Antonia and Bilger [3] but the magnitudes are overpredicted. For instance the experimental data gives the value of $\left(\overline{u'^2}\right)^{1/2} / U_{co}$ approximately equal to 0.28 for x/D > 75 (which corresponds to x > 0.30 m) whereas the *k-ε* model gives the value approximately equal to 0.34 and the RST model gives a much higher value of 0.40.

We observe a good agreement (Figure 9b) between the values of the dissipation rate predicted by both the *k-ε* and RST models and the experimental data [3] who estimated it through the use of the relation $\varepsilon = 15\nu\overline{(\partial u / \partial x)^2}$. The dissipation ε is maximum near the jet inlet and it decreases downstream. This behavior is due to the fully turbulent nature of the flow in the vicinity of the discharge nozzle and afterwards due to the entrainment of the non-turbulent co-flow, the dissipation decreases.

6. Conclusion

The flow configuration of heated turbulent round jet in a cold co-flow was investigated using both the *k-ε* model and the Reynolds stress transport model and the predictions were compared with the experimental data reported in the literature. In general, the agreement between the present computational results and experimental data was good. In almost all the cases the trend was captured well by both the models, although in certain cases the values were not in accordance with the experimental data. The results in the present paper suggest that, although the mean velocity profiles show remarkable similarity at small values of x/D, the turbulence field does not attain a universal self-preserving state. The mean temperature profiles predicted by the *k-ε* model and RST model match well at locations close to the discharge but deviate far downstream. The magnitude of the highest temperature decreases in the streamwise direction indicating the entrainment of co-flow in to the main jet flow.

REFERENCES

[1] J.O. Hinze, *Turbulence*, McGraw-Hill Publishing Co., New York (1975).

[2] A. Dewan, M. Pathak, A.K. Dass, P. Sardar, Prediction of turbulent round jet in a co-flowing ambient using k-ε model of turbulence, *Proc. 18th National & 7th ISHMT-ASME Heat and Mass Transfer Conference,* IIT Guwahati, India (2006).

[3] L.J.S. Bradbury, J. Riley, The spread of a turbulent plane jet issuing into a parallel moving air stream, *J. Fluid Mechanics* 27 (1967) 381-394.

[4] J.R. Cho, M.K. Chung, A *k-ε-γ* equation turbulence model, *J. Fluid Mechanics* 237 (1992) 301-322.

[5] N.R. Panchapakesan, J.L. Lumley, Turbulence measurements in axisymmetric jets of air and helium, Part 1. Air jet, *J. Fluid Mechanics* 246 (1993) 197-223.

[6] D.R. Webster, P.J.W. Roberts L. Ra'ad, Simultaneous DPTV/PLIF measurements of a turbulent jet, *Experiments in Fluids* 30 (2001) 65-72.

[7] R.A. Antonia, R.W. Bilger, An experimental investigation of an axisymmetric jet in a coflowing air stream, *J. Fluid Mechanics 61* (1973) 805-822.

[8] S.B. Pope, An explanation of the turbulent plane jet/round jet anomaly, *AIAA J* 16 (1978) 279-281.

[9] J.F.J. Maczynski, A round jet in an ambient co-axial stream, *J. Fluid Mechanics* 13 (1961) 597-608.

[10] M. Pathak, A. Dewan, A.K. Dass, Computational prediction of a slightly heated turbulent rectangular jet discharged into a narrow channel crossflow using two different turbulence models, *International Journal of Heat and Mass Transfer* 49 (2006) 3914-3928.

[11] F.S. Lien, M.A. Leschziner, Assessment of turbulent transport models including non-linear RNG eddy-viscosity formulation and second-moment closure, *Computers and Fluids* 23 (8) (1994) 983-1004.

[12] B.E. Launder, D.B. Spalding, *Lectures in Mathematical Models of Turbulence*, Academic Press, London, England (1972).

[13] FLUENT 6.2 documentation, Fluent Inc., USA, 2005.

[14] B.E. Launder, M.M. Gibson, Ground effects on pressure fluctuations in the atmospheric boundary layer, *J. Fluid Mechanics* 86 (1978) 491-511.

[15] B.E. Launder, Second-moment closure: present... and future?, *Inter. J. Heat Fluid Flow* 10 (4) (1989) 282-300.

[16] B.E. Launder, G.J. Reece, W. Rodi, Progress in the development of a Reynolds-stress turbulence closure, *J. Fluid Mechanics* 68 (3) (1975) 537-566.

[17] S. Fu, B.E. Launder, M.A. Leschziner, Modeling strongly swirling recirculating jet flow with Reynolds-stress transport closures, *Proc.* Sixth Symposium on Turbulent Shear Flows, *Toulouse,* France (1987).

[18] M. Rokni, T.B. Gatski, Predicting turbulent convective heat transfer in fully developed duct flows, *International Journal of Heat and Fluid Flow* 22 (2001) 381-392.

In: Applied Mathematical Modeling
Editor: Eetu N. Virtanen, pp. 23-87
ISBN: 978-1-60021-975-7

Chapter 1

MATHEMATICAL MODELING APPROACHES FOR OPTIMIZATION OF CHEMICAL PROCESSES

Gabriela Corsano[1], Jorge M. Montagna[2,*], Oscar A. Iribarren[2] and Pío A. Aguirre[1]
INGAR- Instituto de Desarrollo y Diseño-Avellaneda 3657- (3000) Santa Fe-Argentina.
[1]Universidad Nacional del Litoral
[2]Universidad Tecnológica Nacional

ABSTRACT

Mathematical modeling is a powerful tool for solving optimization problems in chemical engineering. In this work several models are proposed aimed at helping to make decisions about different aspects of the processes lifecycle, from the synthesis and design steps up to the operation and scheduling. Using an example of the Sugar Cane industry, several models are formulated and solved in order to assess the trade-offs involved in optimization decisions. Thus, the power and versatility of mathematical modeling in the area of chemical processes optimization is analyzed and evaluated.

1.1. INTRODUCTION

Mathematical modeling is a powerful tool to solve different problems which arise in chemical engineering optimization. Problems as designing a plant, determining the number of units for a specific task, assigning raw materials to different production processes and deciding the production planning or production targets are some of the issues that can be solved through mathematical modeling. In other words, the mathematical formulations are used to make decisions at different levels, from the synthesis and design of the process up to its operation and scheduling.

* Corresponding Author: Tel.: 54-342-4534451. Fax: 54-342-4553439; E-mail: mmontagna@ceride.gov.ar

In this work, different possible mathematical modeling in chemical engineering are developed. Generic formulations are presented and applied to a particular study case of the Sugar Cane industry: the simultaneous production of sugar and derivative products. Through the examples, several models will be posed corresponding to different considered conditions or decisions. Moreover, different scenarios will be considered and trade-offs between decisions will be represented and analyzed.

In general, the aforementioned problems in chemical engineering can be faced either sequentially or simultaneously. The sequential approach deals with each decision separately, one after the other. For example, the plant configuration is first determined, then the unit sizes, after that the processing variables e.g. processing times and flowrates, and lastly the production planning. Sometimes depending on the scenario, when the plant configuration or unit sizes are known, only the operating, scheduling or planning problem are solved separately. On the other hand, modeling all or some decisions simultaneously implies to consider in an overall model all the necessary variables and constraints in order to solve the plant synthesis, design, operation and planning, all together.

Some definitions, concepts and known methodologies of process system engineering are first presented in order to provide a background for the present work. Then, the mathematical models proposed in this work for different scenarios are formulated. Also real-world applications of synthesis, design, operation, scheduling, and planning problems are posed and solved. The optimal solutions are analyzed in each case and the advantages and disadvantages of each formulation are discussed. Finally, desirable future work in this area is briefly addressed.

1.2. Definitions

In this section, some definitions are given in order to introduce the main characteristics of the processes to be modelled.

Batch and Semi-continuous Units

Batch and semi-continuous processes are essential to the chemical processing industry. Products produced in a batch plant range from simple agricultural bulk chemicals to innovative, complex, and high value pharmaceutical products. In a batch unit no simultaneous removal and addition of material streams are performed. Contrarily to the case of continuous processes, where processing tasks are arranged in a sequence and all tasks occur simultaneously, continuously receiving feed and producing product streams. A batch unit operates in a non-continuous manner, receiving feed at certain times and after a processing time, producing products or intermediates. Semi-continuous units are characterized by a processing rate for each product; they are operated continuously with periodic startups and shutdowns.

Batch processes consist of a collection of processing equipment where batches of the various products are produced by executing a set of processing tasks or operations like reaction, mixing or distillation. Every processing equipment can perform only some particular operations. Thus, it is possible to recognize production paths consisting of a sequence of processing equipment, which indicates potential routes that a batch might follow. Processing equipment that can perform the same operations can be grouped in a production stage.

The major classification of batch processes is based on the consideration of the type of production paths followed by the batches of products in the plant.

Single Product, Multiproduct and Multipurpose Batch Plants

In a single product batch plant all the units are totally dedicated to the production of one product. It is the simplest plant operation and no product changeovers are necessary. All units can be designed to exactly fit a batch size and no sequencing has to be performed.

In a multiproduct plant, two or more products are produced following the same production path. When the products are produced through different production paths the plant is a multipurpose plant. Voudouris and Grossmann (1996) introduced a special classification for the multipurpose batch plant. More specifically they divided the multipurpose plants in sequential plants and non-sequential plants (Figure 1.1). In a sequential plant it is possible to recognize a specific direction in the plant floor that is followed by the production paths of all the products. Non-sequential plants are all the remaining cases. Multiproduct plants are used when the production recipes are similar to each other. As the similarities decrease, the plant becomes a multipurpose batch plant. Among these, sequential multipurpose plants are common in industry and hence of practical importance.

When two or more plants are nearly located and there are interconnections between them, the integration scenario is called multiplant complex (MC) integration. The model for a MC frequently considers mass and utility balances between the plants, besides the design, operation, scheduling and planning considerations. In some cases, there is a so called "mother plant", that supplies raw materials and energy to others plants, called "derivative plants". Figure 1.2 shows a MC scenario, with a mother plant and two derivative plants.

An important characteristic of batch processes is the way in which the batches are transferred from one unit to the following one in the production path. In this work, the Zero-wait (ZW) transfer policy is adopted. In the ZW policy the material at any stage will be transferred immediately to the next stage after finishing its processing. In this case, special timing constraints are required in the model. The ZW transfer is commonly used when no intermediate storage is available or when it cannot be held further inside the processing vessel (e.g., due to chemical reaction). The option when storage tanks with finite capacity are available is the finite intermediate storage (FIS) policy. When the material is allowed to hold inside the vessel, the policy is known as non-intermediate storage (NIS) policy.

In a batch plant, all the stages have their own processing time and their own maximum possible batch size. The actual size of a batch being processed in a plant depends on the capacity of a size limiting stage, and the time elapsed between the processing of consecutive batches cannot be shorter than the required by a time limiting stage. The limiting cycle time for a product corresponds to the longest processing time among all the stages, and determines the time between two successive batches. There are two types of bottlenecks: size and time. Units can be added at a stage in order to debottleneck the plant. These units can be operated in-phase or out-of-phase.

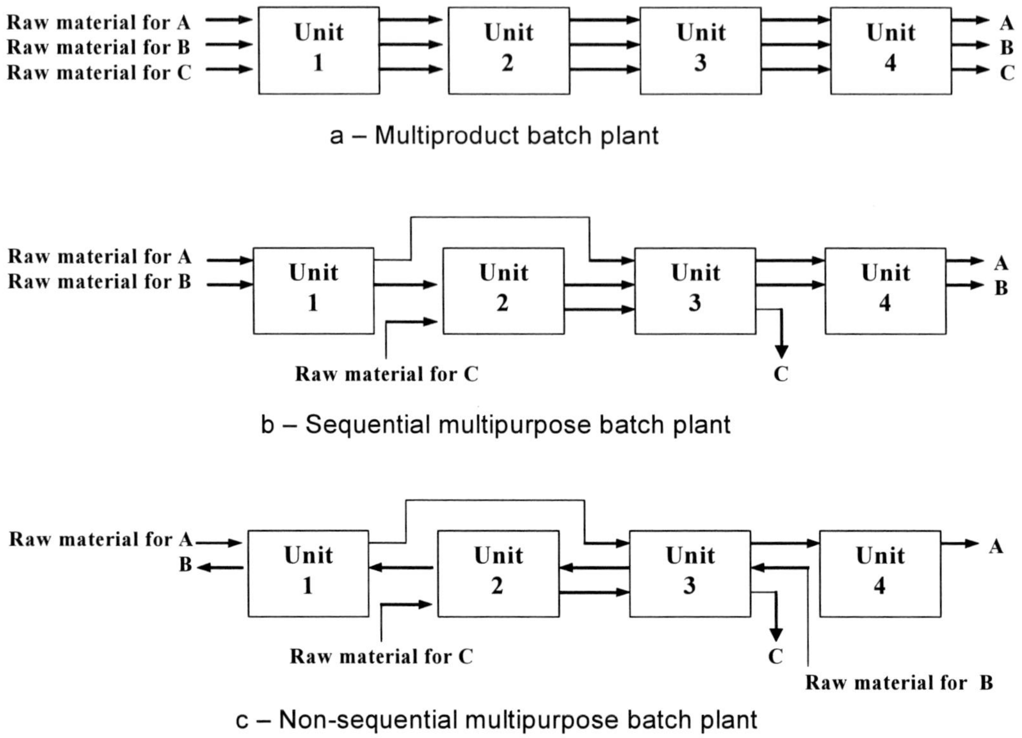

Figure 1.1. Batch plant classification.

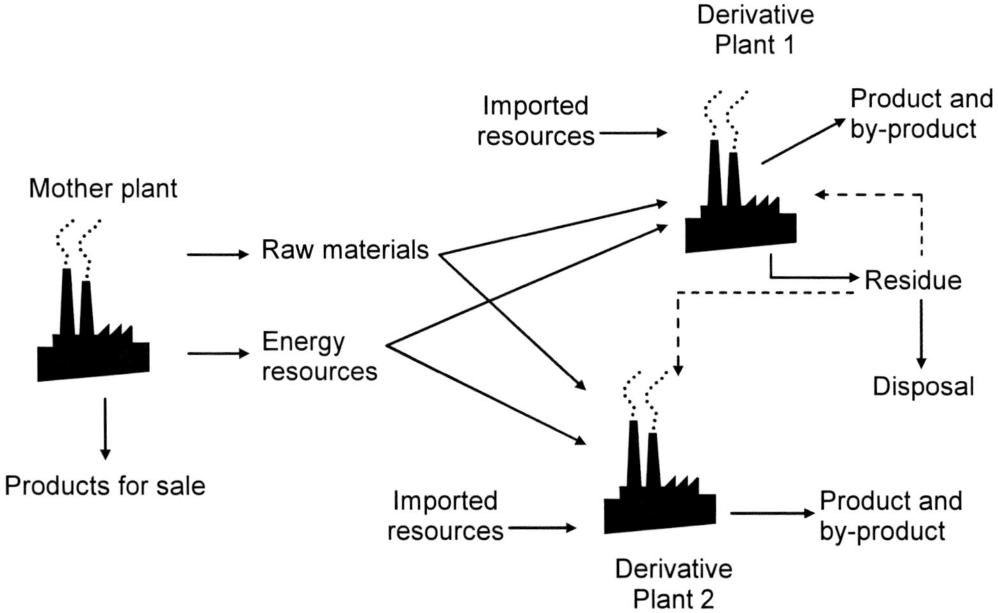

Figure 1.2. Multiplant complex scheme integration.

Parallel units operating out-of-phase are added to remove a time bottleneck in a series of batch stages. With two units operating out-of-phase the stage cycle time is halved compared with only one unit. If a stage is time limiting, then adding parallel units out-of-phase will reduce the limiting cycle time so that the equipment utilization of the remaining units is increased, and thus the required unit sizes are reduced. Figure 1.3 shows the out-of-phase duplication of a stage.

Parallel units operating in-phase are added to remove capacity bottlenecks. The batch from the previous stage is split and assigned to all the in-phase units of that stage. The units process all the material at the same time and upon completion the batches are mixed together and transferred to the next stage. This does not affect the limiting cycle time but the largest batch size that can be processed in the plant is increased.

A batch unit is fed with the material that arrives from the previous unit. Some batch stages require an extra feeding, e.g. the addition of substrate for fermentation stages. For these cases an interesting optimization option is unit duplication in series, where successive equal or not-equal units operate sequentially in order to fulfil certain final product specification. This type of duplication cannot be generalized since it depends on the processing variables. Duplication in series will be treated in this work as a novel process optimization decision.

Optimization Model Decisions: Synthesis, Design, Operation, Scheduling and Planning

In an optimization problem (OP), a decision is made aimed at the maximization or the minimization of a related performance criterion such as elapsed time or cost, by exploiting certain available degrees of freedom under a set of constraints. OPs arise in almost all branches of industry, e.g. in product and process design, production, logistics and even strategic planning. While the word optimization, in non-technical language, is often used in the sense of improving, the mathematical optimization community sticks to the original meaning of the word related to finding the best solution either globally or at least in a local neighbourhood. The OPs presented in this work consider decision making in the synthesis, design, operation, scheduling and planning of production plants.

In the process synthesis the structure of the plant is determined, satisfying a set of constraints and specifications. That is, once the problem inputs (feed volume and composition, for example), tasks and desired products have been specified, the synthesis problem determines the optimal selection of equipment and their interconnections that transform the inputs into final products.

The design problem deals with the sizing of equipment and storage units. This optimization can be done with size factors and operating times as constant parameters in the simplest models or as process dependent variables in more detailed models.

The operation decisions deal with finding the optimal value for process variables as operating times, batch sizes, concentrations, yields, etc. In the simplest models these decision variables are involved in linear equations or algebraic models that describe the units performance (Salomone and Iribarren, 1992). In more detailed models, this kind of decisions is described through differential equations (Bhatia and Biegler, 1996; Corsano, et al., 2004).

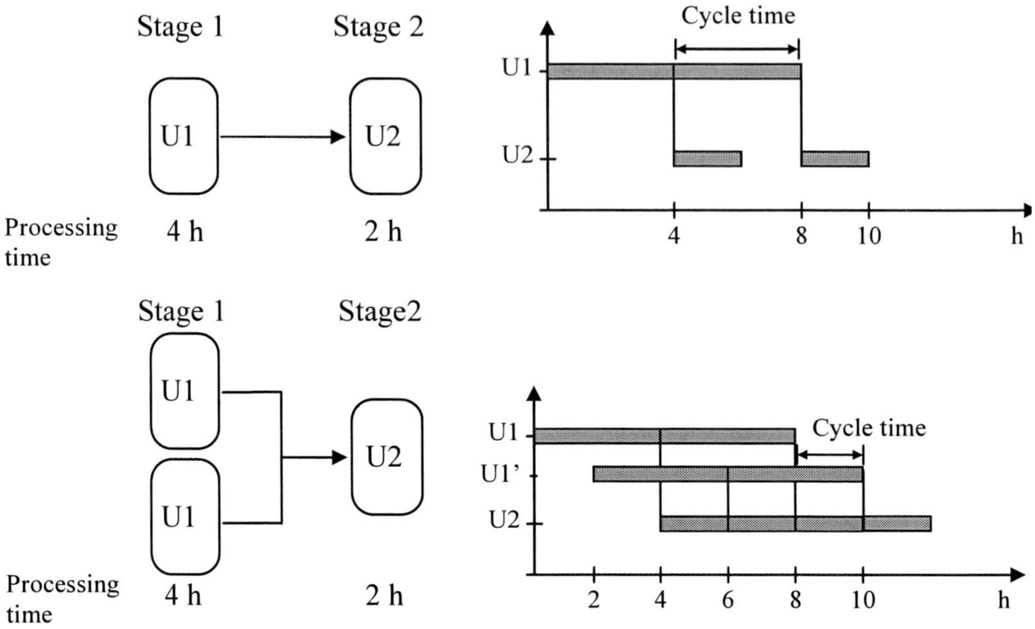

Figure 1.3. Out-of phase duplication. Up, original plant. Bottom, plant with two units in parallel out-of-phase in stage 1.

Production planning determines the optimal allocation of resources within the production facility over a time horizon of a few weeks up to a few months, whereas short-term scheduling provides the feasible production schedules to the plant for day to day operations. Since the boundaries of planning and scheduling problems are not well established and there is an intrinsic integration between these decision making stages, there is a lot of work in the literature addressing the simultaneous consideration of planning and scheduling decisions (Wu and Ierapetritou, 2007).

In multiproduct and multipurpose plant, where two or more products are handled, an important decision to be modelled is the sequence or order in which the products are produced. In single product campaign models all the batches of a product are processed without overlapping with others products. On the other hand, for mixed product campaign models a production sequence has to be determined and the campaign is repeated over the horizon time. From the point of view of the mathematical model, mixed product campaigns raise greater challenges than the formulations for single product campaigns.

Table 1.1 lists the most important synthesis, design, operating and scheduling/planning decisions and the related objective functions in order to have a brief overview of the model complexity and characteristics.

Mathematical Formulations

Mathematical models for optimization usually lead to structured problems such as:

- linear programming (LP) problems,
- mixed integer linear programming (MILP) problems,
- nonlinear programming (NLP) problems, and
- mixed integer nonlinear programming (MINLP) problems.

When the objective function and constraints of the model are linear and the optimization variables are continuous, then the problem is a LP formulation.

If some decision variables are integer, and the model has linear objective function and constraints, then the problem is a MILP formulation.

If the objective function or some constraints are non-linear functions, and the decision variables are continuous, then the problem is a NLP formulation. Finally, if in the last case some variables are integer, the model is formulated through a MINLP programming.

Table 1.1. Problem decisions and objectives

	Synthesis and Design	Operation	Scheduling/Planning
Decisions	Plant configuration Resources utilization Number of units in series Blend and recycle allocation Unit sizes In parallel units duplication Intermediate storage tank allocation Heating and cooling areas Power consumption (vapor and electricity)	Batch blending, batch splitting and batch recycling flow rates Material and energy resources allocation Component concentrations Unit processing times	Cycle time of each production process Units cycle time Units idle time Number of batches Mixed campaign configuration Production
Objectives	Minimizing the investment cost	Minimizing the operative cost Minimizing the utilities costs	Minimizing the makespan Maximizing the profit Minimizing the inventory cost

Literature Review

Since the 70s decade, many authors have contributed with new algorithms and formulations for the mathematical modeling of diverse process optimisation decisions. In general each one focused on a different problem, context or structural alternative in such a way that almost all the relevant alternative structures and options in the batch processes design and operation problems can be considered. However, this progress has occurred as extensions of individual works and not as complete reformulations in the sense of including all the possibilities or at least a significant set of alternatives and decisions. There are no models or formulations in the literature that permit consider and select among all the possible structures and alternatives simultaneously. This is so due to the great effort that the simultaneous resolution of the problems involves. However, in this way, the trade-offs among decisions cannot be assessed. Following, the main references about mathematical modeling in batch processes optimization are cited.

There is abundant literature on batch process synthesis and design. The models range from the simplest formulations as "fixed time and size factors" (Knopf et al., 1982; Yeh and Reklaitis, 1987; Ravemark and Rippin, 1998; Montagna et al., 2000) to the more complex as dynamic models (Bhatia and Biegler, 1996; Corsano et al., 2004, 2006a).

The process performance models are additional algebraic equations describing the time and size factors as functions of the units' process variables. A more detailed description of the performance of batch stages requires that they be modelled with differential equations. First Barrera and Evans (1989) and later Salomone et al. (1994) proposed that this simultaneous optimization should be approached by integrating the batch plant model with dynamic simulation modules for the batch units. Another way of incorporating the units' dynamic models is to discretize the differential equations to convert them into algebraic constraints of the program. This was the approach used by Bhatia and Biegler (1996) for simple process examples, and by Corsano et al. (2004, 2006) for detailed integrated processes.

Several authors have incorporated scheduling constraints into the synthesis problem of multiproduct and multipurpose batch plants. Again, the simplest models are those that take fixed time and size factors (Birewar and Grossmann, 1989) which are expressed through NLP when structured decision are not considered and MINLP otherwise. More detailed formulations where nonlinear task models (processing time, utility usage, unit availability) and nonlinear capital cost functions are considered, a nonconvex MINLP problem will arise (Zhang and Sargent, 1994, 1996). Therefore, the model is very large in size and difficult to solve. In these cases, decomposition approaches are used in order to find "good solutions" (Barbosa-Povoa and Pantelides, 1999).

Whereas significant development has been made in the design, planning, and scheduling of batch plants with single or multiple production routes, the problem of synthesis, design, and operation in batch multiplant complexes has received much less attention. Previous works have been focused on specific decision levels. Tools and models that solve separately the different aspects of a process have been developed. However, to obtain a good preliminary design of a production process, it is important to consider design, synthesis, operation, and scheduling simultaneously. There are few works in the literature dealing with multiplant complex integration (Lee et al., 2000, Kallrath, 2002, Jackson and Grossmann, 2003).

Work Outline

In the following sections, three problems are presented, emphasizing the detailed mathematical formulation and results analysis in each case. In Section 2 a superstructure optimization model is presented, that solves the synthesis, design and operation problem for batch plants through a NLP formulation. In Section 3 a heuristic method is presented for finding the optimal mixed product campaign configuration of multiproduct/multipurpose batch plants. The planning problem is simultaneously solved with the design and operation problem. The third problem deals with the optimal processes integration in a multiplant complex. Different aspects of design, operation, scheduling and planning are analyzed according to diverse operational conditions and environmental issues. MC scenario solutions are compared with multiproduct scenario solutions for a specific industrial studied case. Finally, a summary of the contributions of this work are outlined.

2. NLP Superstructure Modeling for the Optimal Synthesis, Design and Operation in a Batch Plant

2.1. Introduction

The synthesis and design problem of a batch process plant determines the plant structure, the number of units to be used at each stage and the unit sizes. Previously published works on this area resorted to Mixed Integer Non-Linear models (MINLP) to solve these problems. Binary variables allowed contemplating the allocation of units and the different alternatives to organize them at each stage. The various models were characterized by a certain number of operations, which are the necessary steps to elaborate the product according to the previously settled recipe. This problem was initially modelled with this format by Grossmann and Sargent (1979).

All previous works on this area start from a plant where the number of stages is settled by a decision made in a previous step, taking into account that synthesis and design problems are separately solved. Thus, the only structural decision that usually remains is the one related to unit duplication at each previously determined stage. However, there are many operations that pose alternatives as regards the number of stages to be used that cannot be considered with this approach. For example in the case of fermentation, depending on reaction rate, equipment cost, or raw material availability, the number of units to be used and the way in which they should operate (in series or in parallel) may vary. Then these options should be considered in relation to the sizing decisions. As illustrated with this example, a strong relationship exists among the number of stages, the way in which each stage is configured and the operative characteristics of stages (fermentors feeding, processing times, substrate concentration, among others), which has not been posed yet. This is due to the difficulties in solving a problem in which all these elements are optimization variables.

This work intends to solve the aforementioned problem. Firstly, a model with a high level of detail is posed. Operations have been represented through discretized differential equations that describe mass balances. Furthermore, constraints on feeds to each processing unit and equations of interconnections between stages are considered. This level of detail has been posed by few authors. Some exceptions that can be mentioned are Bhatia and Biegler (1996), even though with a simpler model since they work with a predetermined number of stages and they do not admit units duplication.

The option of optimizing the number of stages in series to be considered in this work has not been included in previous general models of batch plants design. The plant structure problems have been solved in previous synthesis models resorting to binary variables. It should be stressed that in many cases, when the level of detail of the operations included in the process was significant (Bhatia and Bieglier, 1996; Barrrera and Evans, 1989) this last option was not even considered. As previously pointed, this formulation leads to difficulties or limits the capacity for solving the model.

This work optimizes the plant superstructure in order to model this problem. This is a systematic method for the synthesis of chemical process networks. For the given raw materials and final products, there are a lot of feasible production paths that can produce the desired products. In a superstructure optimization model, all the possible alternatives for the process are included in the overall model. In order to obtain the optimal solution, the resolution procedure selects the optimal production alternative among the available options.

This work proposes to explicitly include a superstructure that contemplates all possible options for the plant structure with units in series or in parallel. These alternatives can be obtained by means of different mechanisms. Corsano et al. (2004) presented a heuristic procedure through a simplified optimization model that provides an upper bound for the number of stages of each operation. Another option is to pose an exhaustive enumeration of all the alternatives that arise from the upper bounds for the units in series or in parallel. The designer's criteria and his or her experience are also critical to bound the superstructure size by taking into account only the most expensive units with a significant impact on the process performance.

An original approach is developed to solve the model where discrete alternatives are selected without resorting to binary variables. Taking into account that the model is a non-convex NLP due to the equations that describe the plant operations, difficulties of resolution methodologies for MINLP programs are avoided. This kind of models is usually solved through methods like the outer approximation algorithm (Duran and Grossmann, 1986; Viswanathan and Grossmann, 1990; Varvarezos et al., 1992) where a sequence of NLP and MILP problems is solved. The available MINLP algorithms can miss the optimal solution due to the application of linear cuts to non-convex regions. On the other hand, our optimisation approach finds the optimal solution by modeling a superstructure with as options as proposed by the designer. The optimal solution will eventually eliminate units of the network taking their size to be zero. In this way we solve the model as a NLP problem instead of a MINLP problem. With this approach, the user can provide physically meaningful initializations increasing the robustness and usefulness of the optimization models. Available computer codes for solving MINLP are of general purpose and do not use initial solutions as those proposed in this work. Finally, the major advantage of this strategy is that all the alternatives can be explored explicitly because their number is relatively low. Therefore, the work presents a problem representation which is extremely compact and allows for taking into consideration a significant number of alternatives.

The developed model is presented working on the case of a fermentors network. It is a typical case for a number of reasons. First of all, it is often present in agro-industrial and biotechnological plants. In addition, it is necessary to represent the process with a high level of detail, due to its high economic impact on the plant cost. Taking into account the operation characteristics, it is required to duplicate stages both in series and in parallel. On the other hand, the number of options to be contemplated is bounded by operative considerations. It is important to mention that this approach could be easily extended to other kind of reactions. In the next section, a general model for a single product plant is presented. The model can be easily extended to the multiproduct case where many products are made with the same equipment. Then, its application to the case of a fermentors network is described. The resolution of this example allows assessing the potentiality of the proposed approach. Finally, it is compared to a traditional formulation in which the structural decisions of the plant are represented by means of binary variables.

2.2. Model Formulation

We consider a plant that produces only one product and must meet a certain demand Q for that product on the available time horizon H.

To complete the product processing, P operations are required. Each operation is accomplished over several stages j, whose optimal number has to be determined. For each operation p, there is an upper bound C_p on the number of stages to be contemplated for this operation. In this way, the number of stages to be considered for each operation can be modified. Therefore, for each operation p, there is a set of stages j ranging from 1 to C_p, whose utilization must be determined as a solution to the optimization problem.

It should be pointed that this approach is more realistic when the operation of each stage can be represented by means of a detailed model and is not fixed by a size factor as in the first examples referred to in the introduction section. In this way, it is possible to take into account the different tradeoffs that arise when considering different operative conditions. Therefore, the model to solve the batch operations design becomes more appropriate when the description of the stages operation is explicitly contemplated.

For each operation p, alternatives $a = 1, ..., A_p$ are defined. These alternatives can be either automatically generated (through an optimization model, for example) or proposed by the designer, which is more effective taking into account the feasible options for the kind of process they are working with. Each existing alternative a in operation p must be characterized. This implies defining the following elements:

- Number of stages to be included in the alternative.
- Determining the last stage being included in the alternative (basic information to allow for connection between successive operations).
- Number of in-phase (G_{paj}) and out-of-phase (M_{paj}) duplicated units for each stage included in the alternative.
-

The existing j stages in alternative a of operation p may vary between 1 and C_p. For each alternative, the number of stages is predetermined. Each one of these alternatives has structural options due to the duplication of the units included in it. These options are predetermined in each alternative a.

The transfer policy considered in this work is the Zero Wait (ZW) transfer. A stage-configuration option is in series duplication. In this case, the cycle time (problem variable) for a plant is determined as the longest time of all the stages over each operation involved in the production process. This cycle time settles down the time between two successive batches. Therefore, all the units that require less operating time have some idle time. Aiming at reducing this time, out-of-phase parallel units can be included into the stage in which the cycle time is reached. These units operate out-of-phase, thus allowing for reducing the time spent between two successive batches, and consequently the remaining units size is reduced as a result of having less idle time.

The other configuration option for units duplication is in-phase, in which duplicated units operate simultaneously. In this case, when the batch enters that stage, it is split among all units belonging to the stage, and when finishing the operation the exiting batches are added together. In this way, the processing capacity of a stage can be increased, which is important when the unit size reaches the upper bound.

Figure 2.1 shows an example. An operation (P = 1) has C_p = 3 stages, which indicates that any alternative being used in this operation can have at most 3 stages. The problem designer has settled that the A= 8 alternatives shown in Figure 2.1 should be taken into account. The j stages used in the corresponding a alternative are also indicated. The first alternative includes only one stage, whereas the following two have additional stages in series. The other alternatives included out-of-phase duplicated units for various stages, except for the last one (a= 8), which is the only stage in which in-phase duplicated units are considered. This is shown through overlapped units in Figure 2.1.

The model looks for a plant design that allows producing the required quantity Q in time horizon H at the lowest cost. This general presentation takes into account unit costs and operative costs. The objective function is Total Annual Cost (TAC) minimization and is calculated from the following expression:

$$Min \sum_{p=1}^{P} \sum_{a_p=1}^{A_p} \sum_{\mathrm{j_{pa}} \in a_p} \alpha_{\mathrm{p}} M_{paj} G_{paj} V_{paj}^{\beta_p} + OC \tag{2.1}$$

V_{paj} is unit j size in alternative a for operation p. Its cost is calculated from coefficients α_p and β_p that are usually used in this kind of problems (Ravemark and Rippin, 1998). M_{paj} and G_{paj} correspond to the number of out-of-phase and in-phase duplicated units, respectively, for stage j in alternative a for operation p and they are provided by the designer. OC represents operating costs that depend on how each operation is performed and thus it cannot be represented through a general expression.

In the previous expression, all stages j of all existing alternatives for operation p are considered. Taking into account that the objective function minimizes the units cost, only the best structural option will be chosen, driving to zero the size of all units that are not involved in the optimal structure. The simultaneous operation of two structures will always involve a greater cost taking into account that the exponent coefficient β_p is less than one. For that reason, the unit sizes of non-optimal alternatives will be equal to zero.

According to this problem formulation, a set of constraints are developed. First of all, the required demand Q should be satisfied. For that purpose, the plant production rate PR is employed, which is given by:

$$PR = \frac{Q}{H} \tag{2.2}$$

At the last stage of the last operation, the final product is obtained. The sum of the productions of all the defined alternatives must meet the production requirement for the plant:

$$\sum_{a \in p_{last}} PR_{p_{last}, aj_{last}} \geq PR \tag{2.3}$$

where p_{last} corresponds to the last operation of the process and j_{last} to the last stage in each option a. Therefore, $PR_{p_{last}, a}$ represents the production rate achieved at each alternative a in the last operation.

The total amount produced should be at least equal to the plant requirement. Since the model tries to minimize costs, the quantity to be produced will be just PR, and will be reached by using only one alternative in operation p_{last}. This will be so because in case of using two alternatives it will be necessary to use equipment for both of them, which would notably increase the cost.

It is required to determine the plant cycle time TL. This is determined by the longest time required in the stages being used at the plant. Let T_{paj} be the unit operation time at stage j for alternative a in operation p. This value is calculated from the model that describes that operation. Then, considering ZW policy, it should be:

$$TL \geq \frac{T_{paj}}{M_{paj}} \quad \forall p = 1, \ldots, P, a = 1, \ldots, A_p, j = 1, \ldots, C_p \tag{2.4}$$

M_{paj} corresponds to the number of out-of-phase duplicated units that exist at stage j of alternative a in operation p.

Connection balances should be accounted for between successive stages of each alternative of an operation. Let B_{paj}^{ini} and B_{paj}^{fin} be the batch volume that enters and leaves the unit of stage j in alternative a in operation p; then the balances are:

$$B_{paj}^{ini} = B_{pa,j-1}^{fin} \quad \forall p = 1, \ldots, P, a = 1, \ldots, A_P, j_{pa} \geq 2 \tag{2.5}$$

In case of handling several material streams, this kind of connection constraint should be settled for each of them. As will be seen in the example, this balance can also consider adding extra feeds at each stage.

Connection between successive operations must be also assured. For that reason, the last stage of an operation must get in contact with the first stage of the following operation:

$$\sum_{a_p \in p} B^{fin}_{pa,j_{last}} = \sum_{a_{p+1} \in p+1} B^{ini}_{p+1,a,1} \quad \forall p = 1,\ldots,P-1 \tag{2.6}$$

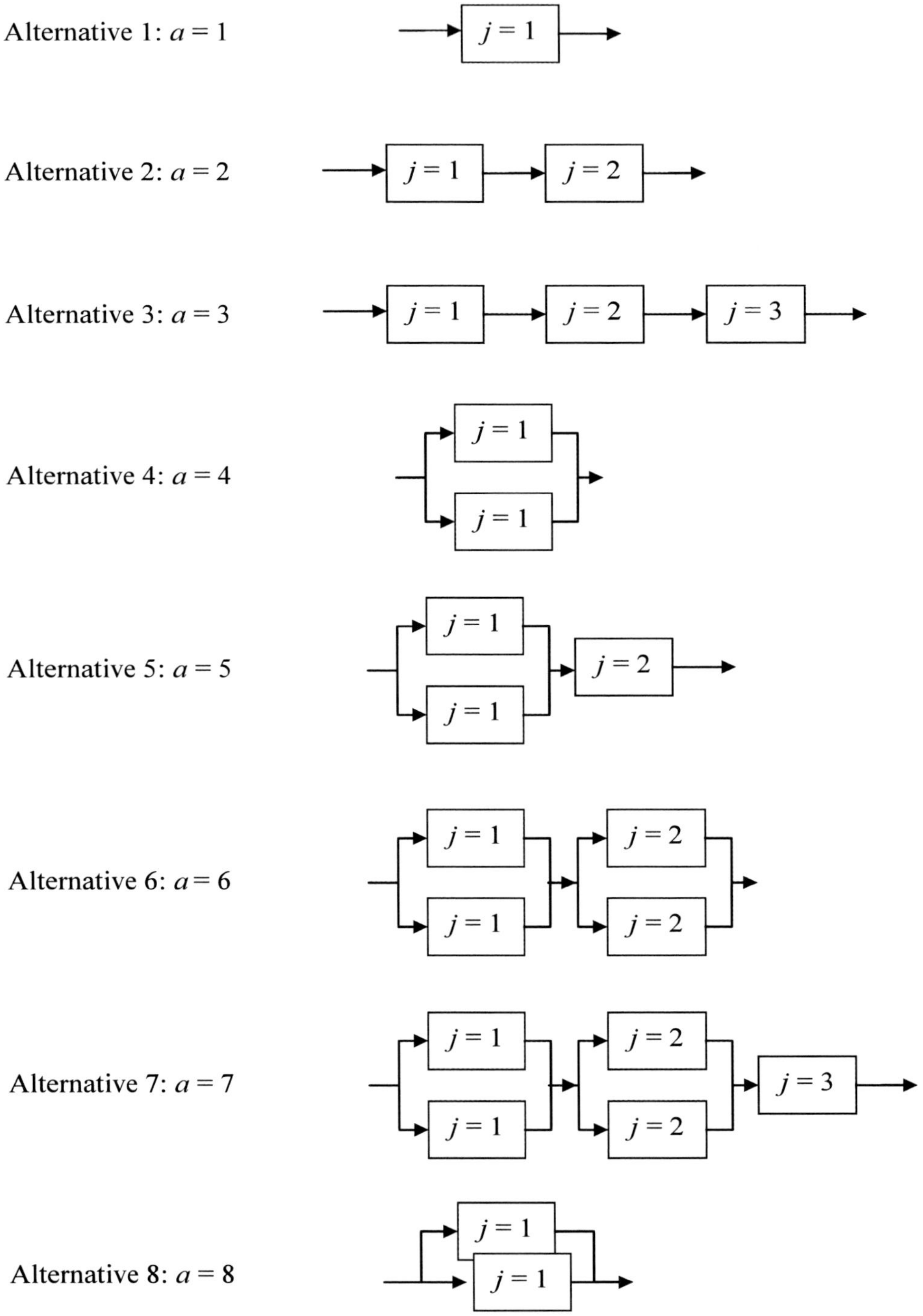

Figure 2.1. Alternatives to be considered for a three-stage operation.

In this case, the total material exiting the last stage j_{last} of all alternatives of operation p, must be equal to all the material entering the first stage of all alternatives of the following operation.

The B_{paj} and T_{paj} values being used must be characterized through appropriate equations. Also the material streams to be considered could be decomposed into several components (substrate, biomass, product, etc.) as it will be shown in the following example. The existing relationship among the material to be processed, the equipment sizes and the time that will be required for processing, arises from the model to be used for describing the involved operation.

Therefore, there is a set of constraints that closely depend on the characteristics of the process being used, and thus it is not possible to formalize them with a general format.

2.3. Fermentation Process for Ethanol Production

In this example corresponding to fermentation for ethanol production, the previously described model is applied to a specific case. The detailed models describing each unit operation are introduced next.

Fermentation for ethanol production consists of two operations, namely: biomass fermentation and ethanol fermentation. At the first operation, only biomass is produced, while at the second both ethanol and biomass are produced, even though the latter is produced at a rate that is lower than that of the previous operation. All stages of both operations can be fed with a mixture of sugar substrates that provide different substrate concentrations. Water can be also added to dilute these substrates. Mass balances of these stages are described by the following differential equations:

$$\textit{Biomass:}\quad \frac{dX_{paj}}{dt} = \mu_{paj} X_{paj} - \upsilon_{paj} X_{paj} \tag{2.7}$$

$$\textit{Substrate:}\quad \frac{dS_{paj}}{dt} = -\frac{\mu_{paj} X_{paj}}{Yx_{paj}} \tag{2.8}$$

$$\textit{Non-Active Biomass:}\quad \frac{dX_{paj}^{dead}}{dt} = \upsilon_{paj} X_{paj} \tag{2.9}$$

$$\textit{Product:}\quad \frac{dE_{paj}}{dt} = \frac{\mu_{paj} X_{paj}}{Ye_{paj}} \tag{2.10}$$

$$\text{where } \mu = \mu_{\max,paj} \frac{S_{paj}}{ks_{paj} + S_{paj}} \tag{2.11}$$

where (2.7) - (2.11) are described for $p = 1, 2$ (biomass and ethanol fermentation respectively); $\forall a_p \in p$; $\forall j_{pa} \in a_p$, X is biomass concentration, S is substrate concentration, X^{dead} is non-active biomass concentration, E is ethanol concentration. For this reason, equation (2.11) is not included in the first operation and μ is growth specific velocity. All these are problem variables. υ represents bacteria death rate, *Ye* is ethanol yield coefficient and *ks* is a substrate saturation constant. These are known parameters for the model. *Yx* is biomass yield and is a function of the feed composition in biomass fermentors and a constant in the ethanol fermentation operation (Corsano et al., 2004).

These equations have been discretized using the trapezoidal method and included in the global model. This model also contains all the constraints presented in the previous section: connections between stages of each alternative of each operation, connections between the last stage of an operation and the first stage of the following operation, constraints that define the time cycle, constraints to meet production requirements, and a set of balances that are similar to those given in equations (2.5) and (2.6) that is performed for each component: biomass, substrate, non-active biomass and ethanol. For example, substrate balances between successive stages are expressed by:

$$V_{paj} S^{ini}_{paj} = \sum_{f \in Feed_{paj}} SF_f V_{f,paj} + V_{pa,j-1} S^{fin}_{pa,j-1} \quad \forall p = 1,2; \forall a \in A_p; \forall j \geq 2 \tag{2.12}$$

where S^{ini}_{paj} represents substrate concentration entering stage j of alternative a of operation p, f are the various materials that constitute the feed for stage j and those that belong to the set *Feed*. In this example we took *Feed*= {*molass, filter juice, vinasses, water*}, each one of them having a substrate concentration that is equal to SF_f and volume equal to $V_{f,paj}$.

$S^{fin}_{pa,j-1}$ is the output substrate concentration of unit j-1 of alternative a. It should be noted that in this case, besides the material coming from the previous stage, other materials from other sources represented by the set *Feed* are also allowed to enter. In a similar way, the balances for the remaining elements and interconnection balances between operations are posed.

Since variable feeds are considered, volume balances have been added:

$$V_{paj} = \sum_{f \in Feed_{paj}} V_{f,paj} + V_{pa,j-1} \quad \forall p = 1,2; \forall a \in A_p; \forall j \geq 2 \tag{2.13}$$

The posed objective is minimizing total annual costs, which are computed as investment cost (given by equipment cost) in addition to operating costs. In this specific case, variable *OC* of expression (2.1) can be posed as shown by expression (2.14). This operating cost is the sum of the cost per m^3 of sugar substrates being used in feed f to stage j of alternative a of operation p. Let γ_f be cost per m^3 of the sugar substrate f being used in feed *Feed*, then the total annual cost can be computed as:

$$Min \left\{ C_{ann} \sum_{p=1}^{P} \sum_{a_p=1}^{A_p} \sum_{j_{pa} \in a_p} \alpha_p M_{paj} G_{paj} V_{paj}^{\beta_p} + \frac{H}{TL} \sum_{p=1}^{P} \sum_{a_p=1}^{A_p} \sum_{j_{pa} \in a_p} \sum_{f \in Feed} \gamma_f V_{f,paj} \right\} \quad (2.14)$$

2.4. Example Resolution

The fermentation model for ethanol production established in the previous section has been solved. Three examples will be presented with different sets of data with the aim of evaluating the optimal design of the plant according to various problem conditions.

This is a problem with $P=2$ operations. For both first examples, the chosen superstructure is shown in Figure 2.2. This figure includes the diverse alternatives that were selected by the designer for each one of the two operations performed at the plant. It should be noted that this superstructure allows considering the combination of the different alternatives that are chosen for each operation.

In the first operation, there are $C_1=2$ stages, so any alternative can have up to two stages. Figure 2.2 also includes $A_1=2$ alternatives that are taken into account in this operation. In the first one, there is only one stage, while in the second one there are two stages in series. For the second operation, $A_2 = 4$ alternatives of up to $C_2 = 2$ stages are considered. In this case, the first alternative consists of only one stage, the second one consists of two stages in series, the third one is the out-of-phase in parallel duplication of the first stage and the last one is the out-of-phase triplication of the first stage. The selection of alternatives for these operations is assumed to be based on the knowledge that the designer has about the problem, regarding its feasibility from an engineering point of view. In this case, as the reaction rate of the biomass fermentation operation is faster than that of the ethanol fermentation operation, the processing time of the first operation is lower than that of the second and therefore the option of in parallel stages duplication in not included for the first operation.

It is necessary to characterize each of the alternatives. Table 2.1 describes the elements of each alternative for biomass and ethanol fermentation. It indicates the number of in-phase and out-of-phase duplicated units at each stage of each alternative, and the last stage of each alternative. This information is used for balances between stages and, in the case of the last one, for determining the process production.

The models have been implemented and solved in GAMS (Brooke et al., 1998) in a CPU Pentium IV, 1.60 Ghz. CONOPT 2 code was used to solve the NLP problem.

The model parameters values for the following examples are shown in Table 2.2.

As a first example, the model of the previously presented superstructure is solved by setting the equipment costs exponent β at 0.43 for both operations. The optimal configuration for producing 100 kg/hr of ethanol corresponds to the first stage of operation 1 and the first stage of operation 2, i.e., alternatives $a = 1$ for $p = 1$ and $a = 1$ for $p = 2$. Figure 2.3 shows the optimal solution. The values of some process and design variables are found in Table 2.3. Sub-index "*final*" of some variables denotes the value of the variable at final time (value corresponding to the last point of discretization). The cycle time of the plant is 16.4 h and the Total Annual Cost is $287,865.

In the second example, the production rate is increased to 500 k/h. The best alternative consists of using a biomass fermentor and two ethanol fermentors in series.

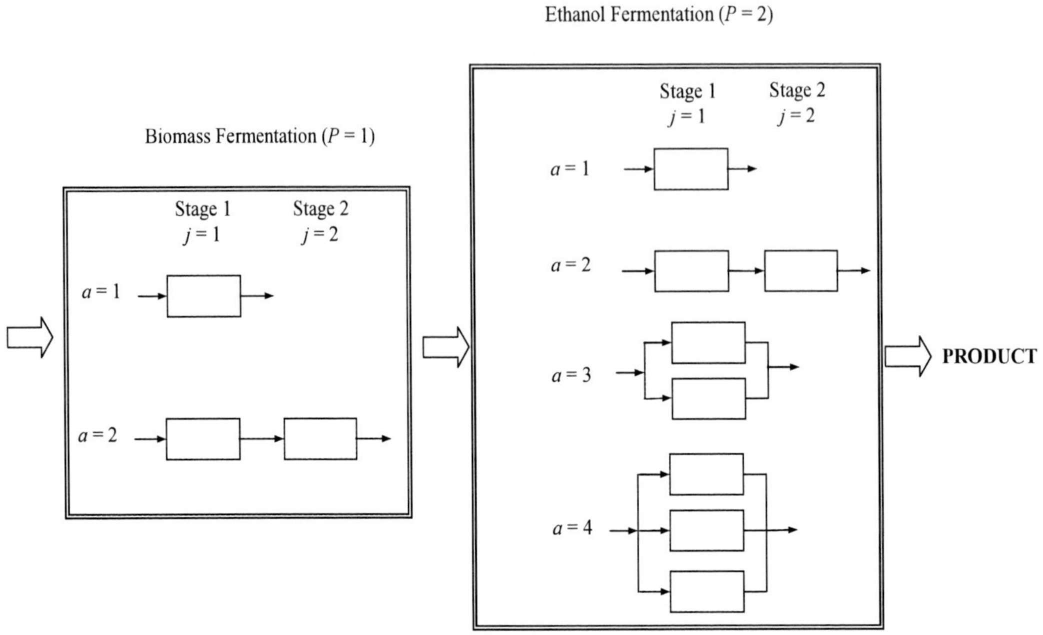

Figure 2.2. Superstructure for the ethanol fermentation model in examples 1 and 2.

Table 2.1. Description of the alternatives of ethanol superstructure for the proposed examples

	Operation: Biomass Fermentation			Operation: Ethanol Fermentation		
	Stage 1	Stage 2	Last Stage	Stage 1	Stage 2	Last Stage
Alternative 1	$M_{111} = 1$	$M_{112} = 0$	1	$M_{211} = 1$	$M_{212} = 0$	1
	$G_{111} = 1$	$G_{112} = 0$		$G_{211} = 1$	$G_{212} = 0$	
Alternative 2	$M_{121} = 1$	$M_{122} = 1$	2	$M_{221} = 1$	$M_{222} = 1$	2
	$G_{121} = 1$	$G_{122} = 1$		$G_{221} = 1$	$G_{222} = 1$	
Alternative 3	NO	NO		$M_{231} = 2$	$M_{232} = 0$	1
				$G_{231} = 1$	$G_{232} = 0$	
Alternative 4	NO	NO		$M_{241} = 3$	$M_{242} = 0$	1
				$G_{241} = 1$	$G_{242} = 0$	

Figure 2.4 shows this solution and Table 2.3 presents the values of some optimal design and process variables. The time cycle is 16 h and the Total Annual Cost is \$ 883,732.

In the third example we decreased the equipment cost exponent of the first operation (β_1) to 0.3 and increased the second ones (β_2) to 1. For this case, we changed the superstructure presented on Table 2.1. Table 2.4 shows the information about the superstructure for both biomass and ethanol fermentation operation, and Figure 2.5 shows this superstructure. The optimal solution consists of the out-of-phase in parallel duplication of the first biomass fermentor followed by one biomass fermentor and two ethanol fermentors in series. This solution corresponds to Alternative 5 of the first operation and Alternative 2 of the second one. Figure 2.6 shows this solution and Table 2.3 presents some of its optimal variables. In-parallel working equipment has the same operative and design characteristics (operation time, size, feeds, flows, etc.). The time cycle is 5.15 h and the Total Annual Cost is $526,822.

Table 2.2. Parameters used in the ethanol production model

Parameter	Value
$\mu_{\max,1aj}$	0.5 h^{-1}
$\mu_{\max,2aj}$	0.1 h^{-1}
α_p	115550
υ_{paj}	0.02 h^{-1}
Yx_{2aj}	0.124
Ye_{2aj}	0.23
H	7500 h $year^{-1}$
ks_{paj}	20 k m^{-3}

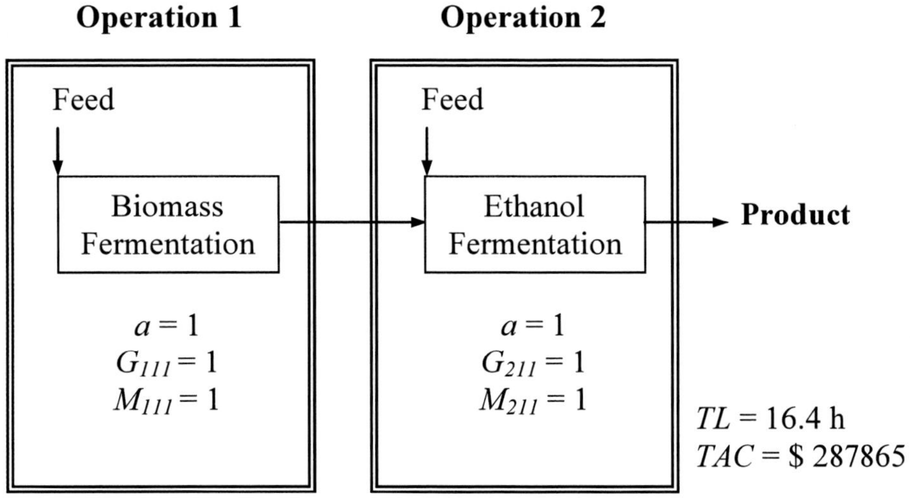

Figure 2.3. Optimal solution for $\beta_p=0.43$ and $PR=100$ KH^{-1}.

2.5. A Comparison with the Traditional Approach

A comparison will be made between the proposed approach, in which the different alternatives of the plant configuration are modelled without resorting to binary variables, and the traditional approach, in which the problem is represented through a MINLP program.

Firstly, it should be highlighted that traditional models do not solve this problem by considering in series stages duplication. Consequently, in order to perform a comparison, we assume that the number of plant stages is fixed, and thus we are facing a problem that is sensibly simpler than the one presented in this work. Therefore, the only way of comparing both approaches is solving a sequence of MINLP problems that contemplate all the alternatives. This could be made in this example because it includes a small number of operations and stages. In larger problems, this task can be extremely burdensome.

It is assumed that each MINLP model contemplates all the previously posed constraints. The difference lies in the fact that the number of stages at each operation is fixed, and the number of out-of-phase and in-phase parallel units for each stage is variable (M_j and G_j). In this case, sub-indexes p and a disappear and we only work with stages j that are included in the plant. M_{paj} and G_{paj}, which were parameters of each alternative, now become variables M_j and G_j, since it is intended to determine the number of units at each stage. All the previously posed equations remain.

Among the previously solved examples, the third example was chosen. The optimal solution obtained there had two biomass fermentation stages where the first one uses out-of-phase parallel duplicated units and two ethanol fermentation stages in series. In order to perform the comparison, four models are solved which contemplate the possible configurations using a predetermined number of units in series for each operation. For this example, up to two stages are used for each operation because this was obtained in the optimal solution of the NLP superstructure model. Then, the number of parallel units and size of each stage are to be determined.

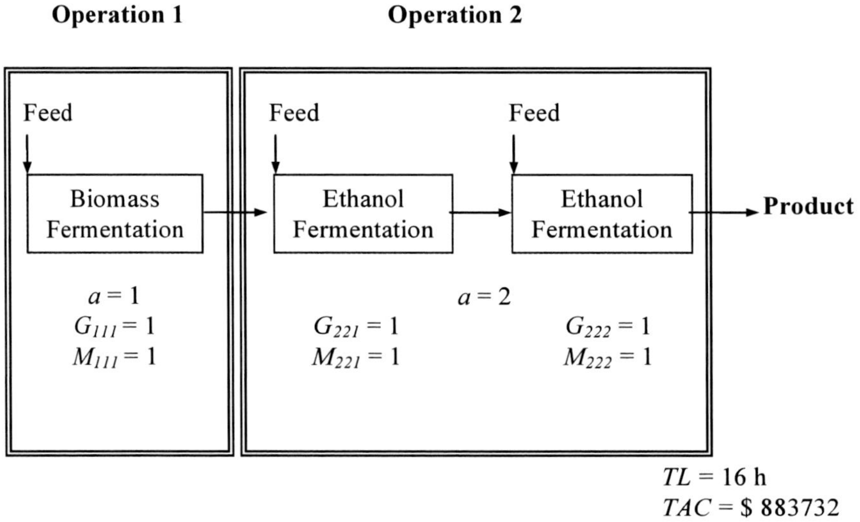

Figure 2.4. Optimal solution for β_p=0.43 and PR =500 kh^{-1}.

Table 2.3. Design and operating optimal solutions

	First Example		Second Example			Third Example			
	Biomass Fermentor	Ethanol Fermentor	Biomass Fermentor	Ethanol Fermentor 1	Ethanol Fermentor 2	Biomass Fermentor 1	Biomass Fermentor 2	Ethanol Fermentor 1	Ethanol Fermentor 2
Time (h)	16.44	16.44	16.02	16.02	16.02	10.33	5.16	5.16	5.16
Size (m^3)	6.72	31.89	11.29	57.84	98.09	4.07	4.17	4.67	5.33
$X_{initial}$ (k m^{-3})	0.1	7.45	0.1	6.74	8.6	0.1	5.75	31.86	34.94
X_{final} (k m^{-3})	35.38	15.39	34.53	14.58	16.43	5.88	35.73	39.8	42.83
X_{final}^{dead} (k m^{-3})	2.7	0.57	2.40	3.94	6.6	0.30	2.1	5.65	9.08
S_{final} (k m^{-3})	2.27	4.4	5.22	8.74	2.37	84.36	17.56	5.52	3.1
E_{final} (k m^{-3})	-------	51.54	-------	49.2	81.64	-------	-------	50.94	96.96

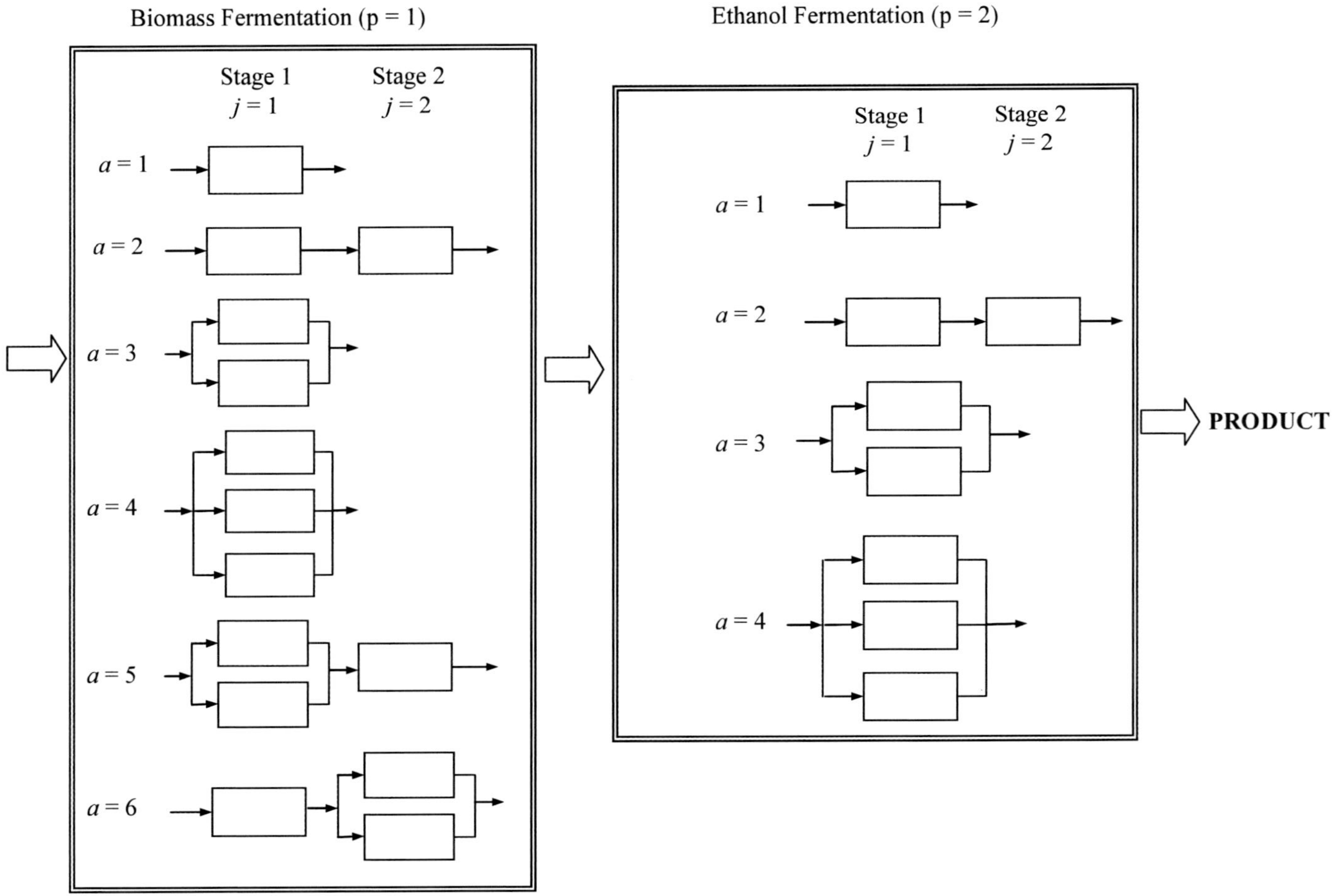

Figure 2.5. Superstructure for the ethanol fermentation model in example 3.

Table 2.4. Description of ethanol superstructure alternatives for example 3

	Operation: Biomass Fermentation			Operation: Ethanol Fermentation		
	Stage 1	Stage 2	Last Stage	Stage 1	Stage 2	Last Stage
Alternative 1	$M_{111} = 1$	$M_{112} = 0$	1	$M_{211} = 1$	$M_{212} = 0$	1
	$G_{111} = 1$	$G_{112} = 0$		$G_{211} = 1$	$G_{212} = 0$	
Alternative 2	$M_{121} = 1$	$M_{122} = 1$	2	$M_{221} = 1$	$M_{222} = 1$	2
	$G_{121} = 1$	$G_{122} = 1$		$G_{221} = 1$	$G_{222} = 1$	
Alternative 3	$M_{131} = 2$	$M_{132} = 0$	1	$M_{231} = 2$	$M_{232} = 0$	1
	$G_{131} = 1$	$G_{132} = 0$		$G_{231} = 1$	$G_{232} = 0$	
Alternative 4	$M_{141} = 3$	$M_{142} = 0$	1	$M_{241} = 3$	$M_{242} = 0$	1
	$G_{141} = 1$	$G_{142} = 0$		$G_{241} = 1$	$G_{242} = 0$	
Alternative 5	$M_{151} = 2$	$M_{152} = 1$	2			
	$G_{151} = 1$	$G_{152} = 1$				
Alternative 6	$M_{161} = 1$	$M_{162} = 2$	2			
	$G_{161} = 1$	$G_{162} = 1$				

This is obviously a much simpler problem but it is included here with the object of assessing the behaviour of this kind of highly non-convex models when handling binary variables explicitly.

The cases modelled with binary variables are:

(i) one biomass fermentation stage and one ethanol fermentation stage
(ii) one biomass fermentation stage and two ethanol fermentation stages
(iii) two biomass fermentation stages and one ethanol fermentation stage
(iv) two biomass fermentation stages and two ethanol fermentation stages.

The upper bounds for the M_j variables are shown in Table 2.5 for each previously mentioned case. On Table 2.6 the results for each studied case are shown, including the operations configuration, operating time, unit sizes, objective function value, number of constraints, number of continuous and discrete variables and CPU time to achieve the solution.

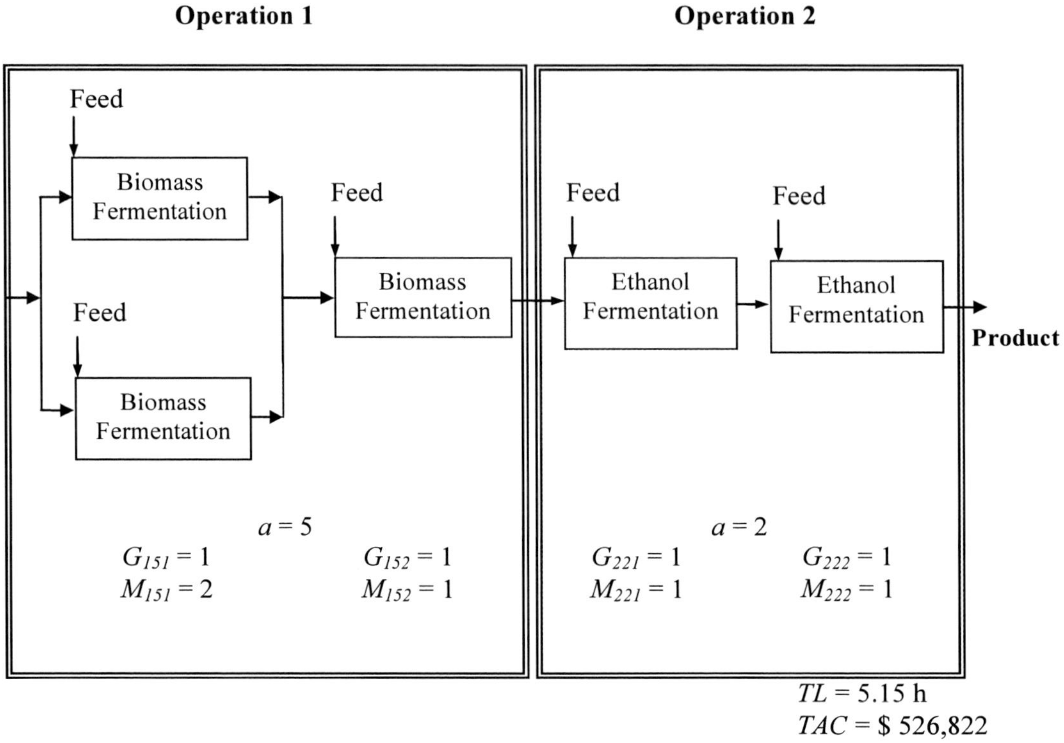

Figure 2.6. Optimal solution for $PR = 100$ kh^{-1}, β_1=0.3 and β_2=1.

The NLP superstructure model of Example 3 has 1707 continuous variables and 1617 constraints and the solution was achieved in 10.5 CPU seconds. As it can be observed, the solutions of the MINLP cases were obtained in a shorter CPU time, but in all cases the number of constraints and variables is smaller and the models are much simpler as it was previously mentioned. The solution of the model (iv) coincides with that obtained in the NLP superstructure model.

A comment on the CPU resolution time is that the differences are not so significant (of the same order of magnitude). Anyway this depends strongly on diverse factors, for example the initialization of variables. Nevertheless, it is necessary to emphasize that, ahead of the comparison of the resolution times, the MINLP model requires a greater effort for the generation of the different options. In this example only four options have been necessary, but in cases with more operations this number can be considerably large, because all possible combinations of alternatives must be accounted for. Another issue to be considered is the need of handling several models with different configurations, which leads to confusions.

2.6. Conclusions and Outlook on the Proposed Superstructure Modeling

This book section presented a novel NLP model for finding the optimal configuration of plants with several operations. The approach is relevant for cases where units can be duplicated either in series or in parallel (in-phase and out-of-phase). Furthermore, all possible options (or as bounded by the designer) are simultaneously considered. This formulation is most appropriate and useful when detailed operation models are posed. The resulting NLP avoids difficulties arising from resolution methodologies of MINLP problems applied to non convex programs.

The model was described as a set of general constraints, but also admits constraints specific for the operation to be designed and optimized. A detailed description for yeast and ethanol fermentation processes can be found in Corsano et al. (2004, 2006a).

For the particular operations on which this modeling technique was applied, the number of alternatives is limited. Indeed, in most cases it is not necessary to consider all the combinations among the various alternatives of each operation because the designer can judge which combinations are feasible for the process that is being optimized.

This model is simple to write, convergence and good solution are guaranteed in reasonable CPU time.

Table 2.5. Out-of-phase parallel units upper bounds for MINLP models

Cases	Upper bound for *M* in biomass fermentation		Upper bound for *M* in ethanol fermentation	
	Stage 1	Stage 2	Stage 1	Stage 2
(i)	2	---	4	---
(ii)	2	---	3	3
(iii)	2	2	4	---
(iv)	2	2	3	3

3. Synthesis and Design of Multiproduct/Multipurpose Batch Plants: A Heuristic Approach for Determining Mixed Product Campaigns

3.1. Introduction

In a multiproduct / multipurpose batch plant, several products are manufactured following the same or different production sequences, sharing the equipment, raw materials and other production resources. The inherent operational flexibility of multiproduct / multipurpose plants gives rise to considerable complexity in the design and synthesis of such plants. In many published case studies, scheduling strategies are not incorporated or well integrated. Usually the simplest scheduling sequence, single product campaign, is considered, which may lead to over-design.

Table 2.6. Optimal solutions of MINLP cases

	Case (i)		Case (ii)			Case (iii)			Case (iv)			
	Bio Fer	Etha Fer	Bio Fer	Etha Fer 1	Etha Fer 2	Bio Fer 1	Bio Fer 2	Etha Fer	Bio Fer 1	Bio Fer 2	Etha Fer 1	Etha Fer 2
Configuration	$M = 2$	$M = 1$	$M = 2$	$M = 1$	$M = 1$	$M = 2$	$M = 1$	$M = 1$	$M = 2$	$M = 1$	$M = 1$	$M = 1$
Time (hr)	14.46	7.23	14.2	7.1	7.1	10.64	5.32	5.32	10.32	5.16	5.16	5.16
Size (m^3)	12.6	13.88	5.84	6.39	7.28	7.76	9.15	10.29	4.07	4.17	4.67	5.33
TAC ($)	643,687		583,410			605,692			526,822			
Constraints	252		387			359			494			
Continuous Variables	270		412			384			526			
Discrete Variables	8		12			12			16			
CPU time (sec)	1.28		2.16			3.71			8.10			

In order to ensure that any resource incorporated in the design can be used as efficiently as possible, detailed consideration of plant scheduling must be taken into account at the design stage. Therefore, in this section, the synthesis, design and operational issues for a sequential multipurpose batch plant are considered simultaneously in an NLP model.

In a sequential multipurpose plant a specific direction in the plant floor is recognized that is followed by the production paths of all the products (Voudouris and Grossmann, 1996). However some processing units are used only by some products. Obviously, the model presented is also valid for the multiproduct batch plant where all the products use all the stages. Besides, alternatives for the number of units in series are introduced. The configuration options are explicitly considered in terms of a superstructure as was presented in the previous section.

The simultaneous optimization of several problems is not an usual approach in the chemical engineering literature. In general, these problems are treated in separate form: first the plant configuration problem, then the sizing problem and lastly the campaign determination problem. This leads to sub-optimal solutions.

The proposed methodology in this work solves in first place a relaxed model where scheduling constraints for mixed campaigns are not considered, with the purpose of obtaining the ratios among the number of batches of the different products considered. With these ratios, it is possible to envisage different campaign configurations. Then, different structures for the mixed campaign are proposed. Taking into account that the approach is applied to problems with a moderate number of products and stages where the model detail is emphasized, the number of campaigns to be planned is manageable.

Once the campaign is defined, the appropriate sequence constraints are added to the relaxed model and thus the design and operation integrated problem of a sequential multipurpose plant is solved. The plant configuration obtained in the relaxed model solution is adopted for the mixed product campaign model, therefore the alternative plant configurations remaining are deactivated in this later model.

An extra plant, or mother plant, is considered in this work to provide the material and power streams that the multipurpose plant requires, so that these resources are bounded. Also, the multipurpose batch plant can produce a by-product, which is a product that is obtained by splitting a batch or processing it in a different manner.

The objective function employed in this formulation is the maximization of the net annual profit as given by the earnings of selling products and savings due to unused resources (those produced and available in the mother plant and not used in the multipurpose plant), minus the annualized investment and operating costs.

Several examples of different mixed campaigns are stated for a Torula Yeast, Brandy and Bakery Yeast production plant to assess the approach proposed.

3.2. Model Assumptions

The problem considered here has the following characteristics:

(i) The plant has batch and semi-continuous units.
(ii) N_p products are processed in the plant.

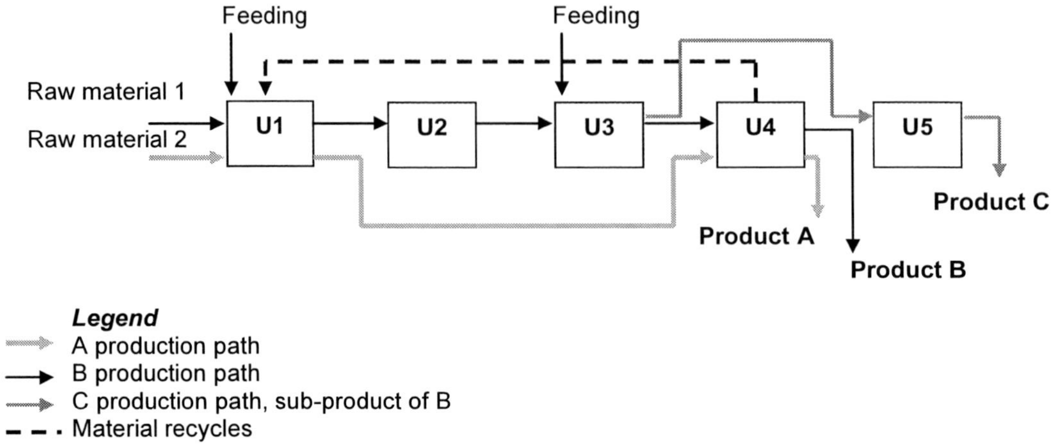

Figure 3.1. Flowsheet of a sequential multipurpose batch plant.

(iii) Not all the products follow the same production path, i.e. a sequential multipurpose batch plant is considered (Figure 3.1).
(iv) The production path for each product is known.
(v) The product demands are upper and lower bounded.
(vi) The processing times are continuous variables and the horizon time is given.
(vii) The mixing, splitting and recycle of batches are allowed.
(viii) The production of by-products is also considered.
(ix) The material and energy resources are bounded.
(x) The unit sizes are continuous variables

The objective is to determine the optimal plant design and operation to meet a specified economic criterion.

Figure 3.1 shows a sequential multipurpose plant where two products, A and B, and a by-product of B, product C are produced. Product A follows the production path U1→ U4 and U1 receives an extra feed (blend of batches) and a recycled batch from U4. Product B follows the path U1→ U2→ U3→ U4. At U3 batches are split to produce product C through U5. U3 also has an extra feed.

3.3. Solution Procedure

The simultaneous optimization of the synthesis, design, operation and scheduling of a sequential multipurpose batch plant, results in a very large size problem, difficult to solve as was pointed in the introduction section. So, a heuristic procedure is proposed in order to solve this simultaneous optimization. The main idea lies in solving first a model without mixed product campaign constraints and then, according to the optimal number of batches of each product obtained in the first model solution, determining the possible campaign configurations. For each campaign configuration proposed, an NLP model is formulated for the optimal plant configuration obtained in the first model solution. The heuristic approach is resumed in the following steps:

(i) First, a model whose constraints consider the design and operation of a multipurpose plant without considering the tasks scheduling constraints is solved. This model is a relaxation of the mixed campaigns problem and is solved as an NLP problem. The model has an embedded superstructure that considers different configuration options for the plant synthesis. The solution of the relaxed model provides the estimated number of batches of each product and the plant configuration.
(ii) Relationships between the number of batches of each product, which are obtained from the relaxed model are established, and the possible sequences of the multiproduct campaigns are selected for the plant synthesis obtained in the relaxed model solution.
(iii) For each proposed campaign configuration, an NLP problem is modelled and solved. In this model, a novel set of tasks scheduling constraints are added to the relaxed model in order to ensure that the production processes of two different products do not overlap in the same unit. In this way, a model for each mixed product campaign is formulated and solved, with the optimal plant configuration obtained from the relaxed model solution. At this step, the plant configuration is fixed and the sizing problem is solved.
(iv) The campaign with the best objective function value is chosen as the optimal solution.
(v) The first model represents a relaxation for the second one. Therefore, the objective function value of the relaxed model solution represents an upper bound for the objective value of the mixed product campaign model. In the studied cases presented below, the gap between these values is very tight, which ensures that the solution obtained for the mixed product campaign is the optimal.

3.4. Mathematical Modeling

A plant with N_j batch units and N_k semi-continuous units is considered. N_p products are manufactured in the plant, which do not necessarily follow the same production path.

Both problems (the relaxed one and the problem with scheduling constraints) include a detailed modeling for all the products and the batch and semi-continuous units.

3.4.1. Relaxed Model

The components and total mass balances at each stage, the connection constraints between stages and the design equations for each stage for each product are considered as a detailed model. If there are recycles or interconnections between the production processes, as it really happens in the study cases, also the balances that correspond to these connections are considered. Mass balances for some units are given by differential equations such as

$$\frac{dC_{xij}}{dt} = h(t,x) \qquad \forall x \tag{3.1}$$

where C_{xij} is the concentration of component x (biomass, substrate, product, etc.), at stage j of the production process of product i. These dynamic equations are discretized and included in the overall model. Note that the discretized equations involve the processing time of the batch item and the time integration step, all of which are considered variables. The number of grid points is a problem data, but as the final processing time is variable, the discretization step length is also an optimization variable determined according to the final time for each unit. For these models, the trapezoidal method was adopted (Atkinson, 1989). For example, if the biomass balance is

$$\frac{dX_{ij}}{dt} = \left(\mu_{ij} - \upsilon_{ij}\right)X_{ij} \tag{3.2}$$

where X represents the biomass concentration, μ the specific growth rate of biomass, and υ the biomass death rate, the corresponding set of algebraic equations is

$$X_{ij}^{(p+1)} = X_{ij}^{(p)} + \frac{l_{ij}}{2}\left(\left(\mu_{ij}^{(p+1)} - \upsilon_{ij}\right)X_{ij}^{(p+1)} + \left(\mu_{ij}^{(p)} - \upsilon_{ij}\right)X_{ij}^{(p)}\right) \tag{3.3}$$

where l is the step length and $p = 1, \ldots, P$ are the grid points.

In addition, for the stages that are shared by several products, the following constraints are considered.

For batch item j and product i:

$$V_j \geq V_{ij} \qquad \forall i = 1,\ldots,N_p, \forall j \in EB_i \tag{3.4}$$

For semi-continuous item k and product i:

$$V_k \geq V_{ik} \qquad \forall i = 1,\ldots,N_p, \forall k \in ES_i \tag{3.5}$$

where V are the batch and semi-continuous unit sizes and EB_i and ES_i represent the set of batch and semi-continuous units in the production path of product i. In both cases the unit size, V_j or V_k, assures to process the required volume V_{ij} or V_{ik}, for all products.

Let t_{ij} be the processing time for product i at stage j, θ_{ik} the processing time for product i at semi-continuous stage k, CT_i the cycle time for the production of product i and Nb_i the number of batches of product i over the horizon time HT, then

$$T_{ij} = \theta_{ik'} + t_{ij} + \theta_{ik''} \qquad \forall i = 1,\ldots,N_p, \forall j \in EB_i \tag{3.6}$$

$$CT_i \geq T_{ij} \qquad \forall i = 1,\ldots,N_p \quad \forall j \in EB_i \tag{3.7}$$

Note that Eq. (3.6) defines the time that the batch unit j will be occupied with product i, which contemplates the material loading ($\theta_{ik'}$) and unloading ($\theta_{ik''}$) time if this unit is located between semi-continuous units. In this approach variables t_{ij} and $\theta_{ik'}$ are assumed to be involved in detailed sub-models, some of them written as differential equations and included in the actual model as was presented in equations (3.1)-(3.3).

Several consecutive semi-continuous units give rise to a semi-continuous sub-train. In this paper, only perfectly synchronized semi-continuous sub-trains are considered, then:

$$\theta_{ik} = \theta_{i,k+1} \qquad \forall i = 1,\ldots,N_p \tag{3.8}$$

where k and $k+1$ belong to the same sub-train.

For products i that share the unit j ($i \in I_j$) the following constraints are considered:

$$\sum_{i \in I_j} Nb_i T_{ij} \le HT \quad \forall j = 1,\ldots,N_j \tag{3.9}$$

In the same way, for all the products i that share the unit k ($k \in I_k$)

$$\sum_{i \in I_k} Nb_i \theta_{ik} \le HT \quad \forall k = 1,\ldots,N_k \tag{3.10}$$

In this way the available time is not exceeded.

If all the products follow the same production path, then Eq. (3.10) becomes redundant because the batch processing time considers the semi-continuous processing times upstream and downstream of the batch unit.

A characteristic of this model is that, for certain batch stages, the number of units in series is a priori unknown. For these stages, a superstructure that contemplates all the possible configurations (or those chosen by the designer as feasible) is modelled and embedded in the global model, as was described in Section 2. Then, if the stage is preceded by a semi-continuous unit, the first unit in the series must include the filling time in its operating time, or the emptying time if this stage has a downstream semi continuous unit (see Figure 3.2).

In this case, the cycle time of stage j is given by

$$T_{ij} = \max_{u}\left\{T_{ij}^{u}\right\} \text{ for each } u \in Nu_j \tag{3.11}$$

or in a continuous formulation:

$$T_{ij} \ge T_{ij}^{u} \text{ for each } u \in Nu_j \tag{3.12}$$

where Nu_j is the set of units in series at stage j.

In order to simplify the results analysis, in this work only units in series are considered as possible configurations for the plant stages. The incorporation of units in parallel on the superstructure model can also be done as proposed in Corsano et al. (2004) and it does not represent a model limitation.

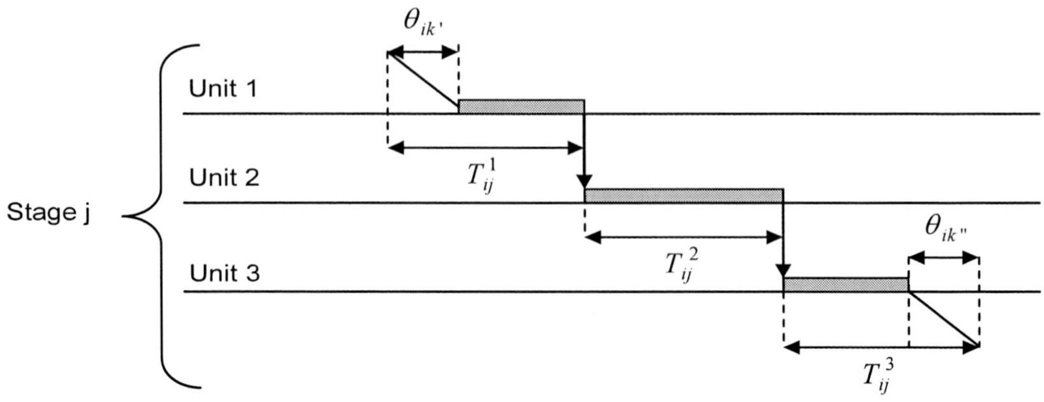

Figure 3.2. In series units configuration for a batch stage.

The material and energy resources s required for each production process can be obtained from another plant that belongs to the same industrial complex called "mother plant" or can be imported from another plant. The unused amount of resource s, i.e. the amount of s that is not consumed by the multipurpose plant, can be sold to other complexes. Let F_s^{prod}, F_s^{imp} and F_s^{ex} be the amounts per hour of produced, imported and exported resource s respectively, and let f_{sij} and f_{sik} be the amount of s consumed for producing product i at the stage j or k respectively, then

$$F_s + F_s^{imp} = \sum_{i=1}^{N_p}\left(\sum_{j=1}^{N_j}\frac{f_{sij}}{CT_i} + \sum_{k=1}^{N_k}\frac{f_{sik}}{CT_i}\right) + F_s^{ex} \tag{3.13}$$

The production rate constraints for each product are

$$Nb_i B_i = Q_i \qquad \forall i = 1,..,N_p \tag{3.14}$$

$$Q_i^{\min} \le Q_i \le Q_i^{\max} \qquad \forall i = 1,..,N_p \tag{3.15}$$

where Q_i is the total production of product i which is bounded by $Q_i^{\min}$ and $Q_i^{\max}$, and B_i is the batch size of product i.

The selected objective function is the maximization of the net annual profit (*NAP*), given by the sum of the earnings from products sales and the exported resources (*SI*) minus the total annualized cost (*TAC*) given by investment (*CInv*) and operating (*CO*) costs. The considered operating costs are the raw material, power resources and disposal costs. Then:

$$NAP = SI - TAC \tag{3.16}$$

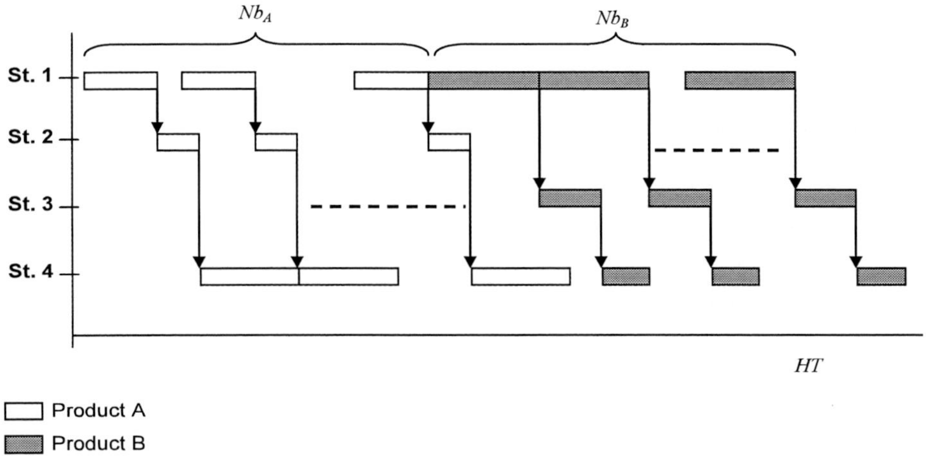

Figure 3.3. Gantt chart for multipurpose plant relaxed model.

$$SI = \sum_{i=1}^{N_p} G_i Q_i + \sum_{s} G_s F_s^{\exp}\ HT \tag{3.17}$$

where G_i represents the product i sale price (\$ ton^{-1}), G_s the resource price of s (\$ ton^{-1}) and $F_s^{\exp}$ the amount of unused resource s (ton).

$$TAC = CInv + CO \tag{3.18}$$

$$CInv = C_{ann}\left(\sum_j \alpha_j V_j^{\beta_j} + \sum_k \alpha_k V_k^{\beta_k}\right) \tag{3.19}$$

$$CO = \sum_{i=1}^{N_p} Nb_i Res_i + HT\left(\sum_s C_s^{prod} F_s^{prod} + \sum_s C_s^{imp} F_s^{imp}\right) \tag{3.20}$$

α_j, β_j, α_k, and β_k are cost coefficients for units, C_s is the cost of resource s, and Res_i the disposal cost of product process i that varies according to the effluent.

The relaxed model solution provides the optimal plant configuration and design and the number of batches of each product to be produced.

It is worth noting that the relaxed model resembles a single product campaign one, in which all the batches of a product are processed without overlapping with others products. The difference is that in the relaxed model, the scheduling constraint of a single campaign model

$$\sum_{i=1}^{N_p} CT_i Nb_i \leq HT$$

is not necessarily satisfied because of Eq. (3.9).

Figure 3.3 schematizes the relaxed model solution of a plant that processes two products with different production paths. Product A follows the path Stage 1 → Stage 2 → Stage 4, while product B follows the path Stage 1 → Stage 3 → Stage 4.

3.4.2. Multiproduct Campaign Model

In many cases, a stream in the production of a product can be recycled to a previous stage in the processing of the same or another product. In such cases the single product campaign is impracticable because the material to be recycled should be stored. In addition single product campaigns require high inventory levels. Furthermore many products cannot be stored because they are degraded. From the point of view of the mathematical model, greater challenges arise than in the formulations for single campaigns. The first decision to be considered is how the multiproduct campaign will be configured. In this work this decision is imposed by the designer from the estimation of the ratio of number of batches of the different products elaborated in the plant.

Let i' be the product with the small number of batches in the relaxed model solution. Let $r_i = round\left(\frac{Nb_i}{Nb_{i'}}\right)$ be the rounding of the relation between the number of batches of each product i and product i', obtained in the relaxed model solution, so that r_i is a model parameter.

Let Nb be the number of times that the mixed campaign is repeated. For the mixed campaign model, the following constraints are imposed:

$$Nb = Nb_{i'} \tag{3.21}$$

$$Nb = Nb_i \, r_i^{-1} \quad \forall i = 1,\ldots, N_p \tag{3.22}$$

Nb is an optimization variable since Nb_i is an optimization variable for each product i.

For example, if three products A, B and C are processed in the sequential multipurpose plant with $N_A = 100$, $N_B = 120$ y $N_C = 310$ in the relaxed model solution, then:

$Nb = N_A$, $r_B = 1$ y $r_C = 3$, i.e. the campaign A-B-C-C-C or some of its permutations is established.

Given that this strategy is here applied to detailed models with a reduced number of products, this is an affordable approach. Therefore, this procedure estimates the proportion among the number of batches of all products. Different campaigns can be proposed by the designer.

Now, new constraints have to be developed to formulate the different conditions that arise when mixed campaigns are used. All the following constraints are posed for a determined mixed campaign.

Let SL_{ij} be the idle time at unit j after processing a batch of product i and before processing the next batch and CT_j be the cycle time for unit j defined by:

$$CT_j = \sum_{i \in I_j} \left(T_{ij} + SL_{ij}\right) \quad \forall j = 1,\ldots, N_j \tag{3.23a}$$

By the same token for semi-continuous units:

$$CT_k = \sum_{i \in I_k} (\theta_{ik} + SL_{ik}) \ \forall k = 1,...,N_k \tag{3.24}$$

If there is more than one batch of some product in the campaign, the processing time and idle time for that product must be added as many times as repetitions occur.

If a stage j has more than one unit in series as a result of the superstructure optimization model performed in the relaxed model, the following constraints have to be added:

$$CT_j^u = \sum_{i \in I_j} (T_{ij}^u + SL_{ij}^u) \quad \forall u \in Nu_j \tag{3.23b}$$

The modeller must establish for each unit the order in which products will be processed, using the relationship between the number of batches previously determined. Next the constraints that must be implemented according to the production path that each product follows are settled down. These constraints are established for two consecutive products, in order to assure that the production processes of two different products do not overlap in a same unit.

Due to the ZW transfer policy adopted,

$$CT_j = CT_{j+1} \quad \forall j = 1,...,N_j \tag{3.25}$$

$$CT_k = CT_{k+1} \quad \forall k = 1,...,N_k \tag{3.26}$$

$$CT_j = CT_k \text{ for some } j = 1, ..., N_j \text{ and } k = 1, ..., N_k \tag{3.27}$$

Suppose that in the production path, the product $i + 1$ is processed in unit j immediately after product i and both follow the path $j \rightarrow j+1$. Three cases are presented.

Case i

if all the products follow the same production path, the following constraints are added to the model with the objective of avoiding tasks superposition at the same processing unit.

$$T_{i+1,j} + SL_{ij} = T_{i,j+1} + SL_{i,j+1} \quad \forall i = 1,...,N_p - 1, \forall j = 1,...,N_j - 1 \tag{3.28}$$

$$CT_j Nb \leq HT \quad \text{for some } j \tag{3.29}$$

Constraints (3.28) were used by Birewar and Grossmann (1990) but with a different definition of the idle time at stage j. In their work they defined SL_{ijk} as the idle time between the batches of products i and k in processing unit j. In that model the processing times and size factors are fixed, the campaign configuration is obtained as a result of the model solution and the model is solved as an MINLP problem.

As Eqns. (3.25) – (3.27) establish that the unit cycle times are equal, the constraint (3.29) written for some j means that it will hold for every unit j.

This first case is shown on Figure 3.4.

It is worth noting that if stage j has more than one unit in series (as a result of the superstructure model optimization), for each of these units the constraints (28) must be satisfied and CT_j is defined by equation (3.23b).

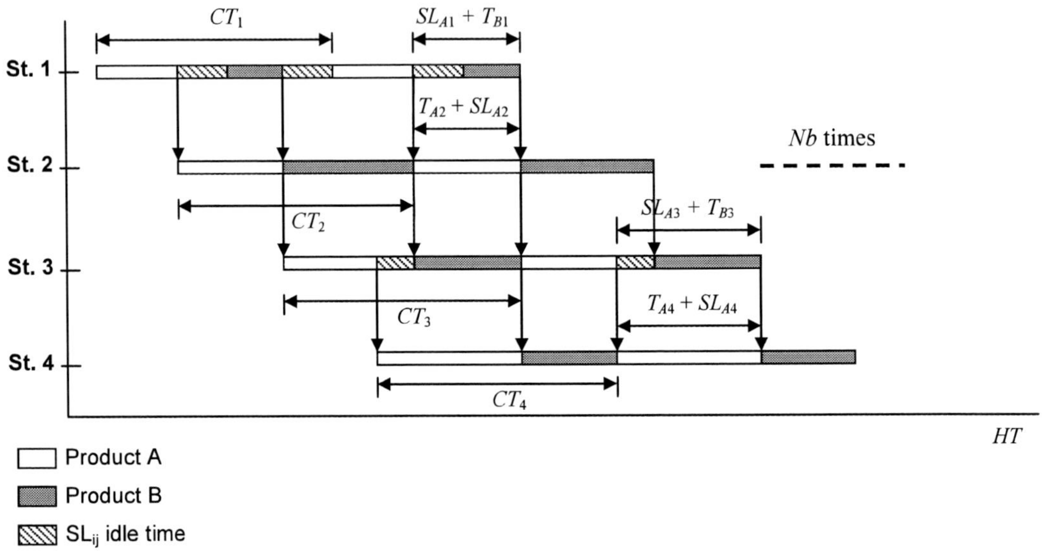

Figure 3.4. Gantt chart for product productions that follow the same path - Case *i*.

Case ii

if a production path is different, the following constraints are imposed for two consecutive products with different production paths. Let A and B be two consecutive products in the production campaign such that A follows the path $l \rightarrow s$ and B the path $m \rightarrow s$ and let J_A and J_B be the set of processing units that are utilized in the process production of A and B respectively, then:

(a) If the processing order is kept at stage s, the following constraint is added in order to avoid the tasks superposition at stage s:

$$T_{Ap} + SL_{Ap} + \sum_{\substack{j \in J_B \\ p \le j \le m}} T_{Bj} - \sum_{\substack{j \in J_A \\ p \le j \le s}} T_{Aj} \ge 0 \tag{3.30}$$

p represent the first stage shared by both products. The two first terms indicate that the B process production at stage p begins after the time $T_{Ap} + SL_{Ap}$.

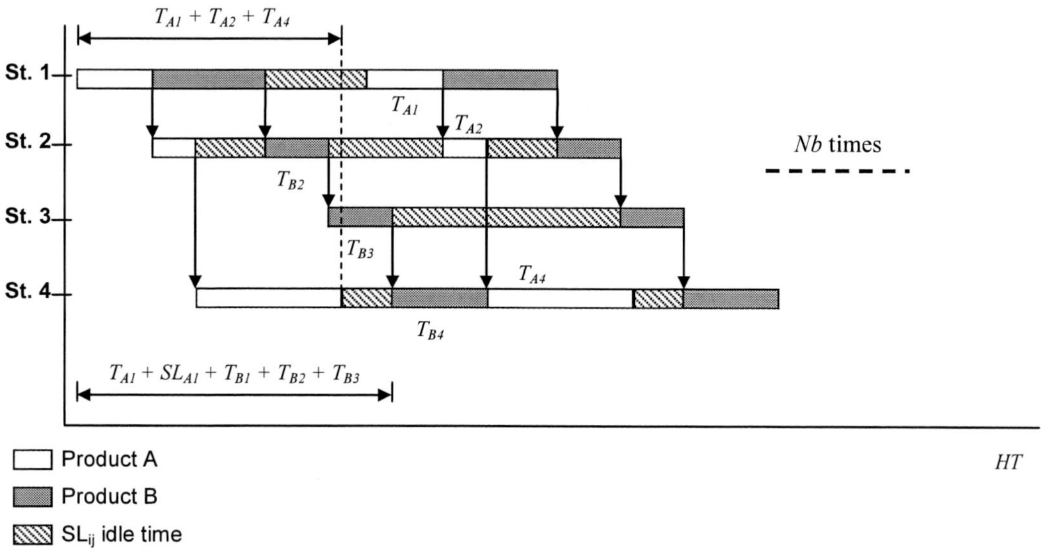

Figure 3.5. Gantt chart for case *ii - a)* without changing the production sequence.

Figure 3.5 shows this situation for the processing of two products, A and B, where the production path of A is $J_A = \{1, 2, 4\}$ and that of B is $J_B = \{1, 2, 3, 4\}$, i.e. $l = 2$, $s = 4$, $m = 3$ and $p = 1$, so the equation that must be added is

$$T_{A1} + SL_{A1} + \sum_{\substack{j \in J_B \\ j \le 3}} T_{Bj} - \sum_{\substack{j \in J_A \\ j \le 4}} T_{Aj} \ge 0$$

Figure 3.6 shows another case where the sequence order is conserved but the first processed product has more stages than the second one and $p \neq 1$. The production path for A is $J_A = \{1, 2, 3, 4\}$ and for B is $J_B = \{2, 4\}$, i.e. $l = 3$, $s = 4$, $m = 2$, $p = 2$ and the constraint added is

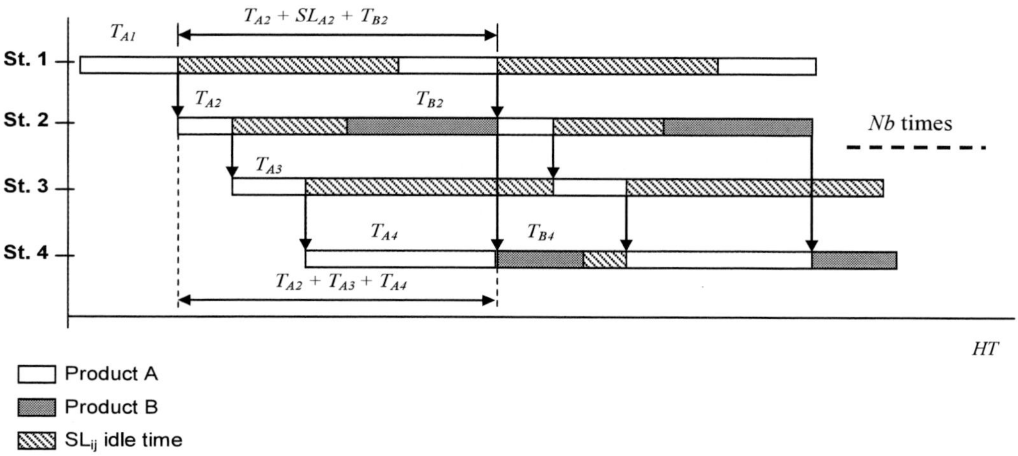

Figure 3.6. Gantt chart for case *ii - a)* with $p \neq 1$.

$$T_{A2} + SL_{A2} + \sum_{\substack{j \in J_B \\ 2 \le j \le 2}} T_{Bj} - \sum_{\substack{j \in J_A \\ 2 \le j \le 4}} T_{Aj} \ge 0 .$$

(b) If the designer chooses to change the production order at stage s the constraint that must be added instead of (3.30) is:

$$\sum_{\substack{j \in J_A \\ p \le j \le l}} T_{Aj} - \left(T_{Ap} + SL_{Ap} + \sum_{\substack{j \in J_B \\ j \le s}} T_{Bj} \right) \ge 0 \qquad (3.31)$$

In this way, as the processing times are variables, the distribution of them will be different. The designer can evaluate both solutions and choose the best economical solution as the optimal solution.

Figure 3.7 shows the scheduling production for two products that change the production order. A follows the production path $J_A = \{1, 2, 3, 4\}$, while for B $J_B = \{1, 2, 4\}$, i.e. $l = 3$, $s = 4$, $m = 2$ and $p = 1$. In stage 1 and 2, product A is processed before product B, while in stage 4 this order is changed.

If stage s is a semi-continuous unit, Eq. (3.30) and (3.31) are valid in each case, taking into account that T_{ij} represents the batch processing time with the loading and unloading times, and therefore the operating times of the semi-continuous units must not be added because they are contemplated on T_{ij}.

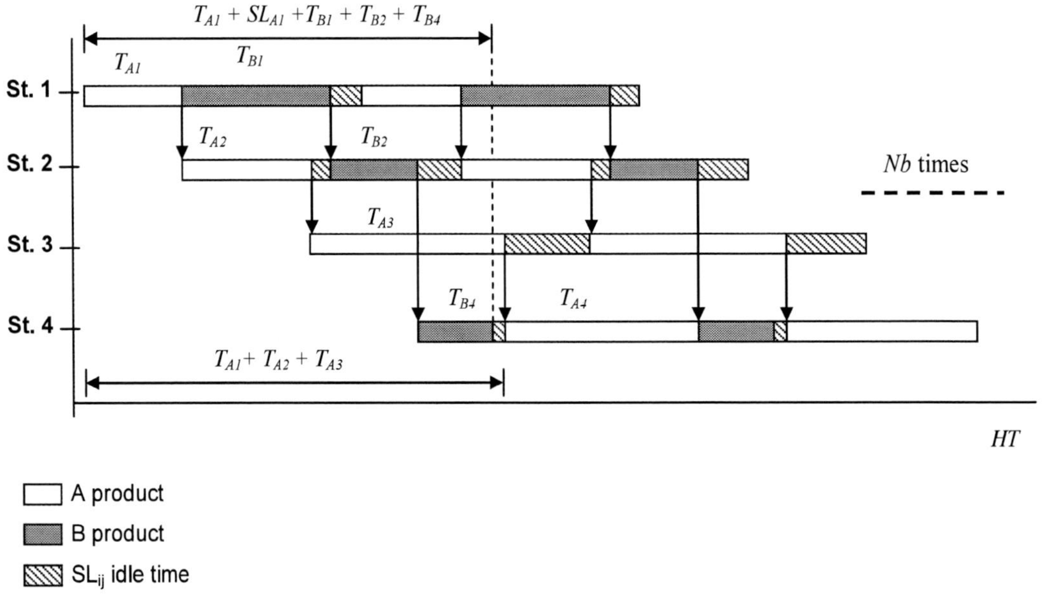

Figure 3.7. Gantt chart for case *ii - b)* changing the production sequence.

For all the situations described for *case ii*, the constraints (3.28) and (3.29) must be added for those products that follow the same path in two consecutive units. For example, for Figure 3.5 the following constraint must be added:

$$SL_{A1} + T_{B1} = T_{A2} + SL_{A2}$$

Case iii

if two consecutive products A and B in the production campaign are such that A follows the path $s \rightarrow l$ and B the path $s \rightarrow m$ no constraints are added, because they follow independent paths.

All the constraints considered in the different cases are added to the previous relaxed model where only the optimal plant configuration is active, that is, the plant structure is fixed.

3.5. Study Case

Sequential Multipurpose Plant: Torula Yeast, Brandy and Bakery Yeast Production Integrated to a Sugar Plant

The integration of several processes into a Sugar Cane Complex is considered. The Sugar plant produces sugar and bagasse for sale and molasses, filter juices, vapor and electricity that are used in the derivatives plant. The derivatives plant is a sequential multipurpose plant with batch and semi-continuous units to produce Torula Yeast, Brandy and Bakery Yeast. The Bakery Yeast is a by-product of the Brandy production that is obtained by the evaporation and drying of the centrifugation residue of this process. Figure 3.8 shows the integration scheme.

The molasses and filter juices produced in the sugar plant serve as sugaring substrates for the biomass and alcohol fermentations. In addition, water and vinasses are added to the fermentation feed. The vinasses are a non-distilled waste of Brandy production. The electricity generated in the sugar plant is used in the centrifuges of the derivatives plant, whereas the fermentors, the evaporator, the spray dryer and the distillation column consume the steam. In addition, if it is necessary, steam can be imported from other power stations with operative cost imputed on the total annual cost. The vapor and the electricity that are not consumed by the derivatives plant can be sold. For the fermentation stages, the superstructure optimization model proposed in the previous section integrated to the overall model is adopted. For a detailed description of the Brandy process model see Corsano et al.(2006b). Therefore, a synthesis, design, operation, and scheduling problem is solved for the sequential multipurpose plant integrated to the Sugar plant as an NLP model.

For the Sugar plant, the model optimizes the amount of extracted filter juices. The sugar plant is considered as an existing mother plant, and the amount of extracted filter juices is a process variable. The production of sugar, molasses, vapor and electricity depends on the amount of filter juice extraction. If more filter juices are extracted, molasses and sugar productions are reduced, so the consumption of vapor and electricity in the Sugar production process is also decreased and therefore the amount of electricity and vapor available for derivatives, and bagasse for sale is increased.

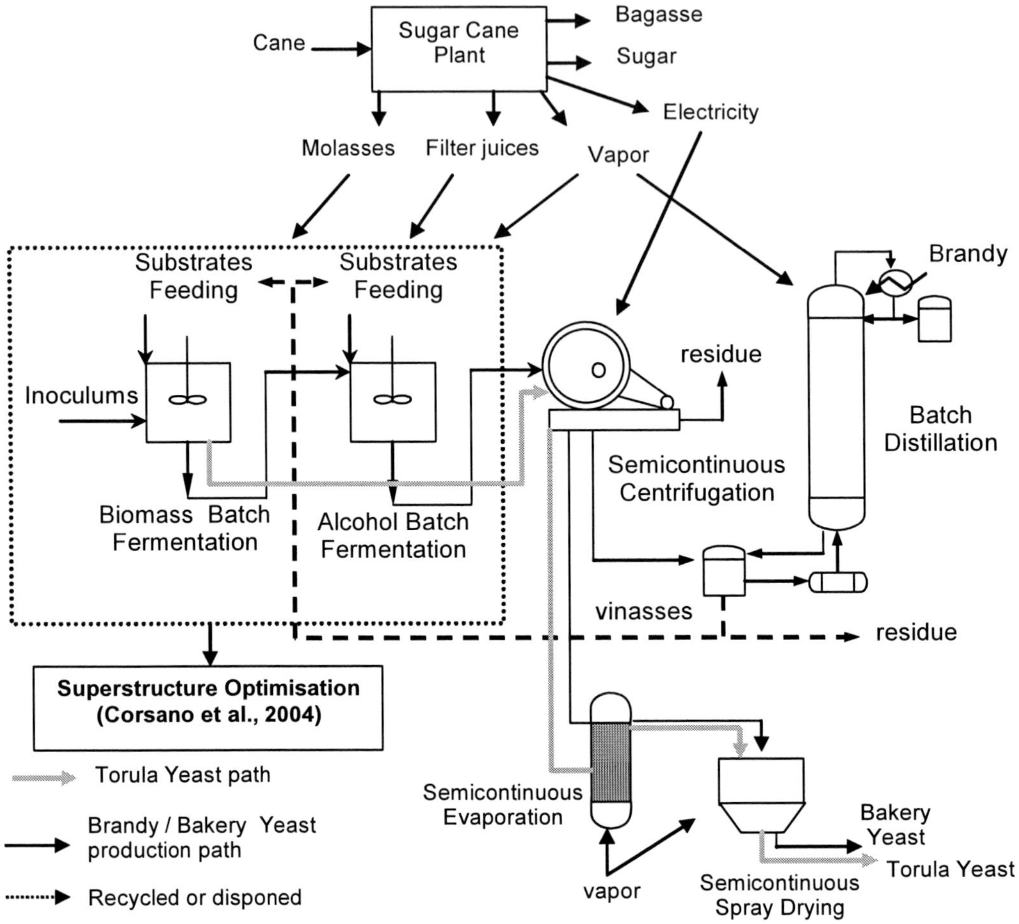

Figure 3.8. Flow sheet for Sugar Cane Complex Integration.

Table 3.1. Decision variables description

Synthesis and Design	Operation	Scheduling
Plant configuration. Number of units in series. Blend and recycle allocation. Unit sizes Heating and cooling areas Power consumption (vapor and electricity) Stage number of distillation column	Batch blending, batch splitting and batch recycling flow rates within the same production process Flow rate recycles from one process to another Material and energy resources allocation from mother plant to the different production processes Production rates Component concentrations Unit processing times for each product	Cycle time of each production process Units cycle time Units idle time Number of batches Mixed campaign configuration

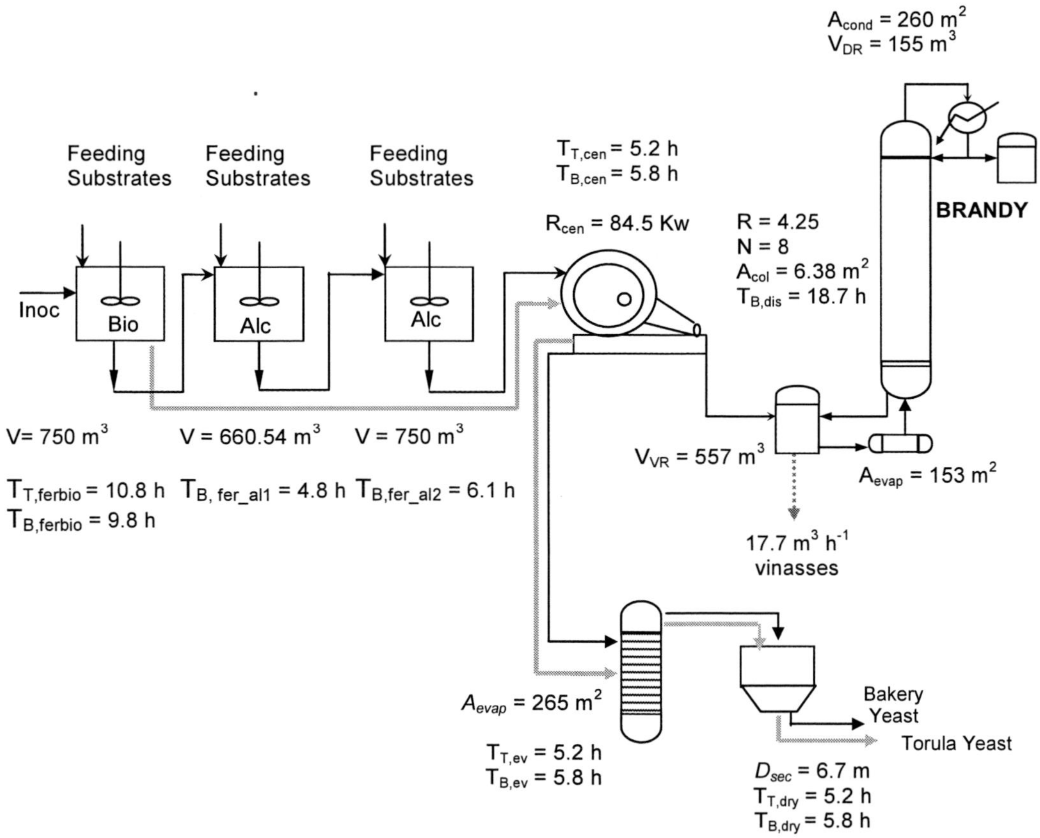

Figure 3.9. Optimal flowsheet for relaxed model plant design.

Table 3.1 lists the most important synthesis, design, operating and scheduling decisions considered in this study case in order to have a grasp of the model complexity, but in the results only some of these variables are reported since the objective of this work is to focus on the heuristic approach and the mathematical models involved.

The models were implemented and solved in GAMS (Brooke, 1998) in a Pentium IV, 1.60 Ghz. The code CONOPT2 was employed for solving the NLP problems.

Firstly the relaxed model is solved. Table 3.2 shows the description and optimal values for some optimization variables. Minimum and maximum productions were fixed for each product at Q_i^{min} = 7500 ton and Q_i^{max} = 37500 ton, while the time horizon was fixed at 7500 h. Figure 3.9 shows the optimal plant synthesis. The fermentation stage configuration consists of one biomass fermentor and two alcohol fermentors in series. The fermentors size is upper bounded by 750 m^3.

As can be observed in Table 3.2, the number of Torula batches is 281 while for Brandy and Bakery Yeast is 305. So, for the mixed campaign model the campaign Brandy-Torula (*B-T*) is proposed, that means, one batch of each product. This is a reasonable campaign in the sense that the vinasses produced in Brandy production would be used in Torula fermentation. The vinasses cannot be stored for long periods of time because of the degradation and inventory considerations, so the campaign *B-T* seems a good option.

Table 3.2. Optimal variables for sequential multipurpose plant relaxed model

Variable	Description	Optimal Value
Q_T	Torula production rate[ton]	14025
Q_{BY}	Bakery Yeast production rate[ton]	10200
Q_B	Brandy production rate[ton]	37500
Nb_T	Torula batches number	281
Nb_B	Brandy and Bakery Yeast batches number	305
CT_T	Cycle time for Torula production [h]	16.0
CT_B	Cycle time for Brandy production [h]	24.5
CT_{BY}	Cycle time for Bakery Yeast production [h]	11.9
NAP	Net Annual Profit [$]	51772500.

Besides this campaign, other alternatives can be assessed adding the corresponding constraints to the relaxed problem, as it is shown below.

According to the proposed methodology, the mixed product campaign model adopts the plant configuration obtained in the relaxed model optimal solution.

Figure 3.10 shows the Gantt chart for the relaxed model solution.

3.5.1. B-T Sequence Campaign for Fermentation Stage and T-B for Semi-continuous Stages (B-T / T-B)

The mixed campaign model consists of the relaxed model plus the constraints corresponding to the *B-T* campaign for the plant synthesis obtained in the relaxed model solution. The sizing problem is solved in this stage. As the Brandy production uses alcohol fermentors that the Torula production does not use, the sequence campaign at the semi-continuous sub-train (centrifuge, evaporator and dryer) is changed. This campaign is denoted by *B-T/T-B*. Then, the following constraints are added to the relaxed model with the plant configuration fixed:

Campaign definition: The number of batches of *B* and *T* must be the same

$$Nb = Nb_B = Nb_T$$

To avoid the task overlapping at the semi-continuous train:

$$(T_{B,ferbio} + T_{B,fer\_al1} + T_{B,fer\_al2} - T_{B,cen}) - (T_{B,ferbio} + SL_{B,ferbio} + T_{T,ferbio}) \geq 0$$

Table 3.3. Optimal variables for *B-T / T-B* model

Variable	Description	Optimal Value
Q_T	Torula production rate [ton]	16500
Q_{BY}	Bakery Yeast production rate [ton]	10350
Q_B	Brandy production [ton]	37500
Nb	Number of times that the mixed campaign is repeated	342
CT_T	Cycle time for Torula production [h]	12.4
CT_B	Cycle time for Brandy production [h]	22.0
CT_{BY}	Cycle time for Bakery Yeast production [h]	9.6
NAP	Net Annual Profit[$]	51675000.

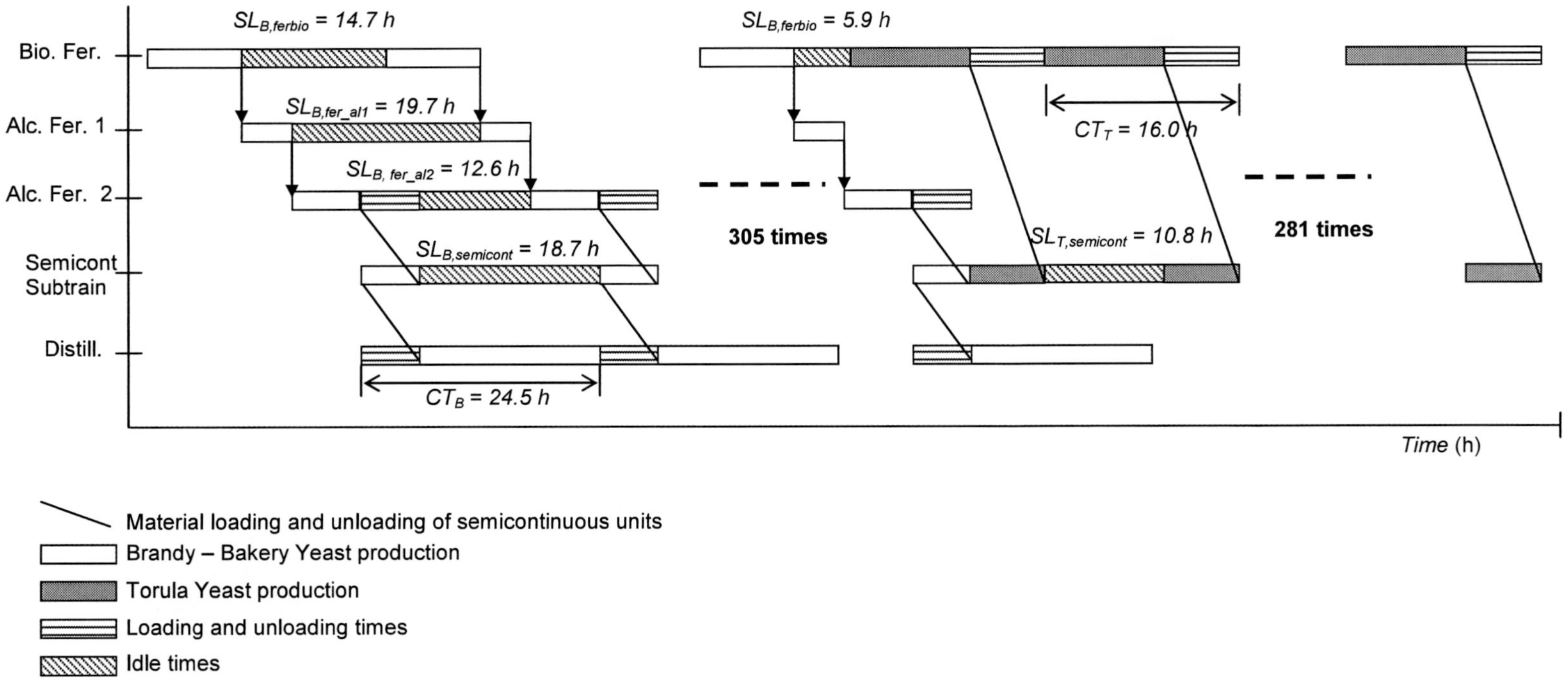

Figure 3.10. Gantt chart for Torula Yeast, Bakery Yeast and Brandy production relaxed model.

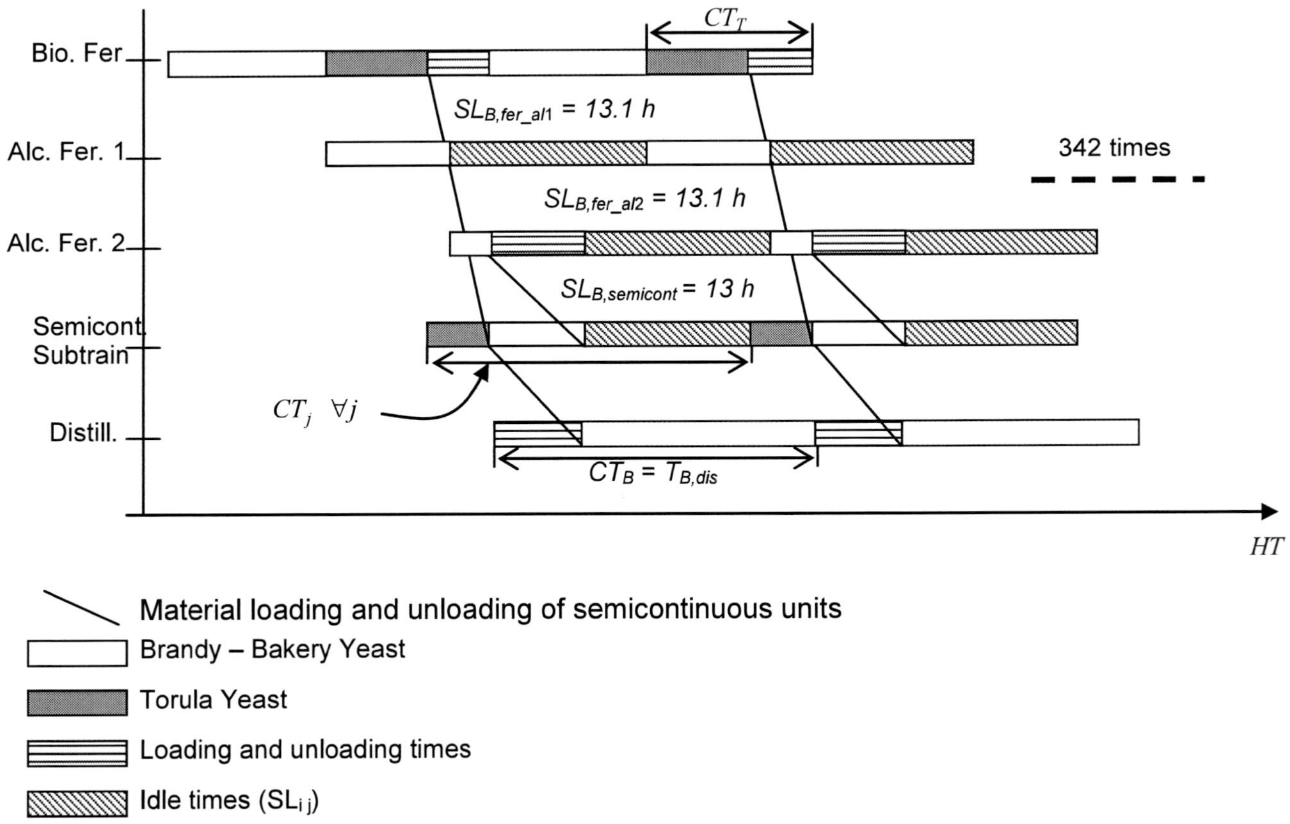

Figure 3.11. Gantt chart for *B-T/T-B* mixed product campaign.

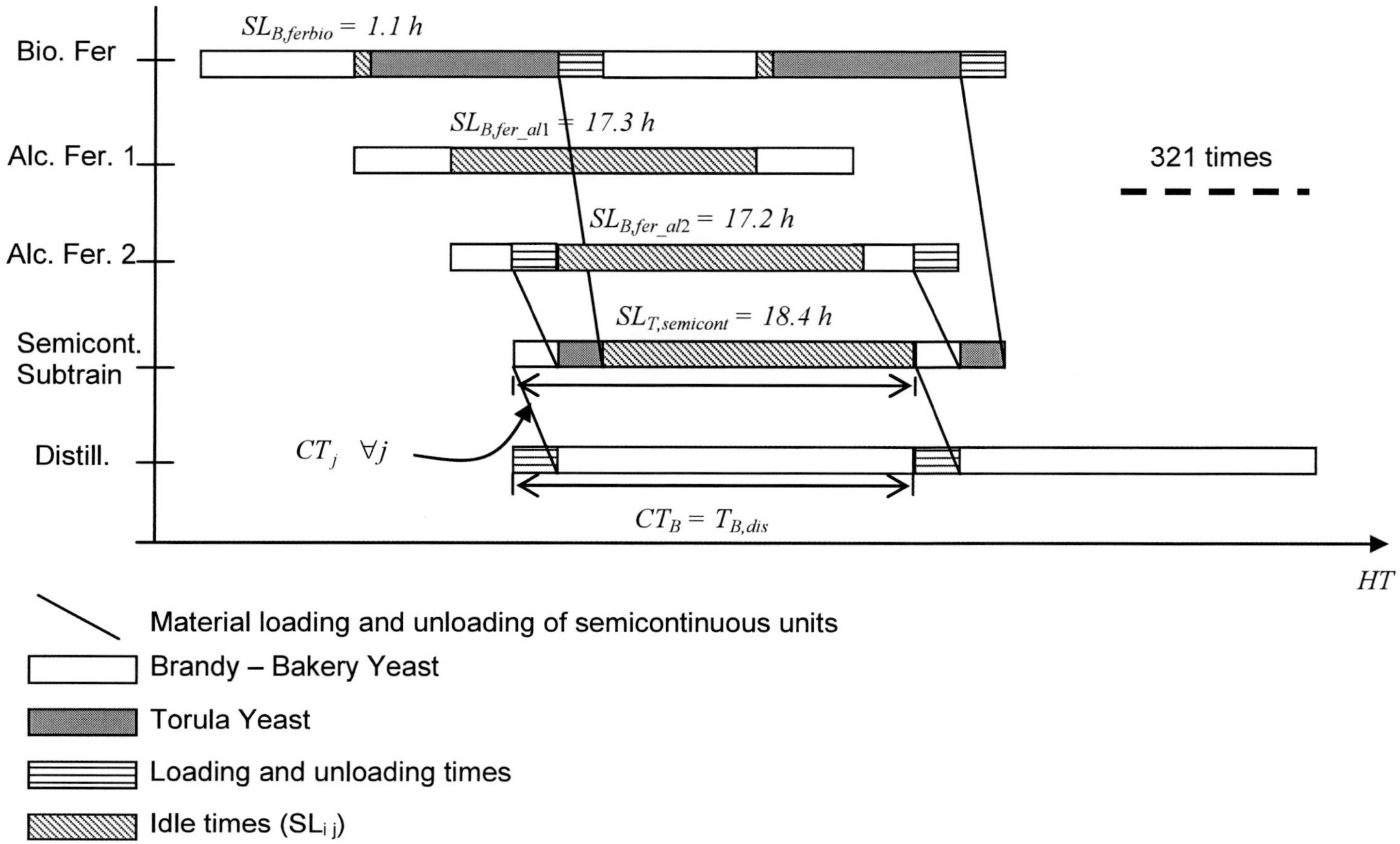

Figure 3.12. Gantt chart for *B-T* mixed product campaign.

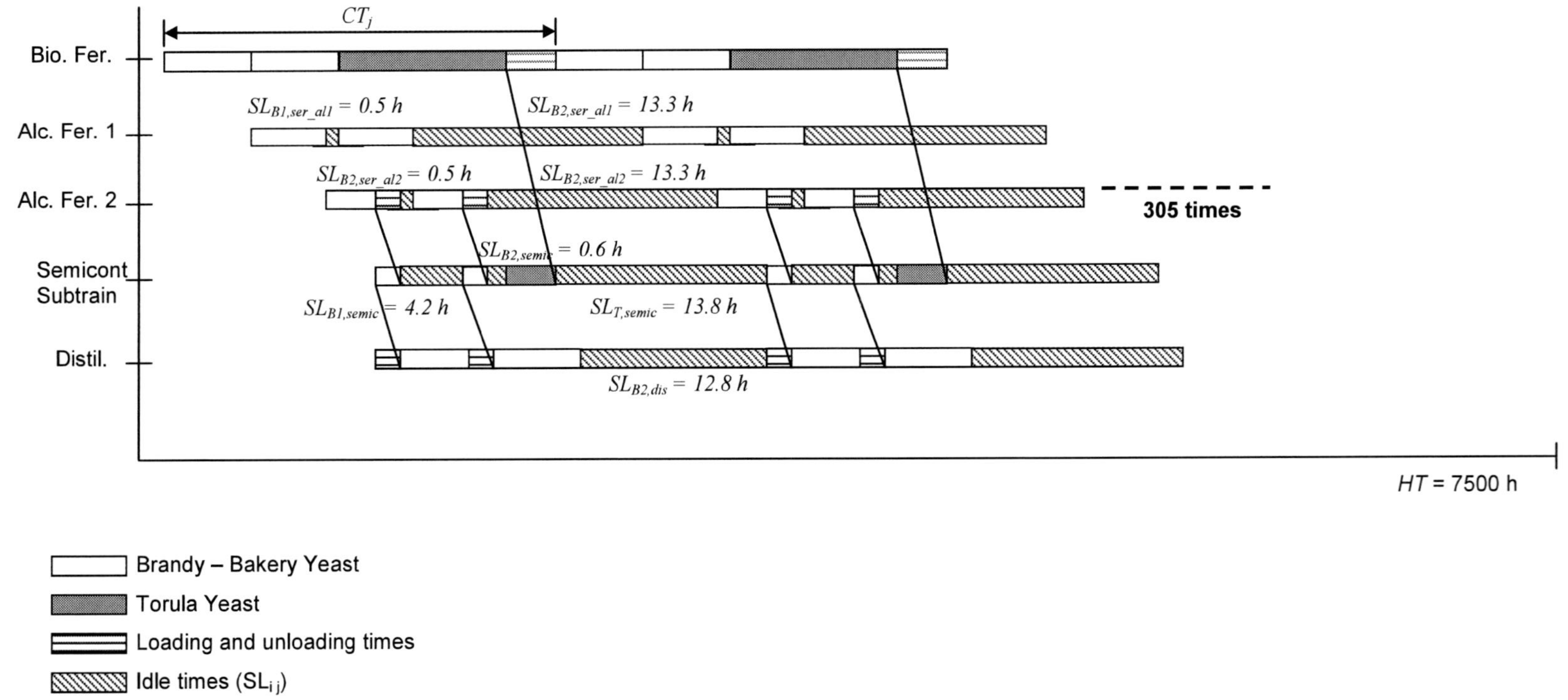

Figure 3.13. Gantt chart for *B-B-T* mixed product campaign.

The subscripts *B* and *T* referrer to the Brandy and Torula production respectively, while *ferbio*, *fer_al*1 and *fer_al*2 represent the biomass fermentor and alcohol fermentors 1 and 2 respectively. The first term in brackets represents the time needed to process the brandy batch up to the alcohol fermentation second unit, while the second parenthesis represents the time at which the Torula batch finished processing on the centrifuge. T_{B,fer_al2} and $T_{T,ferbio}$ consider the centrifuge loading time, so in the first case $T_{B,cen}$ must be subtracted. The units that share both productions are the biomass fermentor and the semi-continuous sub-train, so no additional constraints are added.

The results for some optimal variables are presented in Table 3.3 and the production schedule is displayed in a Gantt chart on Figure 3.11. The optimal design variables are displayed in Table 3.7.

A novel result of this sequential multipurpose plant model with the mixed-product campaign is the fact that, because some units are not used by all products, the operating times of such stages are larger, and therefore, the operating and investment costs of these stages are reduced. This means that a better use of equipment is achieved, as it occurs at the distillation stage. This would not happen if the process adopted a single-product campaign, which is usually the case.

The objective functions of relaxed and mixed product campaign models are not comparable because of the satisfaction of constraints (3.9) and (3.10) in the relaxed model do not imply that the total production of Torula Yeast, Brandy and Bakery Yeast hold in the time horizon. In this case, for the relaxed model solution

$$CT_B\,Nb_B + CT_T\,Nb_T = 24.5 \cdot 305 + 16 \cdot 281 = 11968.5 \geq 7500 = HT$$

However, the total production of Torula Yeast, Brandy and Bakery Yeast is completed in the mixed product campaign model in 7500 hours (*HT*), since for all the stages the cycle time is 22 hours and the campaign is repeated 342 times.

Table 3.4 shows the processing and idle times.

Table 3.4. Processing and idle times for B-T / T-B campaign

	Brandy		Torula	
	Processing Time (h)	Idle Time (h)	Processing Time (h)	Idle Time (h)
Biomass Fermentor	9.6	0	8.7*	0
Alcohol Fermentor 1	8.9	13.1		
Alcohol Fermentor 2	3.6*	13.1		
Semicont. Subtrain	5.3	13.0	3.7	0
Distillation	16.7*	0		

* without considering loading and unloading times.

3.5.2. B-T Sequence Campaign for all the Stages

In this example a change is introduced with respect to the previous one. In the semi-continuous sub-train the same sequence as in the other stages is employed.

For this campaign the following constraints are added to the relaxed model:

Campaign definition:

$$Nb = Nb_B = Nb_T$$

Nb_B and Nb_T are variables.

To avoid the task overlapping at the semi-continuous train:

$$(T_{B,ferbio} + SL_{B,ferbio} + T_{T,ferbio} - T_{T,cen}) - (T_{B,ferbio} + T_{B,fer\_al1} + T_{B,fer\_al2}) \geq 0$$

The first term in brackets represents the time at which the Brandy batch arrives to the centrifuge, while the second one is the time at which the Torula batch leaves the centrifuge.

The model is solved for the same production as for the *B-T / T-B* campaign in order to establish a comparison. The optimal solution for this sequence campaign increases the investment cost in 7% and the idle times in 30%. Table 3.5 shows the processing and idle times and Figure 3.12 the Gantt chart for the *B-T* campaign. The optimal design variables values are displayed on Table 3.7 and economic results on Table 3.8.

As can be seen from comparing Table 3.4 and Table 3.5 the semi-continuous sub-train processing time is shorter than in the previous case, so the equipment sizes are increased and then the investment cost is increased too. The negative terms in Table 3.8 mean that there is unused electricity that is sold in order to increase the profit.

Table 3.5. Processing and idle times for B-T campaign

	Brandy		**Torula**	
	Processing Time (h)	**Idle Time (h)**	**Processing Time (h)**	**Idle Time (h)**
Biomass Fermentor	8.9	1.1	10.8*	0
Alcohol Fermentor 1	5.9	17.3		
Alcohol Fermentor 2	3.6*	17.2		
Semicont. Subtrain	2.4	0	2.4	18.4
Distillation	20.8*	0		

* without considering loading and unloading times.

3.5.3. B-B-T Sequence Campaign for all the Stages

This mixed product campaign does not adjust to the ratio between the number of batches of Torula and Brandy products previously determined. However it is solved in order to assess the effectiveness of the proposed strategy.

For this campaign two consecutive Brandy batches are processed and one Torula batch. Then the superposition of tasks must be avoided in the alcohol fermentors and distillation stage between the two Brandy batches and in the semi-continuous sub-train between Brandy and Torula batches. The following constraints are added to the relaxed model with the plant configuration fixed to the optimal configuration obtained in the relaxed model solution:

To define the campaign:

$$Nb = Nb_T$$

$$Nb = \tfrac{1}{2}\, Nb_B$$

Nb_B and Nb_T are variables.

To avoid the task overlapping at the first alcohol fermentation stage of Brandy:

$$T_{B,ferbio} + SL_{B1,ferbio} = T_{B,fer\_al1} + SL_{B1,fer\_al1}$$

To avoid the task overlapping at the second alcohol fermentation stage of Brandy:

$$T_{B,fer\_al1} + SL_{B1,fer\_al1} = T_{B,fer\_al2} + SL_{B1,fer\_al2} \tag{3.32}$$

To avoid the task overlapping at the distillation stage of Brandy:

$$T_{B,fer\_al2} + SL_{B1,fer\_al2} = T_{B,dis} + SL_{B,dis}$$

To avoid the task overlapping between the second Brandy batch and the Torula batch at the semi-continuous train:

$$\left(2T_{B,ferbio} + SL_{B1,ferbio} + SL_{B2,ferbio} + T_{T,ferbio} - T_{T,cen}\right) - \left(2T_{B,ferbio} + SL_{B1,ferbio} + T_{B,fer\_al1} + T_{B,fer\_al2}\right) \geq 0$$

$SL_{B1,unit}$ and $SL_{B2,unit}$ corresponds to the idle time after the first and second Brandy batches in this campaign respectively at that unit.

It is not necessary to add constraints to avoid task overlapping between the two consecutive Brandy batches at the semi-continuous sub-train because the second alcohol fermentor considers the centrifugation time that represents the fermentor unloading time. Therefore Eq. (3.32) ensures that there will not be task overlapping at the semi-continuous sub-train.

Table 3.6. Processing and idle times for *B-B-T* campaign

	Torula		Brandy		
	Processing Time (h)	Idle Time (h)	Processing Time (h)	Idle Time 1 (h)*	Idle time 2 (h)**
Biomass Ferm.	10.2	0	5.6	0	0
Alcohol Ferm. 1			5.1	0.5	13.3
Alcohol Ferm. 2			3.7	0.5	13.3
Semicont. Subtrain	2.6	13.8	1.4	4.2	0.6
Distillation			4.2	0	12.8

*, ** means idle time after first and second batches in the campaign respectively

For the *B-B-T* campaign, the optimal solution is about a 31% worse than the objective function obtained in example 3.5.1 for the same production. Table 3.6 shows the processing and idle times for this campaign. As can be observed, the distillation processing time is reduced more than three times, so the investment cost is also incremented (Table 3.8). The unit sizes for this campaign can be observed on Table 3.7. The total idle time is increased in 26%. Figure 3.13 shows the Gantt chart for this example.

As the earnings from sales are the same in all cases since the productions are fixed, Table 3.8 compares the investment and operation costs of the first campaign analyzed and this last campaign. In both cases, there is unused electricity that is considered as benefit.

Another disadvantage of this campaign is that the biomass fermentor is sub-occupied in Brandy production (Brandy batch size in biomass fermentor is 308.64 m^3, while the biomass fermentor size is 750 m^3 because it reaches this value for Torula production). This occurs because the same production that was first reached in one batch is now reached after two batches.

Table 3.7. Optimal design variables for sequential multipurpose plant models

Unit	*B-T / T-B* campaign	*B-T* campaign	*B-B-T* campaign
Biomass Fermentor (m^3)	750	750	750
Alcohol Fermentor 1 (m^3)	631.8	655.7	344.4
Alcohol Fermentor 2 (m^3)	724.8	750	391.5
Centrifuge (Kwh)	118.3	184.5	172
Evaporator (m^2)	389	571.4	539.61
Dryer (m of diameter)	7.2	13.	11.3
Distillation			
Condenser Area (m^2)	305.6	272.3	715.7
Evaporator Area (m^2)	180.8	161.1	423.4
Stages Number	9	10	9
Reflux Ratio	5.18	5.5	5.66
Transversal Column Area (m^2)	7.53	6.71	17.64

Table 3.8. Economical comparison between different campaigns

	B-T / T-B campaign	*B-T* campaign	*B-B-T* campaign
Investment costs, ($):			
Biomass Fermentors	512025	512025	512025
Alcohol Fermentors	219150	222675	166425
Centrifuge	154650	209550	199575
Evaporator	172050	210975	204675
Dryer	436950	576750	539850
Distillation	878325	881700	1105125
Operating Costs, ($)			
Inoculums	585825	487800	1125000
Water (fresh and cooling)	234975	249225	251700
Vinasses disposal	124875	130050	190875
Profit for sales of unused electricity, ($)	-80250	- 33000	- 43050
Total ($)	3238575	3447750	4252200

Table 3.9. Studied cases - Computational results

	Relaxed model	*B-T/T-B* campaign	*B-T* campaign	*B-B-T* campaign
Number of variables	2398	761	761	772
Number of constraints	2244	704	705	729
CPU [sec.]	161.4	42.9	60.6	69.9

Only the alcohol fermentation stage is cheaper in this case because the unit sizes are smaller. This occurs because each Brandy batch size is smaller since two batches are processed.

Table 3.7 shows the unit sizes for the different examples studied.

Table 3.9 presents the solutions times and number of variables and constraints of each studied case. The reduction of the number of variables and constraints in mixed product campaign models is because these models the plant configuration is fixed and therefore the superstructure model is not considered.

3.6. Conclusions and Outlook on the Proposed Heuristic Approach for Mixed Product Campaign Model

In this book section a detailed model for the optimal synthesis, design, operation and scheduling of a sequential multipurpose non-continuous plant was developed. The resolution strategy considers two steps. Firstly, a relaxed model is solved to obtain the number of batches of each product and the plant synthesis. Then the designer chooses the mixed campaign based on the ratio between the number of batches. A novel set of tasks scheduling constraints was proposed in order to avoid tasks superposition at processing units.

The problem was formulated as an NLP model and a superstructure model presented in section 2 was embedded in order to solve the synthesis problem. There are not previous published works dealing with the synthesis, design and scheduling problem simultaneously solved as an NLP model.

Another characteristic of these models is the high level of detail reached in the processing unit description, some of them by means of ordinary differentials equations. Batch blending, batch splitting and recycles are allowed as novel components for the multipurpose plant model, decisions taken in this work as optimization variables.

The model was implemented for a Torula Yeast, Brandy and Bakery Yeast production plant. The model was formulated and solved according to the proposed strategy. The optimal solution was compared with different campaign sequences. In all the cases, the investment and operative costs and the idle times were increased. Economic as well as synthesis, design and operational results are also reported. The gap between the objective function value of the relaxed model solution and the objective value of the mixed product campaign model is very tight, which ensures that the solution obtained for the mixed product campaign is the optimal and serves to validate the proposed heuristic methodology.

4. Process Integration: Mathematical Modeling for the Optimal Synthesis, Design, Operation and Planning of a Multiplant Complex

4.1. Introduction

The constant shift towards production of higher-added value products in chemical processing industries has encouraged modeling and optimization studies of batch processes. Whereas significant development has been made in designing, planning and scheduling of batch plants with single or multiple production routes, the problem of synthesis, design, operation, and planning in batch multiplant complexes has received much less attention.

In a multiplant complex (MC), there are two or more interconnected plants that are supplied with raw materials and energy by a mother plant and by external plants. The mother plant produces raw materials and energy resources r (r = 1, …, N_r) that are considered as "produced resources". The derivative plants l (l = 1, …, L) can also receive resources from other not nearby plants (external plants). This type of resources is called "supplied resources". The unused resources, those produced by the mother plant and not used in the derivative plants, can be sold and they are called "exported resources".

The MC considered in this work includes plants that can produce only one product and the production of a by-product is allowed. A by-product is a product which is obtained by splitting a batch or processing it in a different manner.

The derivative plants have batch and semi-continuous units, and mixing, splitting, and recycling batches are allowed. The residue of a derivative plant may become raw material of the same plant or of another derivative plant. If the residue is discarded, a disposal cost is allocated. Recycles amount and concentration are process variables and the allocation or not of a recycled flow to processing unit is also an optimization variable.

In this section a detailed NLP model is developed in order to obtain the optimal synthesis, design, and operation of a MC. The structure of the plant is decided on simultaneously with the design, operating, and scheduling issues. The objective function employed is the maximization of annualized net profit given by the total expected sales price minus the total investment and operation cost, as was presented in the previous section.

The work is focused on providing a tool for generating and evaluating different alternatives of product production. The importance of this approach lies in the integration of synthesis, design, operation, and scheduling decisions working on a multiplant complex where trade-offs must be contemplated among the different processes. When simultaneously approached, these decisions allow considering the relationships and effects between critical elements of the problem that are usually analyzed sequentially.

The example in Figure 1.2 shows that with only two derivative plants, there are interconnections between the derivative plants given by residue recycles and intermediate productions. In our approach, both derivative plants can produce a product and a by-product. Besides producing material and energy resources for derivative plants, the mother plant produces final products to be sold.

The synthesis, design and operation problem for multiplant complex production is defined as follows:

Given N_p derivative products to be processed, their production targets, selling prices, and the production routes for each product, determine the plant configuration for each product including unit sizes, material and energy streams among the plants and the involved processing variables in order to obtain the best economical and environmental solution.

4.2. Model Formulation

In a MC, each product is produced in a different plant and some plants can also produce by-products. Given N_p different derivative products and by-products to be processed in L different plants, the model considers:

Mass balances at every unit of each plant: some of them are given by differential equations which are discretized and included in the global model as algebraic equations, for example

$$\frac{dC_{xlj}}{dt} = h(t,x) \qquad x = 1,\dots,N_x,\ l = 1,\ ..,L,\ j \in E_l \tag{4.1}$$

where C_{xlj} is the concentration of component x (biomass, substrate, product, etc.), at stage j of plant l. E_l represents the set of units of plant l. These dynamic equations are discretized and included in the overall model as algebraic equations in the same way as it was posed in the previous section.

Mass balances between units of the same plant:

$$VS_{lj} = \sum_{r=1}^{N_r} F_{ljr} + VS_{l,j-1} \qquad \forall l = 1,..,L,\ \forall j \in E_l \tag{4.2}$$

$$VS_{lj} C_{xlj}^{ini} = \sum_{r=1}^{N_r} C_{xlj}^{r} F_{ljr} + C_{xl,j-1}^{fin} VS_{l,j-1} \qquad \forall l = 1,..,L,\ \forall j \in E_l \tag{4.3}$$

where superscripts *ini* and *fin* represent the initial and final concentration respectively and VS_{lj} represents the batch volume at stage *j* in plant *l*, and F_{ljr} is the amount of resource *r* used in unit *j* at plant *l*.

Interconnection constraints between the mother plant and derivative plants:

The material and energy required resources for each process production can be obtained from the mother plant (F_r^{prod}) or can be supplied for another plant outside the complex (F_r^s). The unused amount of resource *r*, i.e. the amount of *r* that is not consumed by the derivative plants (F_r^{ex}), can be sold to market. The constraints for the consumed resources are:

$$F_r^{prod} + F_r^s = \sum_{l=1}^{L} \sum_{j=1}^{N_l} \frac{f_{rlj}}{CT_l} + F_r^{ex} \quad \text{for each resource } r \tag{4.4}$$

where f_{rlj} is the amount of *r* consumed at stage *j* in the derivative plant *l*. CT_l represents the cycle time of plant *l* defined by

$$CT_{lt} \geq T_j \quad j \in E_l, l = 1,...,L \tag{4.5}$$

where T_j is the time for processing a batch at stage *j* of plant *l* in time period *t*. Let θ_k be the processing time at semi-continuous stage *k* of plant *l*, then

$$T_j = \theta_{k'} + t_j + \theta_{k''} \qquad for\, j, k', k'' \in E_l \tag{4.6}$$

Note that Eq. (4.6) defines the time that unit *j* will be occupied is the processing time of the batch (t_j), plus the material loading ($\theta_{k'}$) and unloading ($\theta_{k''}$) time if this unit is located between semi-continuous units. It is worth noting that in this approach it is assumed that variables t_j and θ_k are involved in detailed sub-models some of them written as differential equations and included in the actual model (Corsano et al., 2004).

If a product and a by-product are produced in a plant *l*, then CT_l defined as (4.5) is the maximum of all processing time of units *j* used for the production of the product and by-product.

All products must be produced within the production time horizon. In this work, the time horizon is the same for all plants (*HT*); so:

$$Nb_l \, CT_l \leq HT \text{ for each } l = 1, \ldots, L \tag{4.7}$$

If a product and a by-product are produced in plant *l*, then

$$Nb_i \, CT_l \leq HT \text{ for each } i \text{ product produced in plant } l = 1, \ldots, L \tag{4.7'}$$

The interconnection constraints between derivative plants: mass and energy balances and the recycle equations between two different plants and/or from one stage to another in the same production process.

$$F_{lj}^{e} \geq \sum_{l \in L_e} \sum_{j \in EB_e} f_{lj}^{e} \qquad \forall\, e \tag{4.8}$$

where F_{lj}^{e} is the amount of recycle e produced at stage j in plant l, and f_{lj}^{e} is the amount of recycle e consumed at stage j in plant l. L_e and EB_e represent the set of plants and units that consume e respectively. For model simplification purposes, the store cost of raw materials and effluents is not considered in this work.

The design equations for each batch and semi-continuous unit depend on process variables as follows:

$$V_{lj} \geq S_{ij}^{l} B_i \qquad \forall i \text{ produced in } l,\ \forall j \in E_l, \forall l = 1,..,L \tag{4.9}$$

$$V_{lk} \geq S_{ik}^{l} \frac{B_i}{\theta_{ik}} \qquad \forall i \text{ produced in } l,\ \forall k \in E_l, \forall l = 1,..,L \tag{4.10}$$

where S represents size or duty factor of batch and semi-continuous units respectively. These factors are computed as a function of the process variables. In the last two equations, product i and plant l appear simultaneously because of some plants can produce a by-product, and thus they manufacture more than one product. B_i is the batch size of product i, and V_{lj} and V_{lk} are the batch and semi-continuous unit size respectively.

The production constraints of each plant:

For each product i produced in plant l, the required demand has to be satisfied

$$B_i Nb_i = Q_i \quad \forall i \text{ produced in } l, \forall l = 1,..,L \tag{4.11}$$

$$Q_i^{\min} \leq Q_i \leq Q_i^{\max} \tag{4.12}$$

where Q_i is the production of product i which is bounded by $Q_i^{\min}$ and $Q_i^{\max}$.

Just like in the multipurpose plant model, a superstructure model that considers a set of different alternatives configurations is embedded in the overall model so that the model keeps the NLP nature.

The objective function: the maximization of annualized net profits given by the total expected total sales minus the investment and operative costs:

$$Max \sum_{i=1}^{N_p} p_i Q_i + HT \sum_{r=1}^{N_r} q_r F_r^{ex} - \left(\sum_{l=1}^{L} \left(\sum_{j \in E_l} \alpha_{lj} V_{lj}^{\beta_{lj}} + \sum_{k \in E_l} \alpha_{lk} V_{lk}^{\beta_{lk}} + \text{Res}_l \right) + HT \sum_{r=1}^{N_r} \left(c_r F_r^{imp} + c_r^{'} F_r^{prod} \right) \right) \tag{4.13}$$

where Res_l represents the disposal cost for each plant l, p_i and q_r are the selling price of product i and resource r respectively, while c_r and $c_r^{'}$ are the imported and produced resource costs respectively.

The MC model considers constraints (4.1)-(4.13), which are simultaneously optimized and no heuristic or decomposition strategy is applied.

4.3. Multiplant Complex to Produce Derivatives from Sugar Cane

The Cane Sugar Complex consists of three plants interconnected by material and energy flows. The plants that constitute the complex are a Sugar plant (mother plant) and two derivative plants: Torula Yeast and Brandy/Bakery Yeast.

Production processes were described in Section 3.5, with the only difference that in this scenario each product is manufactured in a different plant. Bakery Yeast is a by-product of Brandy plant.

For both derivative plants, the configuration of the fermentation stage is obtained using the superstructure model presented in Section 2 of this chapter embedded in the global integration model, considering only in series units duplication.

Table 4.1. Simultaneous decisions made for each model

Synthesis and Design	Operation	Scheduling
Plant Configuration	Batch blending, batch splitting and batch recycling flow rates within the same production process	Units idle time of units in each plant
Number of units in series for batch stages in each plant		Plant cycle time
Blend and recycle allocation	Flow rate recycles from one process to another	Number of batches of each product
Unit sizes	Material and energy resources allocation from mother plant to the different process production	
Heating and cooling areas		
Stage number of distillation column		
	Component concentrations	
	Processing times of each unit in each plant	
	Power consumption (vapor and electricity)	

The derivatives plants are connected by the vinasses recycles, and the connections between Sugar and derivatives plants are given by material (molasses and filter juices) and energy (vapor and electricity) resources.

For the vinasses recycle equation (4.8) is rewritten as:

$$F^{DV}_{brandy,dist}Nb_{brandy} + Res_{brandy} = \sum_{j\in\{fermentors\}} f^{DV}_{torula,j}Nb_{torula} + \sum_{j\in\{fermentors\}} f^{DV}_{brandy,j}Nb_{brandy}$$

where DV means distillery vinasses and Res_{brandy} represent the residual vinasses.

For the electricity integration between the mother plant and derivative plants, for instance, equation (4.4) is rewritten as

$$F^{prod}_{elec} + F^{s}_{elec} = F_{torula,cent,elec} + F_{brany,centr,elec} + F^{ex}_{elec}$$

where *cent* means centrifuge, the only stage that consumes electricity.

In Table 4.1 the different simultaneous decisions made in this model are described.

Figure 4.1 shows the Sugar Complex integration. This figure shows that the model considers blending, recycling and splitting batches.

4.4. Results and Analysis

The presented model for optimal synthesis, design, operation and scheduling of MC provide a tool for the analysis of different alternative configurations. In addition, the economical tradeoffs among the different contemplated elements can be established and assessed. In this section, the results of the optimization of the model under different operative conditions and their analysis are shown for the Torula Yeast, Brandy and Bakery Yeast processes integrated to a Sugar plant as it was presented in the previous section. This study is based on specific data of product profits, equipment and resources costs, etc. These values can be modified by the manager or the designer, taking into account different contexts or scenarios. In particular, this work tries to show the advantages of considering all these elements simultaneously so as to assess the effects or relationships between them. Regardless of the attained results, which depend on several factors, the analytic capability of this approach is emphasized.

Due to a matter of limited space, only one example is presented here. For more examples and detailed analysis, review the work of Corsano et al. (2006c).

All models were implemented and solved in GAMS (Brooke et al., 1998) in a Pentium IV, 1.60 Ghz. The code CONOPT2 was employed to solve the NLP problems. The MC model has about 1100 equations 1200 variables, and CPU time for its resolution is about 160 sec.

In this section, we show the optimal solution for the MC model with the same model parameters employed in the previous section for Multiproduct Plant (MP) scenario. Then, taking into account the example analyzed in section 3, the both scenarios are compared. Therefore, through two different models, two possible production options are assessed and compared. Thus, we prove the versatility of mathematical models in order to make decisions in different contexts. Not only different options in a specific scenario can be evaluated, but also different scenarios can be compared.

Example

In order to evaluate different production alternatives, the MC model is solved for the same operative condition presented for the case of multiproduct plant model of Section 3, that is, $Q_i^{\min} = 7500$ ton and $Q_i^{\max} = 37500$ ton for Torula and Brandy production, while for Bakery Yeast production no bounds are imposed since this production depends on Brandy production. The time horizon is equal to 7500 h. Table 4.2 shows the optimal design variables. The optimal fermentation configuration for the Torula plant is three units in series while for the Brandy/Bakery Yeast plant is one biomass fermentor and two alcohol fermentors in series. Processing times and unit sizes are displayed in Table 4.3. The ZW transfer policy avoids idle times as shown in Figure 4.2 for both productions. Anyway, it is worth noting that the second alcohol fermentor in the Brandy plant has idle time and this is due to the trade-off that exists between the processing time of this unit and the substrate concentration of vinasses. Longer processing time implies smaller substrate concentration because the substrate is consumed at this stage.

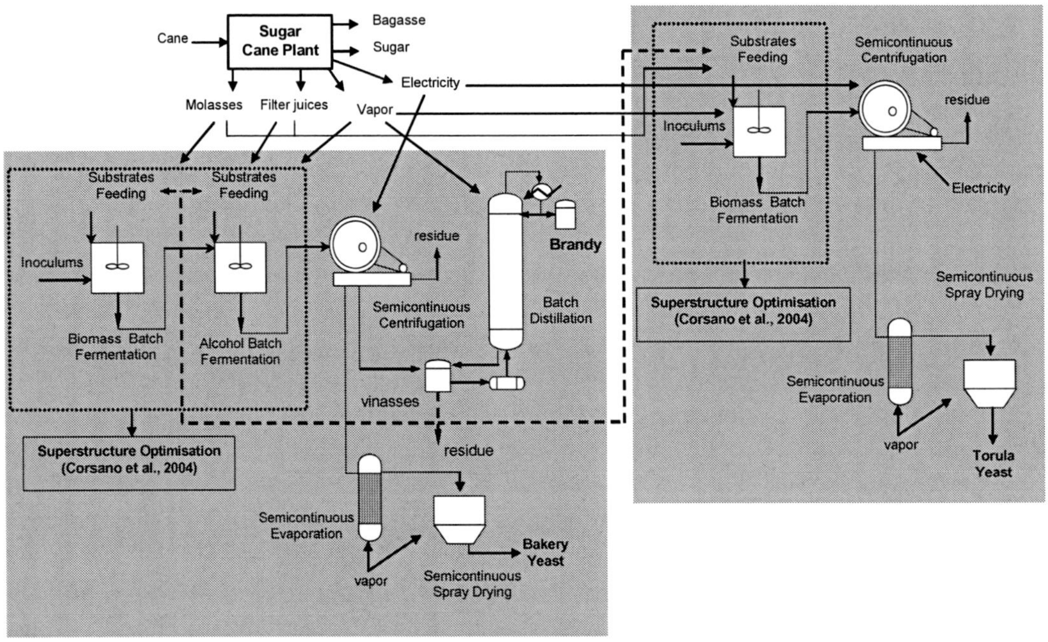

Figure 4.1. Flowsheet for Sugar Cane Complex Integration.

Table 4.2. Processing time and unit sizes of MC model solution

	Torula Plant		Brandy/Bakery Yeast Plant	
	Processing Time (h)	Unit Sizes	Processing Time (h)	Unit Sizes
Biomass Fermentor 1	7.5	9.2 m^3	15.0	140 m^3
Biomass Fermentor 2	7.5	107.2 m^3	-	-
Biomass Fermentor 3	2.8	295.5 m^3	-	-
Alcohol Fermentor 1	-		15.0	518.3 m^3
Alcohol Fermentor 2	-		5.3	593.1 m^3
Centrifuge	4.7	37.1 Kwh	5.8	55.1 Kwh
Evaporator	4.7	124.2 m^2	5.8	152.6 m^2
Dryer (diameter)	4.7	2.5 m	5.8	1.8 m
Distillation Stage number Reflux ratio Distillate tank Still vessel Condenser Evaporator Column area	-	-	9.2	 10 5.1 94.6 m^3 442.4 m^3 373.5 m^2 221 m^2 9.2 m^2

The optimal productions for this scenario are 10,350 t for Torula, 37,500 t for Brandy, and 4,500 t for Bakery Yeast production. As Brandy is the most profitable production, it is produced up to this upper bound; and Torula and Bakery Yeast are produced according to the available resources: molasses, vapor and electricity.

The Net Annual Profit is 51,622,500 $, which is similar to the Net Annual Profits of multiproduct plant. But, since the incomes for Sugar, Bagasse, and Brandy sales are the same in both cases, only the benefits for Torula, Bakery Yeast and electricity sales and the total annualized cost for both scenarios are compared and displayed in Table 4.3 for a detailed analysis. The "partial benefits" means the incomes for Torula, Bakery Yeast and unused electricity sales minus the investment and operative costs. The difference between both total partial benefits is not significant, but the investment cost for the MP plant is 19% smaller than MC. If the operative costs, such as inoculums cost, are not considered, the optimal solution may change. In this case, the operative costs are higher for the MP plant because productions are mutually dependent since all products are produced in the same plant; while at MC plant scenario, processes performances are good and there is an efficient use of units because each product is produced in a separate plant. Therefore, it is very important to simultaneously consider all these characteristics in order to perform a real comparison.

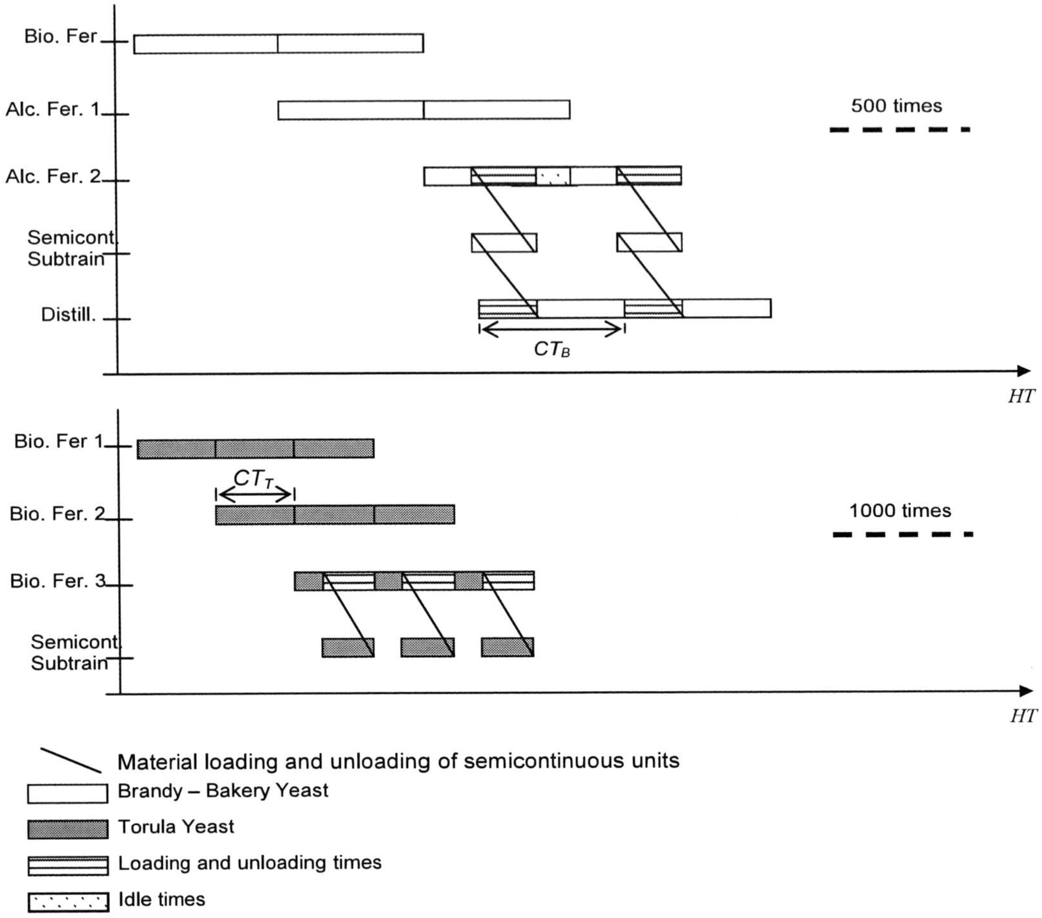

Figure 4.2. Gantt chart for MC scenario.

In addition, human resource demand is higher for the MC scenario than in MP plant. If we assume that one worker is needed for the fermentation stages, one for the semi-continuous units and one for the distillation stage, and considering three work shifts a day, 9 workers will be needed for the MP plant against 15 for MC plant. Human resources demand is not considered in the model.

Since each product is produced in a separate plant in MC scenario, each process production is performed optimally under the integration conditions because the units in each plant are exclusively used for only one product. The biomass fermentation stage performance is better in MC because a good use of units is achieved since there are no idle times and no underused units. The investment cost for this stage is about 30% higher for the MC because the total number of units used in this case is greater than in the MP scenario as a result of production in separate plants. The same difference occurs for semi-continuous units. It can be noted that the inoculums cost is lower for the MC plant due to the good use of fermentative units. The investment cost of alcohol fermentation is lower in the MC plant because unit sizes of this stage are smaller than those in the MP plant solution due to a better yield of biomass fermentation.

Table 4.3. Economic comparison between MC and MP plant models solutions

	MC Plant Model	MP Plant Model
Profit for Torula Yeast sale, ($)	6216000	5152500
Profit for Bakery Yeast sale, ($)	898500	1805250
Profit for electricity sale, ($)	109500	81750
Total profit for sales ($)	7224000	7039500
Investment cost, ($)		
Biomass Fermentors	675750	477000
Alcohol Fermentors	200250	214500
Centrifuge	162000	135750
Evaporator	198750	151500
Dryer	472500	418500
Distillation	901500	791250
Total investment cost	2610750	2188500
Operative cost, ($)		
Inoculums	35250	438750
Water (cooling and diluted)	234000	203250
Vinasses Disposal	0	28500
Imported Vapor	690000	544500
Total operative cost	959250	1215000
Total partial benefits ($)	3654000	3636000

The increment of distillation investment cost is a consequence of the processing time of this stage. A reduction on the processing time implies bigger sizes of transference areas and column diameter of this stage.

The increment of vinasses disposal cost is due to a higher volume and substrate concentration of discarded vinasses. If more vinasses were used in fermentation stages, the unit sizes would be increased and therefore the investment cost of fermentation stages would be also increased. So, there is a trade-off between vinasses use and fermentation investment cost.

Therefore, by means of a detailed analysis of economic results (Table 4.3) it can be concluded that both scenarios solutions are similar. If the synthesis, design and operation items are not taken into account simultaneously, as it is usually the case in the literature, solutions and conclusions might be erroneous. Also, the different achieved conclusions in the last paragraphs emphasize the trade-offs among the involved elements of the problem.

4.5. Conclusions and Outlook on the Proposed Multiplant Integration Model

Detailed models for optimal process synthesis, design and operation were proposed in this work in order to evaluate different production alternatives for a multiplant complex.

One of these models' characteristics is the high level of detail reached in the processing unit description, some of them being attained by means of ordinary differential equations. Batch blending, batch splitting and recycles are allowed as novel components for this type of models, and these decisions are considered as optimization variables in this work. Also, the recycles streams have been considered in order to reduce the environmental impact. All these options are very challenging taking into account the MC scenario, where the connections among all the plants must be assessed.

For a rigorous analysis, several decisions must be considered simultaneously. The proposed models integrate synthesis, design, operating and scheduling decisions.

This formulation was implemented for a Torula Yeast, Brandy and Bakery Yeast productions integrated to a Sugar plant. With the aim of evaluating different production scenarios, the solution of MC model was compared with that obtained in the previous section for a MP model. For the study case, several results were analyzed; some tradeoffs between process and design variables were presented together with the economic impact in each case. Finally the conclusions for each comparison were reported.

Further on these specific results, this work emphasizes the significant value of formulations that take into account all involved decisions simultaneously so as to reach effective solutions.

The idea presented in this work can be applied to other process productions, and the analysis can be performed according to different variations on the operative conditions.

5. General Summary and Suggestions for Further Reading

This chapter has been prepared in order to present different applications of mathematical modeling in chemical engineering. Really it is a powerful tool to solve significant problems in the different stages of the chemical process lifecycle. We have chosen a particular context, and modelled different problems related to batch processes. The obtained conclusions have allowed us to present the valuable results that can be attained when complete and comprehensive models have been developed.

The detailed models provided in this chapter have been focused on the simultaneous optimization of the synthesis, design, operation, and scheduling of chemical processes. These models include decisions of plant configuration, material and energy balances, design equations, product and by-products specifications constraints, environmental considerations, integration processes decisions, blending and recycles constraints, economics objective functions, among others. Thus, we have posed models that show one of the main advantages of mathematical modeling: decisions can be made taking into account all the elements involved and the trade-offs among them. Usually, different actors, areas, stages, etc. may have different and opposite goals that must be related and coordinated in order to achieve the optimal results for the global system.

In the first section, a concise overview of basic concepts and definitions was provided, in order that the reader can be introduced in process engineering models and some methodologies used in this area. For further reading, a classic text of this area is due to Biegler, Grossmann, and Westerberg (1997).

In the second section, a novel superstructure model is presented in order to obtain the optimal plant configuration simultaneously with the optimal plant design and operation. The main characteristic of this approach remains in the fact that discrete decisions are taken without resort to integer optimization variables. The details of this approach and more examples can be found in Corsano et al. (2004, 2006a).

In the third section a heuristic procedure for finding the optimal mixed product campaign for a multipurpose/multiproduct plant is presented. The approach consists in two steps: first a relaxed model is solved in order to obtain the plant configuration and the relationship between the numbers of batch of each product. Then, according to that relationship, a mixed product campaign model is proposed and solved. Sometimes, in this model, the designer criteria or knowledge can influence in the campaign sequence selection and it can be taken into account. For further details of this methodology see Corsano et al. (2007). For others approaches for mixed product campaign models see Birewar and Grossmann (1989, 1990), Xia and Macchietto (1997), Lin and Floudas (2001), and Ierapetritou et al. (1999).

In the fourth section a model for the processes integration in a multiplant complex is developed. The model stresses the optimization of all the production processes embedded in an overall model, emphasizing the relations among all them. The use of recycles streams in order to reduce the environmental impact is considered. The presented model for optimal synthesis, design, operation and scheduling of MC provide a tool for the alternatives analysis of different configurations. In addition, the economical tradeoffs between each production scheme were established and assessed as well as a comparison with a multiproduct plant configuration.

In all the sections, numerical examples corresponding to the Sugar Cane industry and its derivative productions were posed. Although the Sugar Cane example serves as good case of processes integration, the proposed models can be modified to other industrial and engineering examples where several decisions and different alternatives can be analyzed.

References

Atkinson, K. E. *An Introduction to Numerical Analysis*. Second Edition. John Wiley & Sons; New York, 1989.

Barrera M. D.; Evans L. B. Optimal Design and Operation of Batch Processes. *Chem. Engng. Comm.* 1989, *82,* 45.

Bhatia, T. K.; Biegler, L.T. Dynamic Optimisation in the Design and Scheduling of Multiproduct Batch Plants *Ind. Eng. Chem. Res.* 1996, *35*, 2234.

Biegler, L. T.; Grossmann, I. E.; Westerberg, A. W. Systematic Methods of Chemical Process Design. Prentice Hall International Series in the Physical and Chemical Engineering Sciences; New Jersey, 1997.

Birewar, D. B.; Grossmann, I. E. Incorporating scheduling in the optimal design of multiproduct batch plant. *Comp. & Chem. Eng.* 1989, *13*, 141.

Birewar, D. B.; Grossmann, I. E. Simultaneous synthesis, sizing, and scheduling of multiproduct batch plants. *Ind. Eng. Chem. Res.* 1990, *29*, 2242.

Brooke A.; Kendrick D.; Meeraus A.; Raman R. *GAMS, A User Guide*. Scientific press, Calif,1998.

Corsano, G.; Iribarren, O. A.; Montagna, J. M.; Aguirre, P.A.. Batch Fermentation Networks Model for Optimal Synthesis, Design and Operation. *Ind. Eng. Chem. Res.* 2004, *43*, 4211.

Corsano, G., Montagna, J.M, Aguirre, P.A., Iribarren, O.A. Design and Operation Issues using NLP Superstructure modeling. *Applied Mathematical Modeling* 2006a, 30 (9), 974.

Corsano, G.; Iribarren, O. A.; Montagna, J. M.; Aguirre, P.A.; González Suárez, E. Economic Tradeoffs involved in the Design of Fermentation Processes with Environmental Constraints. *Chemical Engineering Research and Design* 2006b, *84 (10)*, 1.

Corsano, G.; Montagna, J. M.; Iribarren, O. A.; Aguirre, P.A. Processes Integration: Multiplant Complex vs Multipurpose Plant Assessment. *Ind. Eng. Chem. Res.* 2006c, *45*, 4253.

Corsano, G., Montagna, J.M, Aguirre, P.A., Iribarren, O.A. A Heuristic Method for the Optimal Synthesis and Design of Batch Plants Considering Mixed Product Campaigns. *Ind. & Eng. Chem. Res.* 2007, 46 (9), 2769.

Duran, M. A.; Grossmann, I. E. Outer-Approximation Algorithm for a Class of Mixed-Integer Nonlinear Programs. *Mathematical Programming* 1986, 36 (3), 307.

Grossmann, I. E.; Sargent, R. W. H. Optimum design of multipurpose chemical plants. *Industrial and Engineering Chemistry Process Design and Development* 1979, *18 (2)*, 343.

Ierapetritou, M. G.; Floudas, C. A.; Vasantharajan S.; Cullick A. S. Optimal Location of Vertical Wells: Decomposition Approach . *AIChE Journal* 1999, *45(4),* 844.

Jackson, J.R.; Grossmann I.E. Temporal decomposition scheme for nonlinear multisite production planning and distribution models. *Ind. Eng. Chem. Res.* 2003, *42*, 3045.

Kallrath, J. Combined strategic and operational planning: An MILP success story in chemical industry. *OR Spectrum* 2002, *24*, 315 – 341.

Knopf F. C.; Okos M. R.; Reklaitis G. V. Optimal Design of Batch / Semi continuous Processes. *Ind. Eng. Chem. Process Des. Dev.* 1982, *21*, 79.

Lee, H-K.; Lee I-B.; Reklaitis, G.V. Capacity expansion problem of multisite batch plants with production and distribution. *Comp. & Chem. Engn.* 2000, *24*, 1597.

Lin, X.; Floudas, C. A. Design, synthesis and scheduling of multipurpose batch plants via an effective continuous-time formulation. *Comp. & Chem. Engn.* 2001*, 25*, 665.

Montagna J. M.; Vecchietti A. R.; Iribarren O. A.; Pinto J. M.; Asenjo J. A. Optimal Design of Protein Production Plants with Time and Size Factor Process Models. *Biotechnol. Prog.* 2000, *16*, 228.

Ravemark D. E.; Rippin D. W. T. Optimal Design of Multiproduct Batch Plants. *Comput. Chem. Engng.* 1998, *22*, 177.

Salomone H. E.; Iribarren O. A. Posynomial Modeling of Batch Plants. *Comput. Chem. Engng.* 1992, *16*, 173.

Salomone H. E.; Montagna J. M.; Iribarren O. A. Dynamic Simulations in the Design of Batch Processes. *Comput. Chem. Engng.* 1994, *18*, 191.

Varvarezos, D. K.; Grossmann, I. E.; Biegler, L. T. An Outer-Approximation Method for Multiperiod Design Optimization. *Ind. Eng. Chem. Res.* 1992, 31, 1466.

Viswanathan, J.; Grossmann, I. E. Combined penalty function and outer-approximation method for MINLP optimization. *Comp. & Chem. Engn.* 1990, 14 (7), 769.

Voudouris, V. T.; Grossmann, I. E. MILP model for scheduling and design of a special class of multipurpose batch plants. *Comp. & Chem. Engng.* 1996*, 20 (11)*, 1335.

Wu, M. D. ; Ierapetritou, M. Hierarchical approach for production planning and scheduling under uncertainty, *Chem. Eng. Process* 2007, in press.

Xia, Q.; Macchietto, S. Design and synthesis of batch plants—MINLP solution based on a stochastic method. *Comp. & Chem. Engn.* 1997, *21*, S697.

Yeh N. C.; Reklaitis G. V. Synthesis of Batch / Semicontinuous Processes. *Comput. Chem. Engng.* 1987, *11*, 639.

Zhang, X.; Sargent, R. W. H. The optimal operation of mixed production facilities- a general formulation and some solution approaches for the solution. In *Proceeding of 5th Int. Symp. Process Systems Engineering* 1994, 171. Korea.

Zhang, X.; Sargent, R. W. H. The optimal operation of mixed production facilities - extensions and improvements. *Comp. & Chem. Engn.* 1996, *S20*, 1287.

In: Applied Mathematical Modeling
Editor: Eetu N. Virtanen, pp. 89-143

ISBN: 978-1-60021-975-7

Chapter 2

ON MODELLING SUBCOOLED BOILING FLOWS: A REVIEW

***G. H. Yeoh*[1,*] *and J. Y. Tu*[2]**

[1]Australian Nuclear Science and Technology Organisation (ANSTO)
PMB 1, Menai, NSW 2234, Australia
[2]School of Aerospace, Mechanical and Manufacturing Engineering
RMIT University, Vic. 3083, Australia

ABSTRACT

In the context of computational fluid dynamics (CFD), modeling low-pressure subcooled boiling flow is of particular significance. A review is provided in this article of the various numerical modeling approaches that have been adopted to handle subcooled boiling flow. The main focus in the analysis of such a challenging problem can be broadly classified according into two important categories: (i) Heat transfer and wall heat flux partitioning during subcooled boiling flow at the heated wall and (ii) Two-phase flow and bubble behaviors in the bulk subcooled flow away from the heated wall. For the first category, details of both empirical and mechanistic models that have been proposed in the literature are given. The enhancement in heat transfer during forced convective boiling attributed by the presence of both sliding and stationary bubbles, force balance model for bubble departure and bubble lift-off as well as the evaluation of bubble frequency based on fundamental theory depict the many improvements that have been introduced to the current mechanistic model of heat transfer and wall heat flux partitioning. For the second category, details of applications of various empirical relationships and mechanistic model such as population balance model to determine the local bubble diameter in the bulk subcooled liquid that have been employed in the literature are also given. A comparison of the predictions with experimental data is demonstrated. For the local case, the model considering population balance and improved

* Mailing address: Dr. Guan Heng Yeoh, B40, ANSTO, Private Mail Bag 1, MENAI, NSW 2234, Australia. Email: Guan.Yeoh@ansto.gov.au Tel: +61-2-9717-3817; Fax: +61-2-9717-9263.

wall heat partition shows good greement with the experimentally measured radial distributions of the Sauter mean bubble diameter, void fraction, interfacial area concentration and liquid velocity profiles. Significant weakness prevails however over the vapor velocity distribution. For the axial case, good agreement is also achieved for the axial distributions of the Sauter mean bubble diameter, void fraction and interfacial area concentration profiles. The present model correctly represents the plateau at the initial boiling stages at upstream, typically found in low-pressure subcooled boiling flows, followed by the significant rise of the void fraction at downstream.

NOMENCLATURE

a_{if}	interfacial area concentration
A_B	bubble area
A_C	cross-sectional area of boiling channel
A_q	fraction of heater area occupied by bubbles
Bo	boiling number = Q_w/Gh_{fg}
c_f	constant in equation (44)
C_1, C_2	constants defined in equation (67)
C_{bw}	constant in equation (43)
C_D	drag coefficient
C_L	shear lift coefficient
C_{TD}	turbulent dispersion coefficient
C_p	specific heat
C_s	constant defined in equation (63)
C_{w1}, C_{w2}	wall lubrication constants
C_μ	turbulent constant = 0.09
$C_{\mu b}$	bubble induced turbulent constant
C_v	acceleration coefficient
d	vapor bubble diameter at heated surface
d_{bw}	bubble departure diameter
$d_{b,max}$	maximum allowed bubble diameter
d_i, d_j	daughter particle diameters
d_{ij}	equivalent diameter
d_w	surface/bubble contact diameter
D	average bubble diameter
D_b	departing bubble diameter
D_d	bubble departure diameter
D_l	bubble lift-off diameter
D_s	Sauter bubble diameter
D_{B1}, D_{B2}	reference bubble diameters
D_B	death rate due to break-up
D_C	death rate due to coalescence
f	bubble departure frequency
f_i	scalar fraction related to the number density of the discrete bubble classes

F	degree of surface cavity flooding
F_b	buoyancy force
F_h	force due to the hydrodynamic pressure
F_{du}	unsteady drag force due to asymmetrical growth of the bubble
F_s	surface tension force
F_{sL}	shear lift force
F_x	forces along the *x-direction*
F_y	forces along the *y-direction*
F_{lg}	action of interfacial forces from vapor on liquid
F_{gl}	action of interfacial forces from liquid on vapor
F_{lg}^{drag}	drag force
F_{lg}^{lift}	lift force
$\mathrm{F}_{lg}^{lubrication}$	wall lubrication force
$\mathrm{F}_{lg}^{dispersion}$	turbulent dispersion force
g	gravitational constant
$\vec{g}$	gravitational vector
G	mass flux
G_s	dimensionless shear rate
h	interfacial heat transfer coefficient
h_0	initial film thickness
h_f	critical film thickness at rupture
h_{fg}	latent heat of vaporization
H	enthalpy
Ja	Jakob number based on wall superheat
Ja_{sub}	Jakob number based on liquid subcooling = $\rho C_{pl}\left(T_{sat}-T_l\right)/\rho_g h_{fg}$
k	turbulent kinetic energy
K	projected area of bubble
l_s	sliding distance
$\vec{n}$	normal to the surface
n_i	number density of the discrete bubble *i*th class
n_j	number density of the discrete bubble *j*th class
N	number of bubble classes
N_a	active nucleation site density
P_B	production rate due to break-up
P_C	production rate due to coalescence
Q_w	wall heat flux
Q_c	heat transfer due to forced convection
Q_e	heat transfer due to evaporation
Q_{tc}	heat transfer (transient conduction) due to stationary bubble
Q_{tcsl}	heat transfer (transient conduction) due to sliding bubble
r	bubble radius at heated wall or flow spacing within annular channel
r_c	cavity radius at heated surface
r_r	curvature radius of the bubble at heated surface

Re	flow Reynolds number = Gd_h / μ_l
Re_b	bubble Reynolds number
R_f	ratio of the actual number of bubbles lifting off to the number of active nucleation sites
R_i	radius of inner heated wall
R_o	radius of outer unheated wall
R_{ph}	source/sink term due to phase change
s	spacing between nucleation sites
S_i	additional source terms due to coalescence and breakage
St	Stanton number
t	thermo-fluid time scale
t_g	bubble growth period
t_{ij}	coalescence time
t_{sl}	bubble sliding period
t_w	bubble waiting period
T	temperature
T_{sub}	local subcooling temperature = $T_{sat} - T_l$
T_{sub1} T_{sub2}	Reference subcooling temperatures
ΔT	difference in temperature
P	pressure
u	velocity
$\vec{u}$	velocity vector)
u_t	velocity due to turbulent collision
u_τ	friction velocity
v_i	specific volume of discrete bubble ith class
V_b	bubble velocity
V_B	bubble volume
x	Cartesian coordinate along x
x^+	non-dimensional normal distance from heated wall
y_w	adjacent point normal to the wall surface

Greek Letters

α	advancing angle
α_g	vapor void fraction
α_l	liquid void fraction
β	receding angle or measured constant in equation (44)
χ	coalescence rate
δ_t	Thermal boundary layer thickness
ε	turbulent dissipation rate
ϕ_{WN}	bubble nucleation rate
ϕ_{COND}	bubble condensation rate
η	thermal diffusivity

λ	thermal conductivity or size of an eddy
μ	viscosity
μ^{eff}	effective viscosity
θ	bubble contact angle
θ_i	inclination angle
θ_{ij}	turbulent collision rate
ρ	density
σ	surface tension
τ_{ij}	bubble contact time
ξ	size ratio between an eddy and a particle in the inertial sub-range
ξ_H	heated perimeter of boiling channel
Γ_{lg}	interfacial mass transfer from vapor to liquid
Γ_{gl}	interfacial mass transfer from liquid to vapor
Ω	break-up rate

Subsripts

axial	axial distribution
eff	effective
g	vapor
inlet	channel entrance
l	liquid
local	local distribution
ONB	onset of nucleate boiling
s	surface heater
sat	saturation
sub	subcooling
Tb	bubble induced turbulence
Tl	liquid shear-induced turbulence
w	wall

INTRODUCTION

During the last five decades, numerous numerical models have been developed to predict the fluid flow and heat transfer characteristics of subcooled boiling flow. is the boiling process by nature is inherently complex, which still makes the analysis extremely challenging. Subcooled boiling flow can usually be characterized by the presence of thermodynamic non-equilibrium between the liquid and vapor phases. A *high-temperature* two-phase region exists near the heated wall whilst a *low-temperature* single-phase liquid prevails away from the heated surface. Consider the schematic drawing of a subcooled boiling flow accompanied by a typical boiling curve describing the void fraction distribution in Figure 1. Heterogeneous bubble nucleation occurs within the small pits and cavities designated as nucleation sites on the heated surface. These nucleation sites are activated when the temperature of the surface exceeds the saturation liquid temperature at the local pressure. Away from the wall, the temperature of the bulk fluid remains below saturation and is, by definition, subcooled. At a point called the onset of nucleate boiling (ONB), boiling occurs and bubbles remain attached to the heater surface. Further downstream, as the bulk temperature liquid temperature increases, the bubbles grow larger and begin to detach from the heater surface. The void fraction increases sharply at a location called the net vapor generation (NVG), which indicates the transitional point between two regions: low void fraction region followed by another region in which the void fraction increases significantly thereafter. This review will mainly focus on subcooled boiling flow especially at low-pressure applications.

Modeling low-pressure subcooled boiling flow in the context of computational fluid dynamics (CFD) can be broadly classified according to two categories: (i) Heat transfer and wall heat flux partitioning during subcooled boiling flow at the heated wall and (ii) Two-phase flow and bubble behaviors in the bulk subcooled flow away from the heated wall.

The first category generally entails the development of models to predict the heat transfer rate during subcooled nucleate boiling flow by appropriately partitioning the wall heat flux. The various approaches that have been adopted are empirical correlations for wall heat flux, empirical correlations for the partitioning of the wall heat flux and mechanistic models for wall heat flux and partitioning.

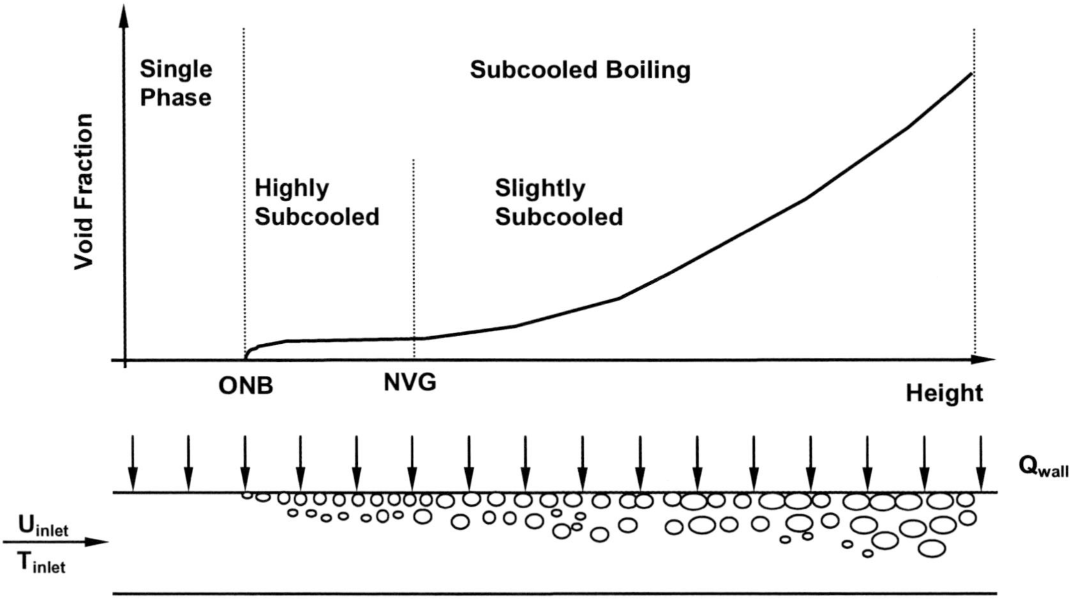

Figure 1. A schematic illustration of a subcooled boiling flow in a heated channel.

On the basis where empirical correlations for wall heat flux are employed, these correlations are usually limited to the prediction of total wall heat flux for a particular boiling flow situation. These empirical heat transfer models attempt to predict the wall heat flux given the wall superheat ΔT_w for a particular flow condition in the regions covering the singles-phase forced convection, partial nucleate boiling and full-developed boiling. In the partial nucleate boiling region, which is the transition region between single-phase forced-convection and fully-developed nucleate boiling, the correlations of Bergles and Rohsenow [1], Mikic and Roshsenow [2], Chen [3] and Kandlikar [4] are widely used. Several correlations have also been proposed for the fully-developed nucleate boiling of which specific relationships developed by McAdams et al. [5], Thom et al. [6] and Kandlikar [4] are popular expressions. Despite their extensive usage, they are still nevertheless merely correlations of experimental data and as such they preclude the modeling of heat transfer mechanisms and are unable to provide any information regarding the partitioning of the wall heat flux between the liquid and vapor phases.

A number of correlations for wall heat flux partitioning have also been proposed in the literature. These correlations differ from above, namely, what heat transfer mechanisms are considered, and what fraction of the total heat flux is transferred by each of these mechanisms. Various processes occurring at the heater surface such as the information required for the vapor generation and condensation rates are distinguished. The primary goal of the correlations is to predict the bulk void fraction; empirical correlations are formulated to provide information on how the wall heat flux is to be partitioned but not of the total wall heat flux itself. Useful models such as proposed by Griffith et al. [7], Zuber and Findlay [8], Lahey [9], Chatoorgoon et al. [10] and Zeitoun [11] are noted for their significant contributions to the study of subcooled boiling flow.

Mechanistic models based on relevant heat transfer mechanisms occurring during the boiling process can however be readily employed for both the prediction of the wall heat flux as well as the partitioning of the wall heat flux between the liquid and vapor phases. Most CFD investigations on subcooled boiling flow are mainly based on the use of these mechanistic models. In essence, the wall heat flux can be partitioned into three heat flux components: single-phase, transient conduction and evaporation. Single-phase heat transfer is assumed to occur in areas of the heated surface unaffected by the bubbles. Transient conduction (quenching heat flux component) occurs over the area of the heater surface under the influence of bubbles while the single-phase heat flux component acts over the area unaffected by bubbles. The model consists of three unknown parameters, which should be modeled accurately. They are: active nucleation site density (N_a), departing bubble diameter (D_b) and bubble frequency (f).

The second category involves the handling of the fluid flow in the bulk subcooled liquid. The two-fluid model based on the Eulerian framework represents the most detailed and accurate macroscopic formulation of the thermal and hydrodynamic characteristics of any two-phase systems. In the two-fluid model, the field equations are expressed by two sets of six conservation equations of mass, momentum and energy. For subcooled boiling flow, one set is dedicated to resolve the gas bubble phase while the other is solved for the liquid water phase. Through an appropriate averaging of the local instantaneous balance equations, phasic interaction terms appear in each of the averaged balance equations. These interfacial terms represent the mass, momentum and energy transfers through the interface between the phases. The existence of these interfacial transfer terms is rather significant as they determine the rate of phase changes, and the degree of mechanical and thermal non-equilibrium between phases. Most importantly, they provide the necessary closure relations to the two-fluid model formulation. In the mass conservation equations, mass transfer is accounted between phases due to gas bubbles being condensed in the bulk subcooled liquid. In the momentum conservation equations, the important interfacial effects between the gas and liquid phases due to the drag force as well as other possible so-called non-drag forces in the form of lift, wall lubrication and turbulent dispersion are incorporated. In the energy conservation equations, the interfacial heat transfer accounts for the phase change due to condensation.

Also, the prediction of the local bubble sizes in the subcooled liquid flow is expected to be strongly influenced by the complex bubble behaviors in the two-phase flows occurring near the heated wall. A detailed mathematical analysis indicates that empirical correlations can at best only provide a macroscopic description of the boiling phenomenon. In reality, complex mechanistic behaviors of bubble coalescence and bubble collapse due to condensation (microscopic in nature) exist in the two-phase subcooled boiling flow. Physical insights suggest that the numerical solutions could be significantly improved by accounting these bubble mechanistic behaviors especially in determining the local bubble size such as the Sauter bubble diameter. In the two-fluid model, the Sauter bubble diameter appears in the interfacial transfer terms of the mass, momentum and energy. An accurate determination of the Sauter bubble diameter is crucial as the bubble size influences the inter-phase heat and mass transfer through the interfacial area concentrations and momentum drag terms. Because of the many successes in applying the population balance approach towards better describing and understanding complex industrial flow systems [12–16], the potential to implement and extend the modeling of subcooled boiling flow to predict the non-uniform bubble size distribution is of enormous significance. It will be demonstrated later that the specific development of the population balance approach for subcooled boiling flows has contributed to a considerable improvement of the two-fluid boiling model formulation.

A review of the various models/correlations developed for the prediction of heat transfer and wall heat flux partitioning as well as the formulation of conservations equations for the two-fluid boiling model and population balance model is given in the following sections.

Conservation Equations for the Two-Fluid Boiling Model

The two-fluid model treating both the vapor and liquid phases as continua solves two sets of conservation equations governing mass, momentum and energy which can be expressed as:

Continuity Equation of Liquid Phase

$$\frac{\partial \rho_l \alpha_l}{\partial t} + \nabla \cdot \left(\rho_l \alpha_l \vec{u}_l \right) = \Gamma_{lg} \tag{1}$$

Continuity Equation of Vapor Phase

$$\frac{\partial \rho_g \alpha_g}{\partial t} + \nabla \cdot \left(\rho_g \alpha_g \vec{u}_g \right) = \Gamma_{lg} \tag{2}$$

Momentum Equation of Liquid Phase

$$\frac{\partial \rho_l \alpha_l \vec{u}_l}{\partial t} + \nabla \cdot (\rho_l \alpha_l \vec{u}_l \vec{u}_l) = -\alpha_l \nabla P + \alpha_l \rho_l \vec{g} + \nabla \cdot \left[\alpha_l \mu_l^{eff} \left(\nabla \vec{u}_l + (\nabla \vec{u}_l)^T \right) \right] + \left(\Gamma_{lg} \vec{u}_g - \Gamma_{gl} \vec{u}_l \right) + \mathrm{F}_{lg} \tag{3}$$

Momentum Equation of Vapor Phase

$$\frac{\partial \rho_g \alpha_g \vec{u}_g}{\partial t} + \nabla \cdot (\rho_g \alpha_g \vec{u}_g \vec{u}_g) = -\alpha_g \nabla P + \alpha_g \rho_g \vec{g} + \nabla \cdot \left[\alpha_g \mu_g^{eff} \left(\nabla \vec{u}_g + (\nabla \vec{u}_g)^T \right) \right] + \left(\Gamma_{gl} \vec{u}_l - \Gamma_{lg} \vec{u}_g \right) + \mathrm{F}_{gl} \tag{4}$$

Energy Equation of Liquid Phase

$$\frac{\partial \rho_l \alpha_l H_l}{\partial t} + \nabla \cdot (\rho_l \alpha_l \vec{u}_l H_l) = \nabla \cdot \left[\alpha_l \lambda_l \nabla T_l + \frac{\mu_{Tl}}{Pr_{Tl}} \nabla H_l \right] + + \left(\Gamma_{lg} H_g - \Gamma_{gl} H_l \right) \tag{5}$$

Energy Equation of Vapor Phase

$$\frac{\partial \rho_g \alpha_g H_g}{\partial t} + \nabla \cdot (\rho_g \alpha_g \vec{u}_g H_g) = \nabla \cdot \left[\alpha_g \lambda_g \nabla T_g + + \frac{\mu_{Tg}}{Pr_{Tg}} \nabla H_g \right] + \left(\Gamma_{gl} H_l - \Gamma_{lg} H_g \right) \tag{6}$$

In subcooled boiling flow, the source term Γ_{lg} in equation (1) represents the mass transfer rate due to condensation in the bulk subcooled liquid. It can be expressed by:

$$\Gamma_{lg} = \frac{h a_{if} \left(T_{sat} - T_l \right)}{h_{fg}} \tag{7}$$

where h is the inter-phase heat transfer coefficient determined from Ranz and Marshall [17] Nusselt number correlation and a_{if} is the interfacial area between phases per unit volume. For mass conservation, it is noted that the source term of the mass transfer rate Γ_{gl} in equation (2) is the negative of Γ_{lg} in equation (1), *viz.*, $\Gamma_{gl} = -\Gamma_{lg}$. The wall vapor generation rate Γ_{wg} is modeled in a mechanistic way derived by considering the total mass of bubbles detaching from the heated surface as:

$$\Gamma_{wg} = \frac{Q_e}{h_{fg} + C_{pl} T_{sub}} \tag{8}$$

where Q_e is the heat transfer due to evaporation. This wall nucleation rate is usually accounted as a specified boundary condition for the continuity equation of vapor phase.

Interfacial transfer terms in the momentum and energy equations represented by Γ_{lg} and F_{lg} denote the transfer terms from the gas phase to the liquid phase. The mass transfer Γ_{lg} for subcooled boiling flow is already given in equation (7) whilst the total interfacial force F_{lg} considered in the present study includes the following forces:

$$\mathrm{F}_{lg} = \mathrm{F}_{lg}^{drag} + \mathrm{F}_{lg}^{lift} + \mathrm{F}_{lg}^{lubrication} + \mathrm{F}_{lg}^{dispersion} \tag{9}$$

The terms on the right-hand side in equation (9) represent the drag force, lift force, wall lubrication force and turbulent dispersion force respectively. This force strongly governs the distribution of the liquid and gas phases within the flow volume. The interfacial momentum transfer between gas and liquid due to drag force F_{lg}^{drag} is a result of the shear and form drag of the fluid flow. It is modeled according to Ishii and Zuber [18] as:

$$F_{lg}^{drag} = -F_{gl}^{drag} = \frac{1}{8} C_D a_{if} \rho_l \left| \vec{u}_g - \vec{u}_l \right| \left(\vec{u}_g - \vec{u}_l \right) \qquad (10)$$

Owing to the horizontal velocity gradient, bubbles rising in a liquid are subjected to a lateral lift force. Following Drew and Lahey [19], the lift force in terms of the slip velocity and the curl of the liquid phase velocity is described by:

$$F_{lg}^{lift} = -F_{gl}^{lift} = \alpha_g \rho_l C_L \left(\vec{u}_g - \vec{u}_l \right) \times \left(\nabla \times \vec{u}_l \right) \qquad (11)$$

In addition to the lift force, a lateral force is also formed to prevent bubbles attaching on the solid walls due to surface tension. The wall lubrication force, which is in the normal direction away from the heated wall and decays with distance from the wall, can be modeled as [20]:

$$F_{lg}^{lubrication} = -F_{gl}^{lubrication} = -\frac{\alpha_g \rho_l \left(\vec{u}_g - \vec{u}_l \right)}{D_s} \max\left(0, C_{w1} + C_{w2} \frac{D_s}{y_w} \right) \vec{n} \qquad (12)$$

To avoid the emergence of attraction force, the force is set to zero for large y_w.

For turbulence assisted bubble dispersion, turbulence induced dispersion taken as a function of turbulent kinetic energy and gradient of the liquid void fraction can be expressed in the form of [21]:

$$F_{lg}^{dispersion} = -F_{gl}^{dispersion} = -C_{TD} \rho_l \kappa \nabla \alpha_l \qquad (13)$$

More detail descriptions of these forces can be found in Anglart and Nylund [22] and Lahey and Drew [23]. The drag coefficient C_D in equation (10) has been correlated for several distinct Reynolds number regions for individual bubbles according to Ishii and Zuber [18]. For the lift coefficient C_L, Lopez de Bertodano [20] and Takagi and Matsumoto [24] suggested a value of $C_L = 0.1$. Drew and Lahey [19] proposed $C_L = 0.5$ based on objectivity arguments. Tomiyama [25] developed however an Eötvos number dependent correlation that allows negative coefficients to emerge if the bubble diameter is larger than 5.5 mm (i.e. for air-water system). This results in a negative lift force forcing the large bubbles to be emigrated towards the tube centre. For the range of flow conditions considered in this paper, the bubbly flow regime persisted in the subcooled boiling flow; only positive lift forces appeared for small bubbles. The constant of $C_L = 0.01$ as suggested by Wang et al. [26] has been found to be appropriate for the wide range of boiling flows investigated. The wall lubrication constants $C_{w1} = -0.01$ and $C_{w2} = 0.05$ as suggested by Antal et al. [21] are employed while the recommended value for $C_{TD} = 0.1$ according to Kurul and Podowski [27] is adopted for the turbulent dispersion force.

Unlike single phase fluid flow problem, no standard turbulence model has been tailored for two-phase turbulent flow. In majority of two-phase flow applications, the standard two-equation k-ε turbulence model is employed to resolve the turbulent flow associated with the continuous liquid and dispersed vapor phases. For simplicity, the standard k-ε model has been employed throughout the various subcooled boiling investigations herein. It is nevertheless noted that Menter's k-ω based shear stress transport (SST) [28] could also be alternatively employed to resolve the two-phase turbulent subcooled boiling flow. In the recent work by Cheung et al. [29] and numerical calculations performed by Frank et al. [30], the "wall" peaking behavior of the void fraction distribution in an isothermal bubbly flow was seen to be better captured through the SST turbulent model. This could be attributed to a more realistic prediction of the turbulent dissipation close to the wall through the k-ω formulation. Further development should be focused on two-phase flow turbulence modeling in order to better understand or improve existing models.

On the basis of the eddy viscosity hypothesis, the shear-induced viscosity in the momentum liquid phase and vapor phase equations can be taken as the sum of the molecular viscosity and turbulent viscosity of each phase. By the inclusion of Sato's [31] model for bubble-induced turbulence, the effective viscosity is thus considered as the total of the shear-induced turbulent viscosity and the bubble-induced turbulent viscosity. The viscosity of the liquid phase can be expressed as:

$$\mu_l^{eff} = \mu_l + \mu_{Tl} + \mu_{Tb} \tag{14}$$

where the liquid turbulent viscosity is given by

$$\mu_{Tl} = \rho_l C_\mu \frac{k_l^2}{\varepsilon_l} \tag{15}$$

and the extra bubble induced turbulent viscosity is evaluated according to

$$\mu_{Tb} = \rho_l C_{\mu b} \alpha_g D_s \left| \vec{u}_g - \vec{u}_l \right| \tag{16}$$

The constant $C_{\mu b}$ has a value of 0.6. Effective viscosity in the vapor phase can now be simply evaluated as

$$\mu_g^{eff} = \mu_l^{eff} \frac{\rho_g}{\rho_l} \tag{17}$$

APPLICATION OF EMPIRICAL RELATIONSHIPS TO MODELING SUBCOOLED BOILING FLOW

Mechanistic Models and Closure through Various Empirical Relationships

A number of mechanistic models have been developed for the prediction of wall heat flux and partitioning. Del Valle and Kenning [32] concentrated on the formulation of a mechanistic model for subcooled nucleate flow boiling by taking into consideration the bubbly dynamics at the heated wall. This model employed some of the concepts developed initially by Graham and Hendricks [33] for wall heat flux partitioning during pool nucleate boiling. The mechanistic model by Kurul and Podowski [27], adapted by Judd and Hwang [34] on the premise of a model also for wall heat flux partitioning during pool nucleate boiling, is still widely employed in the CFD commercial code CFX4.4 and ANSYS-CFX11. In the paper by Steiner et al. [35], a modified Chen-type superposition model was used to compute the wall heat flux in subcooled boiling flow. More recently, a new approach to the partitioning of the wall heat flux has been proposed by Basu et al. [36,37] and reviewed in Warrier and Dhir [38]. The fundamental idea of the model is to consider all the energy from the wall is transferred to the liquid adjacent to the heated wall. Thereafter, a fraction of the energy is transferred to the vapor bubbles by evaporation while the remainder goes into the bulk liquid.

The mechanistic model of Kurul and Podowski has been applied with much success in many CFD investigations of subcooled boling flows at low pressures. In essence, this model entails the partitioning of the wall heat flux into three heat flux components. They are: (i) Heat transferred by conduction to the superheated layer next to the wall (nucleate boiling or surface quenching), Q_q; (ii) Heat transferred by evaporation or vapour generation, Q_e; and (iii) Heat transferred by turbulent convection, Q_c. The wall heat flux partitioning can be written as

$$Q_w = Q_q + Q_e + Q_c \tag{18}$$

The surface quenching heat flux is determined through the relationship:

$$Q_q = \left[\frac{2}{\sqrt{\pi}} \sqrt{k_l \rho_l C_{pl}} \sqrt{f} \right] A_q \left(T_w - T_l \right) \tag{19}$$

where T_w is the wall temperature, A_q is the fraction of wall area subjected to cooling by quenching and f is the bubble departure frequency.

Heat flux due to vapour generation at the wall in the nucleate boiling region can be simply calculated from Bowring [39]:

$$Q_e = N_a f \left(\frac{\pi}{6} d_{bw}^3 \right) \rho_g h_{fg} \tag{20}$$

The heat flux according to the definition of local Stanton number St for turbulent convection is given as:

$$Q_c = St\rho_l C_{pl} u_l \left(T_w - T_l\right)\left(1 - A_q\right) \tag{21}$$

where u_l is the local tangential liquid velocity adjacent to the heated surface. The evaluation of the area of the heater surface influenced by bubbles A_q is given by

$$A_q = K\frac{\pi d_{bw}^2}{4}N_a \tag{22}$$

where the empirical constant K is used to account for the area of the heater surface influenced by the bubble. A value of $K = 4$ is often recommended [27]. However, Kenning and Del Valle [40] have found values ranging between 2 and 5. Judd and Hwang [34] ascertained that a lower value, $K = 1.8$, best fitted their experimental data. In most of our studies [41-43], we have incorporated a Jacob number (Ja_{sub}) based on liquid subcooling dependence as suggested by Kenning and Del Valle [40] according to

$$K = 4.8\exp\left(-\frac{Ja_{sub}}{80}\right) \tag{23}$$

To determine each respective component on the right hand side of equation (18), a bisection algorithm, an iterative procedure, is employed to evaluate the wall superheat that satisfies the applied wall heat flux. This algorithm begins with a guess of the wall superheat, and thereafter calculates each component of the heat flux. The difference between the computed total wall heat flux and the actual applied wall heat flux provides a new wall superheat estimate for the next step in iterative procedure. The iteration continues until the difference error between the applied and calculated wall heat flux falls below a prescribed criterion (i.e. $< 10^{-4}$) of the applied heat flux.

A number of studies examining bubble growth and detachment have resulted in a number of different empirical correlations for bubble departure. We focus on the selected relationships that are relevant to low-pressure subcooled boiling flow. Tolubinsky and Kostanchuk [44] proposed a simple relationship which evaluated the bubble departure as a function of the subcooling temperature as:

$$d_{bw} = \min\left[0.0006\exp\left(-\frac{T_{sub}}{45}\right), 0.00014\right] \tag{24}$$

Incidentally, Kurul and Podowski [27] also employed a similar expression to Tolubinsky and Kostanchuk [44] in calculating the bubble departure diameter. The correlation, based on Unal's experimental data [45], is:

$$d_{bw} = 0.00014 + 10^{-4}T_{sub} \tag{25}$$

On the basis of the balance between the buoyancy and surface tension forces at the heating surface, Fritz [46] proposed a correlation which includes the contact angle of the bubble, *viz.*,

$$d_{bw} = 0.0208\theta\sqrt{\frac{\sigma}{g\left(\rho_l - \rho_g\right)}} \tag{26}$$

where θ is the contact angle. The above expression was modified by Kocamustafaogullari and Ishii [47] for low pressure as

$$d_{bw} = 2.5\times10^{-5}\left(\frac{\rho_l - \rho_g}{\rho_g}\right)\theta\sqrt{\frac{\sigma}{g\left(\rho_l - \rho_g\right)}} \tag{27}$$

A more comprehensive correlation proposed by Unal [45] which included not only the effect of subcooling but also on the convection velocity and heater wall properties is given by:

$$d_{bw} = \frac{2.42\times10^{-5}\,p^{0.709}a}{\sqrt{b\Phi}} \tag{28}$$

where

$$a = \frac{(Q_w - hT_{sub})^{1/3}k_l}{2C^{1/3}h_{fg}\sqrt{\left(\pi k_l/\rho_l c_{pl}\rho_g\right)}}\sqrt{\frac{k_w\rho_w c_{pw}}{k_l\rho_l c_{pl}}};\quad b = \frac{T_{sub}}{2\left[1-\left(\rho_g/\rho_l\right)\right]}$$

$$C = \frac{h_{fg}\mu_l\left[c_{pl}/\left(0.013h_{fg}\,\mathrm{Pr}^{1.7}\right)\right]^3}{\sqrt{\frac{\sigma}{g\left(\rho_l - \rho_g\right)}}};\qquad \Phi = \begin{cases}\left(\dfrac{u_l}{0.61}\right)^{0.47} & \textit{for}\quad u_l \ge 0.61\,m/s\\ 1.0 & \textit{for}\quad u_l < 0.61\,m/s\end{cases}$$

The stated range for this correlation is

Pressure	$0.1 < p < 17.7$ MPa
Wall heat flux	$0.47 < Q_w < 10.64$ MW / m^2
Liquid velocity	$0.08 < u_l < 9.15$ m/s
Liquid subcooling	$3.0 < T_{sub} < 86$ °C

For the bubble departure frequency, most correlations have been derived from the consideration of the bubble departure diameter. Cole's correlation [48] which was derived assuming a balance between buoyancy and drag (drag coefficient constant) for pool nucleate boiling is a popular expression. It is used as a default expression in the CFD commercial code CFX4.4 in the form of:

$$fd_{bw}^{0.5} = \sqrt{\frac{4g(\rho_l - \rho_g)}{3\rho_l}} \tag{29}$$

In the hydrodynamic region, Ivey [49] has suggested a bubble frequency relationship, which can be used for coalesced bubbles. The correlation is simply

$$fd_{bw}^{0.5} = 0.9\sqrt{g} \tag{30}$$

A slightly more complicated expression by Stephan [50], which included the effect of surface tension, may also be employed for the low-pressure subcooled boiling flow. It is given as:

$$fd_{bw} = \frac{1}{\pi}\sqrt{\frac{g}{2}\left(d_{bw} + \frac{4\sigma}{\rho g d_{bw}}\right)} \tag{31}$$

Similar expressions of the bubble frequency correlation that include the surface tension effect in the form of equation (31) are also noted. Peebles and Garber [51] observed the velocity rise in a gravitational field as

$$V_b = 1.18\left[\frac{\sigma g(\rho_l - \rho_g)}{\rho_l^2}\right]^{0.25} \tag{32}$$

On the basis of equation (32), Jakob [52] proposed a bubble frequency by assuming the waiting time to be equivalent to the growth time thus resulting in

$$fd_{bw} = 0.59\left[\frac{\sigma g(\rho_l - \rho_g)}{\rho_l^2}\right]^{0.25} \tag{33}$$

The Lemmert and Chwala's [53] active nucleation site density relationship is commonly used in the wall heat flux partitioning model of which has been correlated based on Del Valle and Kenning [32] data. It can be determined simply from the local wall superheat as

$$N_a = \left[m(T_{sat} - T_w)\right]^n \tag{34}$$

According to Kurul and Podowski [27], the values of m and n are 210 and 1.805 respectively. Recently, Končar et al. [54] demonstrated that better predictions were obtained by reducing the value of m to 185. Other relationships of the active nucleation site density that may also be applicable to low-pressure subcooled boiling flow are subsequently described. Kocamustafaogullari and Ishii [47] correlated existing active nucleation site density data by means of parametric study. They have assumed that the active nucleation site density in pool boiling by both surface conditions and thermo-physical properties of the fluid. They also postulated that the active nucleation site density developed for pool boiling could be used in forced convective system by the use of an effective superheat rather than the actual wall superheat. The active nucleation site density, N_a, can be expressed as:

$$N_a = \frac{1}{d_{bw}^2}\left[\frac{2\sigma T_{sat}}{\Delta T_{eff}\rho_g h_{fg}}\right]^{-4.4} f(\rho^*) \tag{35}$$

where $\rho^* = (\rho_l - \rho_g)/\rho_g$ and the function $f(\rho^*)$ is a known function of a density ratio described by: $f(\rho^*) = 2.157\times10^{-7}\rho^{*-3.2}(1+0.0049\rho^*)^{4.13}$. ΔT_{eff} represents the effective superheat, which is given by $\Delta T_{eff} = S\Delta T_w$ where $\Delta T_w = T_{sat} - T_w$ and S is the suppression factor. Basu et al. [55] proposed an alternative empirical correlation that included the effect of contact angle θ on the active nucleation site density, which is given by:

$$\begin{aligned} N_a &= 0.34\times10^4(1-\cos\theta)\Delta T_w^{2.0} \quad \Delta T_{ONB} < \Delta T_w < 15K \\ N_a &= 3.4\times10^{-1}(1-\cos\theta)\Delta T_w^{5.3} \quad 15K \le \Delta T_w \end{aligned} \tag{36}$$

Very recently, Hibiki and Ishii [56] modeled the active nucleation site density relationship mechanistically by the knowledge of the size and cone angle distributions of cavities. In accordance with Basu et al. [55] correlation, they have also formulated the nucleation site density as a function of contact angle. The correlation is given by:

$$N_a = 4.72\times10^5\left\{1-\exp\left(-\frac{\theta^2}{4.17}\right)\right\}\left[\exp\left\{2.5\times10^{-6} f(\rho^+)\frac{\Delta T_w \rho_g h_{fg}}{2\sigma T_{sat}}\right\}-1\right] \tag{37}$$

where $\rho^+ = \log_{10}((\rho_l - \rho_g)/\rho_g)$ and the function $f(\rho^+)$ is a function described by: $f(\rho^+) = -0.01064 + 0.48246\rho^+ - 0.22712\rho^{+^2} + 0.05468\rho^{+^3}$.

Model Applications

The three most important parameters in the wall heat flux partitioning model are the bubble departure diameter d_{bw}, bubble departure frequency f and the active nucleation site density N_a. It is recognized that not one single dedicated correlation for each of the parameters is expected to be universally valid for all types of flow conditions. A survey in the literature has revealed that the mechanistic model of Kurul and Podowski is a widely used heat transfer closure model for modeling the thermal-hydraulics in nuclear reactor geometries with different correlations of d_{bw}, f and N_a. Some pertinent investigations on the application of various empirical relationships for subcooled boiling flow are presented below.

Anglart and Nylund [22] performed a subcooled boiling flow investigation to predict the void fraction distribution in two-phase bubbly flows in fuel rod bundles. Special attention was devoted into the phenomena which govern the lateral void fraction distribution in a channel. The two-fluid boiling model with the closure of appropriate relationships for the interfacial mass, momentum and energy terms and the standard k-ε turbulence model with the additional consideration of Sato's bubble-induced turbulent viscosity was employed. Assuming spherical bubbles, the interfacial area concentration can be defined as:

$$a_{if} = \frac{6\alpha_g}{D_s} \tag{38}$$

Equation (38) requires the evaluation of the local distribution of the Sauter bubble diameter D_s. In Anglart and Nylund [22], the bubble diameter has been modeled as a function of the local water subcooling T_{sub} from the following equation:

$$D_s = \begin{cases} D_{B1} = 0.15\,mm & for \quad T_{sub} > T_{sub,1} = 13.5\,\mathrm{K} \\ \dfrac{D_{B1}\left(T_{sub} - T_{sub,2}\right) + D_{B2}\left(T_{sub,1} - T_{sub}\right)}{\left(T_{sub,1} - T_{sub,2}\right)} & for \quad T_{sub,2} \le T_{sub} < T_{sub,1} \\ D_{B2} = 1.5\,mm & for \quad T_{sub} \le T_{sub,2} = 0\,\mathrm{K} \end{cases} \tag{39}$$

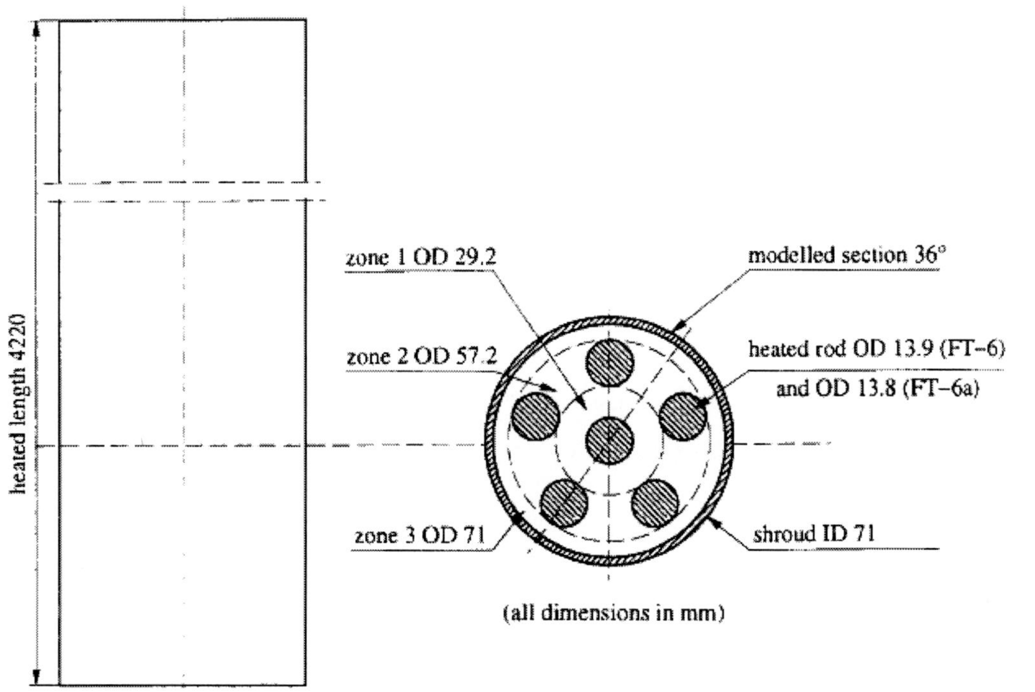

Figure 2. A schematic illustration of the test section with six rods (after Anglart and Nylund [22]).

For the wall heat flux partitioning model, closure was achieved through correlation of equation (25) for the bubble departure diameter d_{bw}, Cole's relationship for the bubble departure frequency f and the Lemnart and Chwala's expression with $m = 210$ and $n = 1.805$ for the active nucleation site density N_a. Void fraction measurements were performed by Nylund et al. [57] in the six-rod fuel bundles as shown in Figure 2.

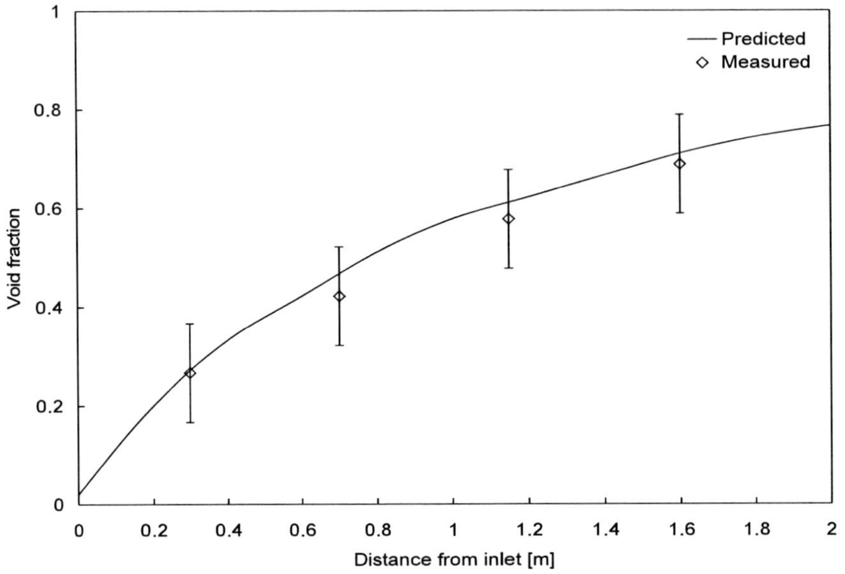

Zone 1.

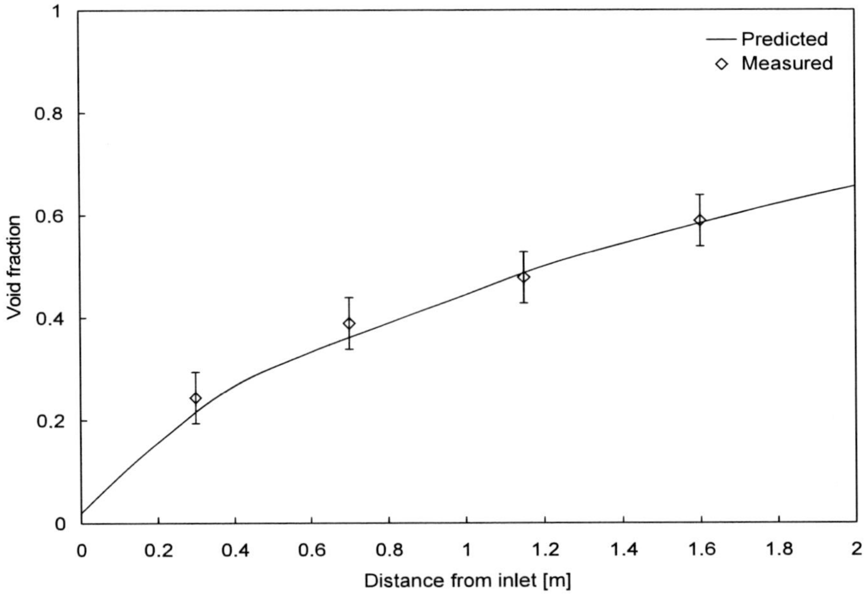

Zone 2.

Figure 3. Axial void distributions in test section (after Anglart and Nylund [22]).

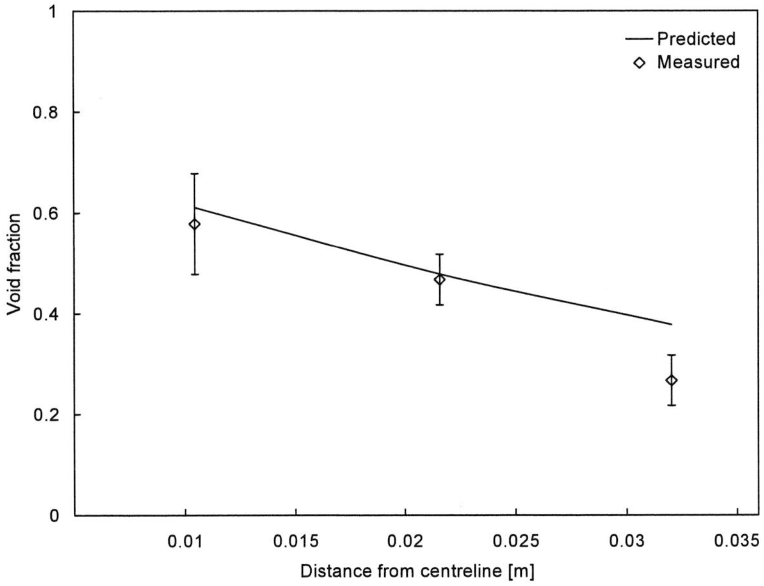

1148 mm from inlet.

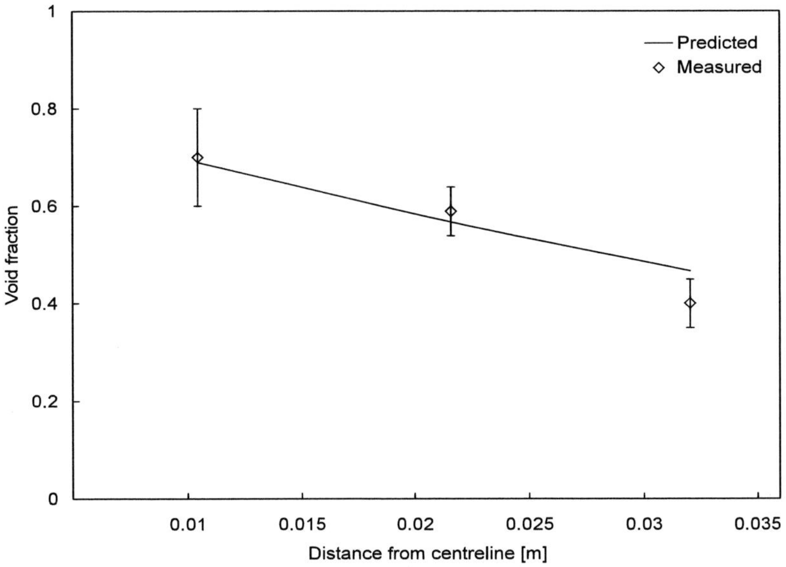

1598 mm from inlet.

Figure 4. Lateral void distributions in test section (after Anglart and Nylund [22]).

Numerical calculations were performed for one-tenth of the cross-section, utilizing the symmetry property of the bundle geometry. Comparisons of the some results of the axial and lateral distributions of the computed and measured void fraction are illustrated in Figures 3 and 4. Satisfactory agreement that had been achieved between the predicted and measured void fractions at specific locations of the test section clearly highlighted the model's potential of resolving more complex fuel bundle geometries.

On the basis of the parametric study focusing on different closure empirical relationships that were carried out in our early work [41], we have established the use of Unal's bubble departure relationship (equation (28)) as the preferred model for bubble departure d_{bw}. We further advocated the use of a more sophisticated expression proposed by Zeitoun and Shoukri [58,59] for evaluating the Sauter bubble diameter D_s instead of the linear functional expression of equation (39), which is given by:

$$\frac{D_s}{\sqrt{\sigma / g\Delta\rho}} = \frac{0.0683(\rho_l / \rho_g)^{1.326}}{Re^{0.324}\left(Ja_{sub} + \dfrac{149.2(\rho_l / \rho_g)^{1.326}}{Bo^{0.487} Re^{1.6}}\right)} \tag{40}$$

where Re is the flow Reynolds number, Bo is the boiling number and Ja_{sub} is the Jakob number based on the liquid subcooling. To close the wall heat flux partitioning model, Cole's relationship for the bubble departure frequency f and the Lemnart and Chwala's expression with m = 210 and n = 1.805 for the active nucleation site density N_a were employed as in Anglart and Nylund [22].

Numerical predictions of the above model were examined for a range of experimental data covering a range of heat flux, subcooling and flow conditions at low pressures. For the purpose of illustration, results of axial void fraction distributions compared against experimental measurements of Donveski and Shoukri [60] and Dimmick and Selander [61] are presented herein. The data of Donveski and Shoukri [60] were obtained for a vertical 306 mm long concentric annular test section with an outer diameter of 25.4 mm. Dimmick and Selander's [61] data were obtained for subcooled boiling flow inside a tube of 12.29 mm inside diameter. These configurations are typical geometries that can be found in nuclear applications. Figures 5 and 6 illustrate the comparison between the measured and predicted void fraction. In all the cases, the agreement was found to be generally very good.

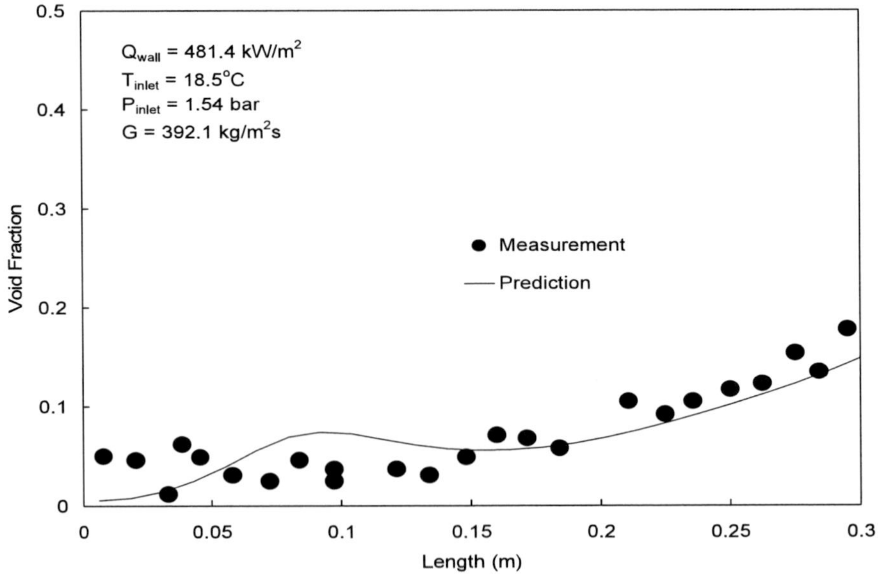

Figure 5. (Continued on next page)

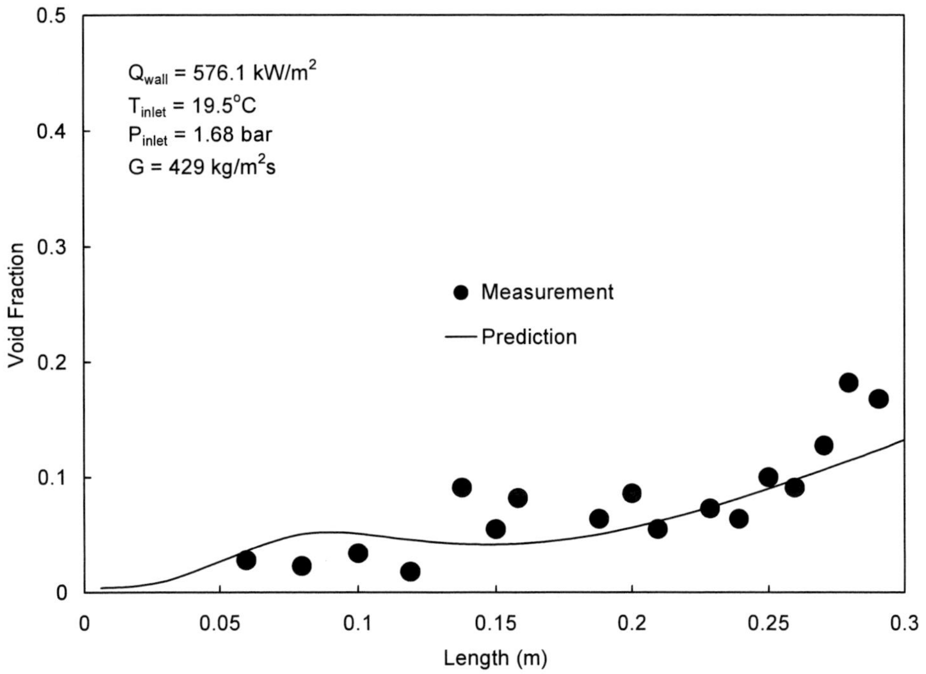

Figure 5. Comparison of predicted void fraction against Donveski and Shoukri's [60] data.

Nevertheless, further applications of the above boiling model in comparing the local radial measurements of a low-pressure subcooled boiling annular channel flow carried out by Yun et al. [62] and Lee et al. [63] revealed significant weakness, predominantly in the radial prediction of the Sauter bubble diameter [64]. Numerical calculations were compared against measurements performed for a test channel of a 2.376 m long vertical concentric annulus with a heated inner tube. The inner tube of 19 mm in outer diameter was composed of a heated section of 1.67 m in length and two unheated sections. A transparent glass tube placed in the vicinity of the measuring plane at 1.61 m allowed visual observations of the boiling flow. Comparison of the bubble size distribution evaluated from the model employing the Zeitoun and Shoukri's relationship as described in equation (40) and calculations based on different reference diameters for the linear bubble function based on equation (39) against experimental measured data of different heat fluxes, inlet subcooling temperatures and flow conditions are shown in Figure 7. We have assumed for the case of "Linear 1" the local bubble diameters were evaluated between D_{B1} = 0.15 mm and D_{B2} = 4.0 mm while for the case of "Linear 2" they are determined between D_{B1} = 0.15 mm and D_{B2} = 7.0 mm. We further assumed that both the reference diameters corresponded to identical reference subcooling temperatures of $T_{sub,1}$ = 13.0 K and $T_{sub,2}$ = −5 K. The closure for the wall heat flux partitioning model was achieved through Kocamustafaogullari and Ishii's modified expression of Fritz for the bubble departure diameter d_{bw}, Cole's relationship for the bubble departure frequency f and the Lemnart and Chwala's expression with m = 210 and n = 1.805 for the active nucleation site density N_a. As demonstrated in Figure 7, the locally predicted bubble diameter distributions determined through the empirical relationships were not consistent with those measured during the experiments. The failure of model was predominantly attributed by the absence of the mechanistic behavior of bubble coalescence and collapse due to condensation not appropriately accounted in the model. Evidently, the bubble size distribution was not strictly dependent on only the local subcooling effect.

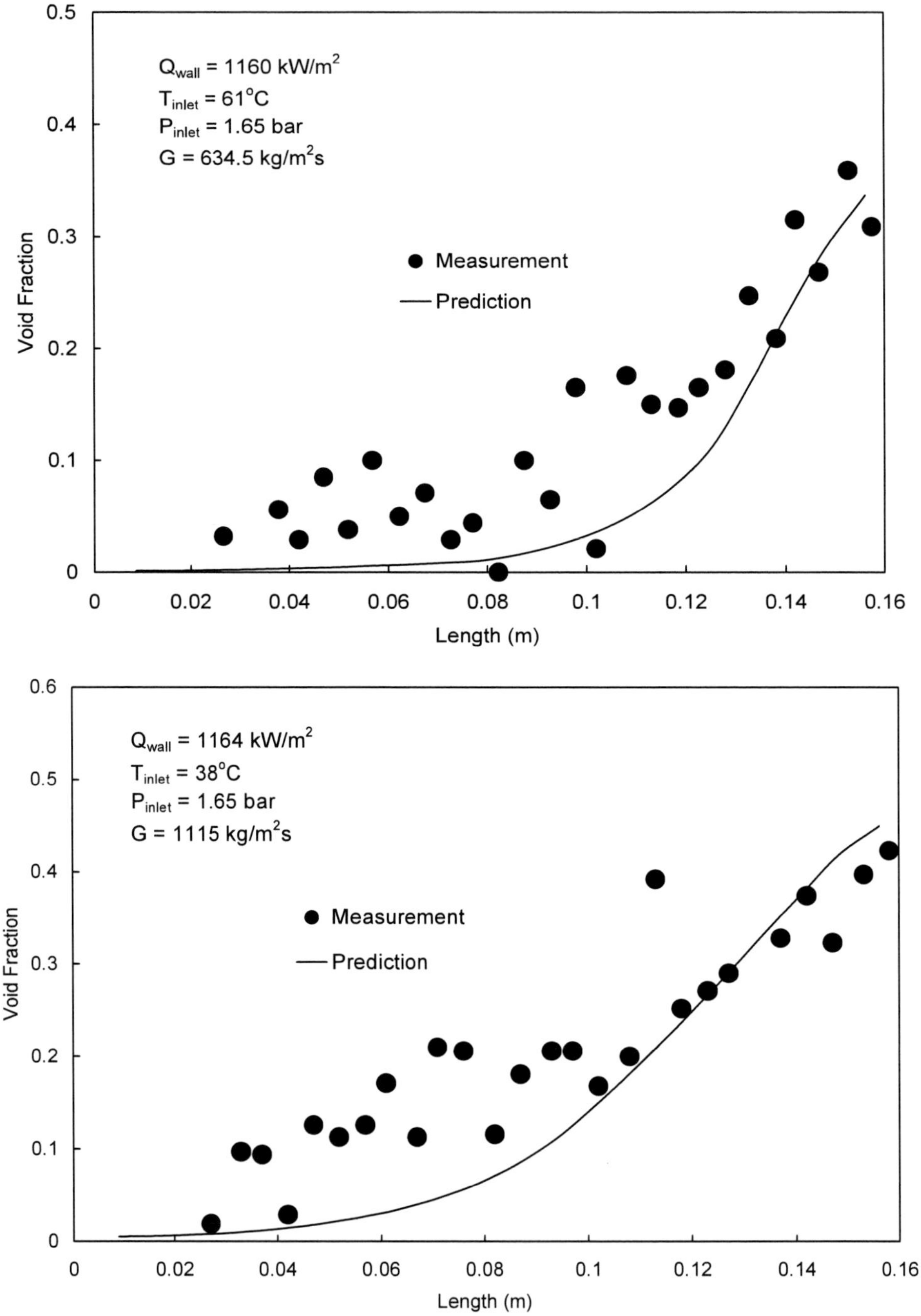

Figure 6. Comparison of predicted void fraction against Dimmick and Selander's [61].

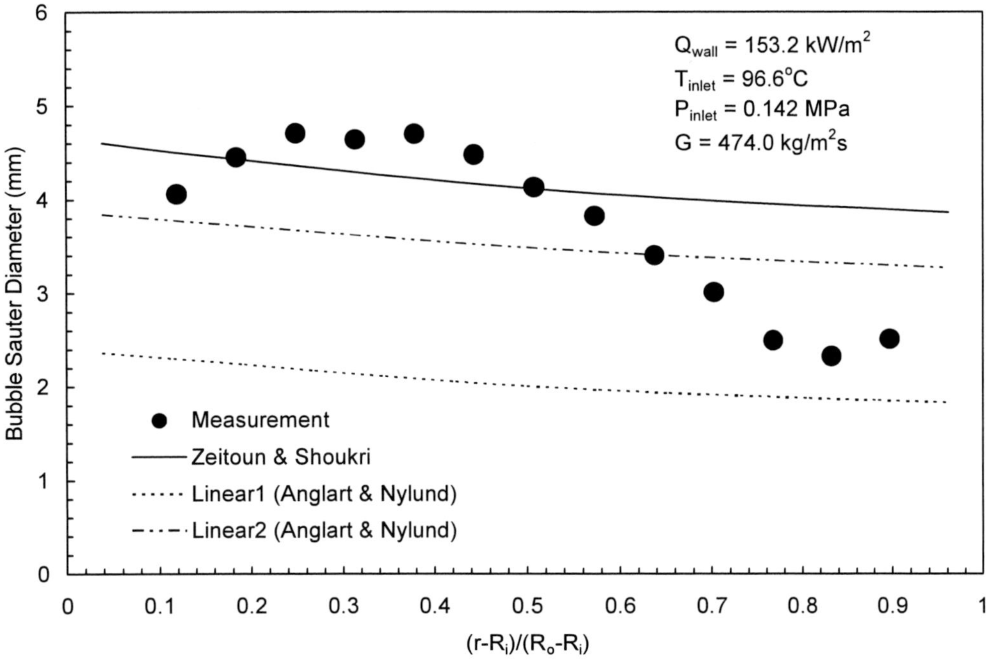

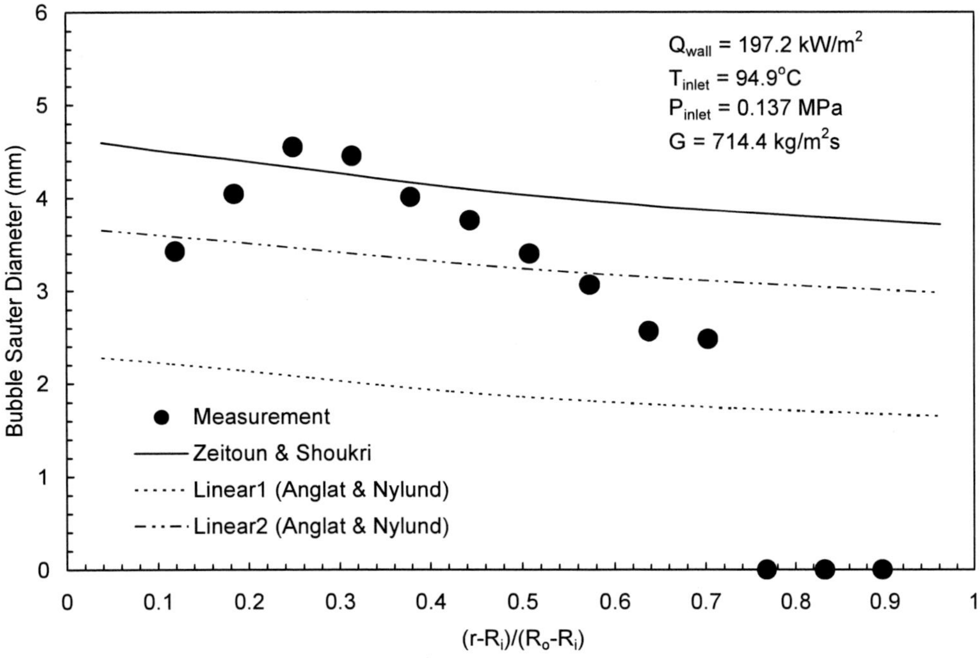

Figure 7. (Continued on next page)

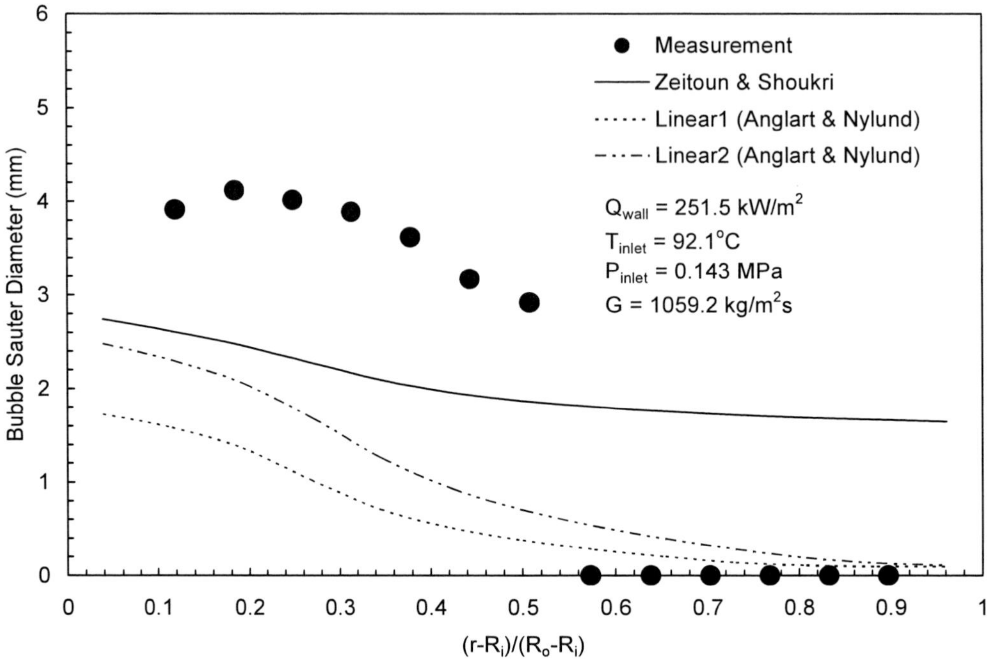

Figure 7. Comparison of local mean radial profiles of Sauter bubble diameter between predictions of empirical bubble diameter and measurements at the measuring plane.

Similar numerical investigations on the evolution of cross-sectional distribution of the local Sauter bubble diameter were also performed in Končar et al. [54]. A simple model to determine the radial distribution of the Sauter bubble diameter was proposed. The shift of the maximum bubble size away from the heated wall was modeled by a linear evolution in the radial direction as:

$$D_s^* = \min\left(d_{bw} + y_w, d_{b,\max}\right) \tag{41}$$

where y_w is the radial distance from the near-wall cell center. The maximum allowed bubble diameter $d_{b,\max}$ was prescribed as $2d_{bw}$. Multiplying equation (41) by a relation describing condensation in the subcooled flow field yields the relationship of the local Sauter bubble diameter as:

$$D_s = D_s^* \exp\left(-\frac{T_{sub} - T_{sub,w}}{2 \cdot T_{sub,w}}\right) \tag{42}$$

where $T_{sub,w}$ represents the local subcooling in the near-wall cell. To evaluate the bubble departure diameter d_{bw}, Končar et al. [54] also applied the correlation proposed by Unal [45] as was applied in our early work [41]. The correlation was nevertheless modified by adding a coefficient C_{bw} to describe large bubbles at low-pressure conditions, *viz.*,

$$d_{bw} = C_{bw} \frac{2.42 \times 10^{-5} p^{0.709} a}{\sqrt{b\Phi}} \tag{43}$$

The default value of the coefficient C_{bw} was set to 2 but they have however found the need to adjust the coefficient for their numerical predictions in some experimental cases.

Of particular significance, comparison of their model predictions was made against the measurements of Lee et al. [63]. The flow conditions of the four experimental cases considered are shown in Table 1. In d_{bw}, the default value of C_{bw} was used. Predicted radial distributions of the void fraction for cases stipulated in Table 1 are illustrated in Figure 8. It is noted that in all the figures presented, the dimensionless parameter $(r-R_i)/(R_o-R_i) = 1$ indicates the inner surface of the unheated flow channel wall while $(r-R_i)/(R_o-R_i) = 0$ indicates the surface of the heating rod in the annulus channel. Although the measurements of local void fraction were only performed at one axial location ($z = 1.61$ m), simulations at other locations were presented to demonstrate the axial evolution of the subcooled boiling flow. Satisfactory agreement between predictions and measurement was achieved except for Case 2 where the local void fraction was significantly under-predicted. It should be noted that no comparison of their model was made especially on their predicted local Sauter bubble diameter distribution through equation (42) against local measurements.

In the recent paper by Krepper et al. [65], the mechanistic model of Kurul and Podowski [27] through closure relationships using equations (24), (29) and (34) and determination of local bubble size through linear equation (39) were investigated for their capability to contribute towards fuel assembly design. Despite the heat transfer closure achieved through these relationships, the application of CFD with the subcooled boiling model for the simulation of a hot fuel channel was shown to at least provide some means of better evaluating different geometrical designs of the spacer grids and assessing the integrity of fuel rods within a fuel assembly. They have however indicated that further improvements of the microscopic models are necessary and the model for the bubble size at departure has to be replaced by a more mechanistic model. The simulation of the bubble size in the bulk subcooled liquid has to also consider the prevalence of coalescence and fragmentation of bubbles.

Table 1. Experimental conditions of subcooled boiling flow taken from Lee et al. [63]

Experiment No.	Pinlet (MPa)	Qw (kW/m2)	G (kg/m2 s)	Tinlet (oC)	Tsub,inlet (oC)
1	0.115	169.76	478.14	83.9	19.6
2	0.121	232.59	718.16	84.0	21.2
3	0.130	114.78	476.96	95.6	11.5
4	0.125	139.08	715.17	93.9	12.0

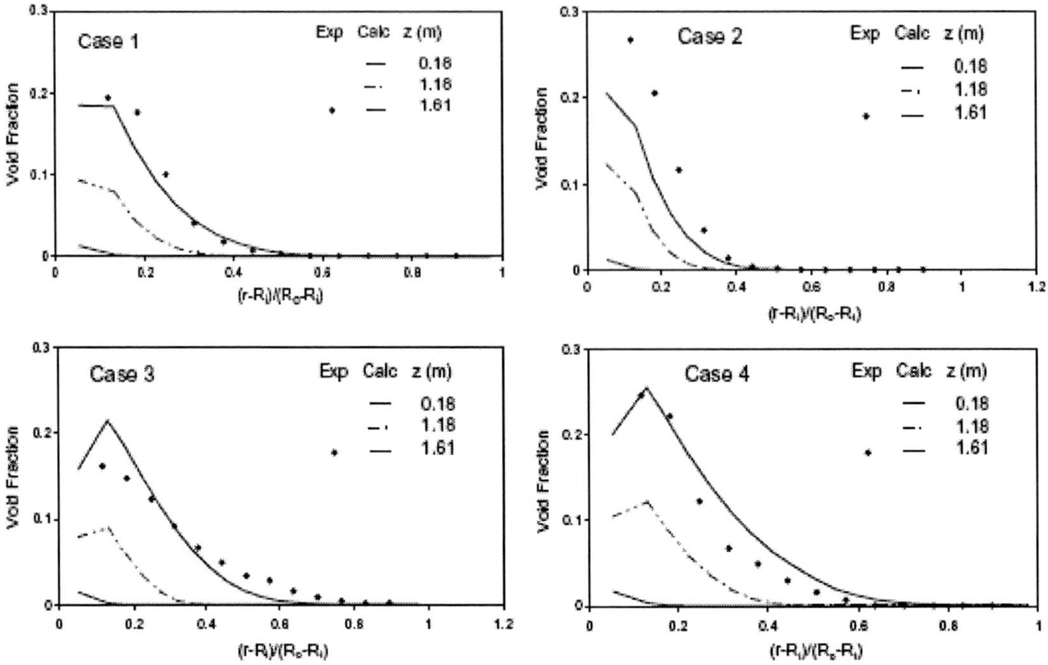

Figure 8. Comparison of predicted radial distributions of void fraction against measurements of Lee et al. [64] (after Končar et al. [54]).

PHENOMENOLOGICAL OBSERVATIONS AND POPULATION BALANCE MODEL

Empirical correlations as exemplified in the previous section can at best only provide a macroscopic description of the boiling phenomenon. It is therefore not surprising that these relationships are unable to adequately represent some of the many important complex mechanistic behaviors of bubbles such as bubble coalescence and condensation (microscopic in nature) that have been observed in experiments. The importance to appropriately consider these behaviors has been clearly demonstrated by the incorrect prediction of the local Sauter bubble diameter distributions as shown in Figure 7.

As observed in experiments by Lee et al. [63] and Tu et al. [64], high-speed photography in the vicinity of the heated wall as shown in Figure 9 clearly confirmed the presence of large bubbles away from the heated wall. More importantly, the bottom vapor bubble at upstream can be seen to be sliding along the heated surface before being impeded by the downstream bubble (see Figure 9(a)). Such observation was also confirmed by experiments performed by Bonjour and Lallemand [66] and Prodanovic et al. [67], which clearly indicated the presence of bubbles sliding shortly after being detached from the heated cervices before lifting into the liquid core. In their experiments, they have also seen that these upstream bubbles traveling closely to the heated wall have the tendency of significantly colliding with any detached bubbles downstream and subsequently forming bigger bubbles due to bubble merging or coalescence of adjacent bubbles. The supposition of larger bubbles being present due to bubble coalescence was evidenced away from the heated wall as shown in Figure 9(b). Focusing on the local bubble mushroom region as depicted in Figure 9(b) and tracking its development through time, the region increased in size along the heated wall downstream, confirming the significant coalescence of bubbles. On the other hand, very few bubbles were present in the bulk subcooled liquid further away from the heated wall and near the unheated wall. In this flow region, the effect of condensation gradually caused the bubbles to decrease in size due to the subcooling temperature of the liquid as they migrated towards the opposite end of the unheated wall of the flow channel. This was further confirmed by experimental observations of Gopinath et al. [68] (see Figure 10), which illustrated a bubble gradually being condensed in a subcooled liquid away from the heated surface.

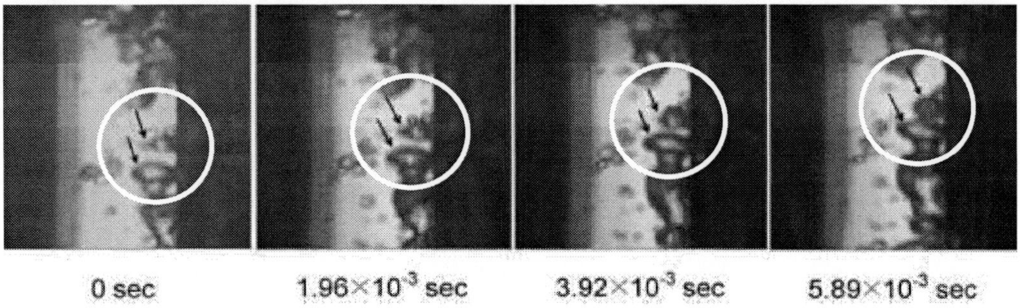

(a)

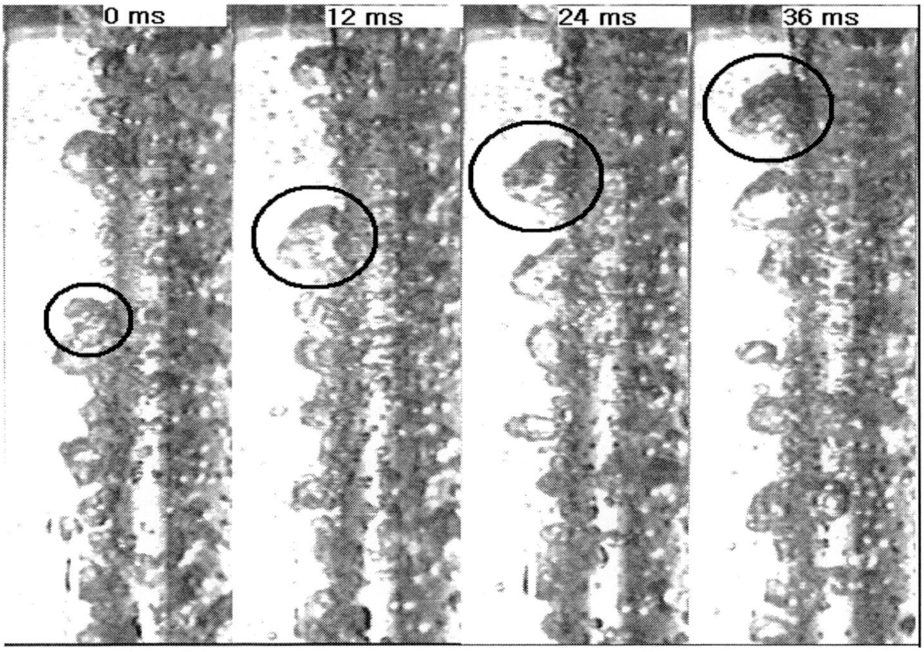

(b)

Figure 9. Experimental observations: (a) bubble sliding and collision with downstream bubble and (b) significant bubble coalescence as indicated by the bubble mushroom region near the heated wall of the annular flow channel (after Lee et al. [63] and Tu et al. [64]).

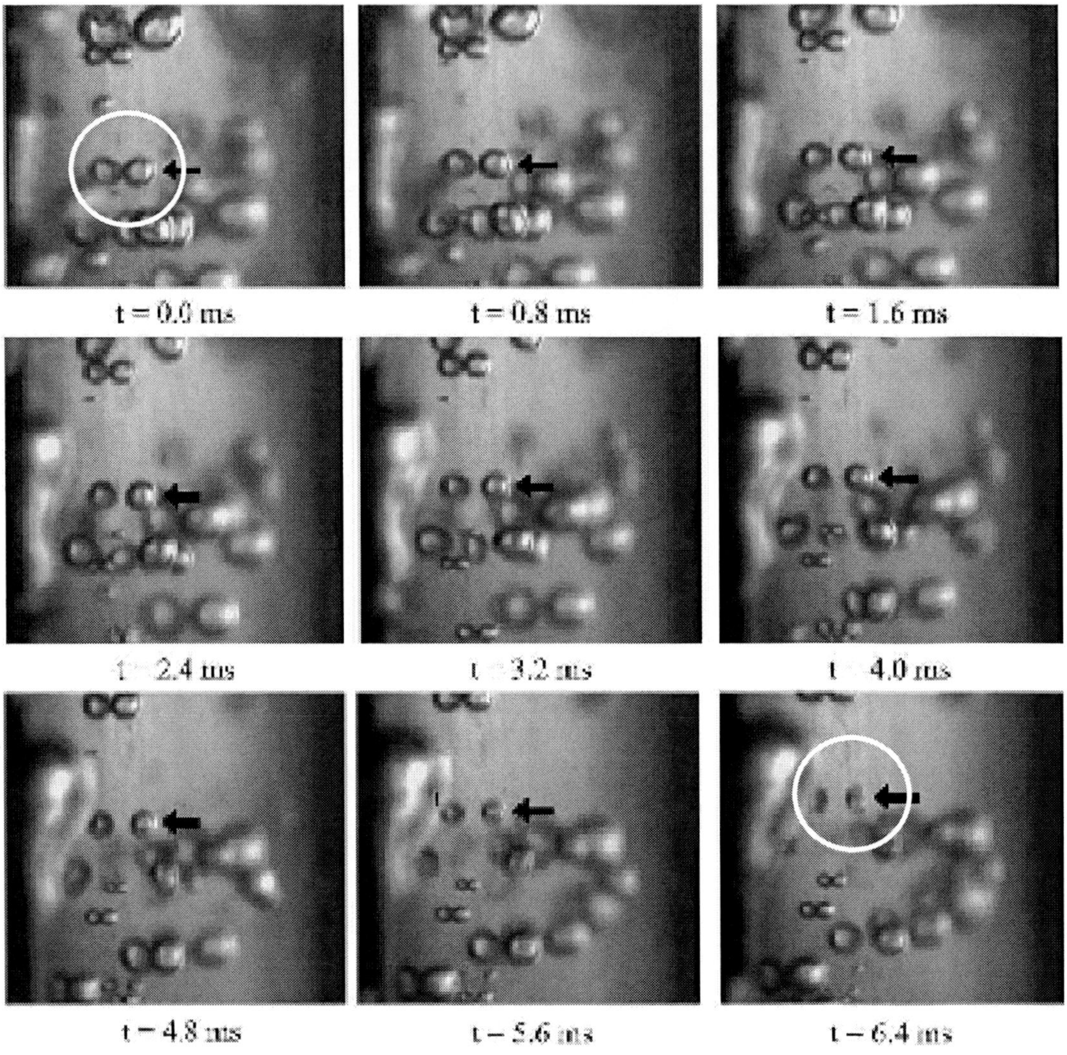

Figure 10. Bubble undergoing condensation in the bulk subcooled liquid (after Gopinath et al. [68]).

On the basis of the important phenomenological observations of the complex mechanistic behaviors of bubble coalescence near the wall and bubble collapse due to condensation in the single phase fluid flow as observed above, the population balance model which can accurately account for the coalescence and breakage characteristics of bubbles presents one of the many viable approaches in predicting the non-uniform bubble distribution in subcooled boiling flow. Phenomenological models developed by Luo and Svendsen [69] and Prince and Blanch [70] and have allowed detailed description of the mechanisms for break-up and coalescence of intermittent bubbles, which are illustrated in Figure 11.

For the consideration of the fragmentation of bubbles, the model developed by Luo and Svendsen [69] is employed for the break-up of bubbles in turbulent dispersions. In this model, binary break-up of the bubbles is assumed and the model is based on the theories of isotropic turbulence. The break-up rate of bubbles of volume v_j into volume sizes of v_i can be obtained as:

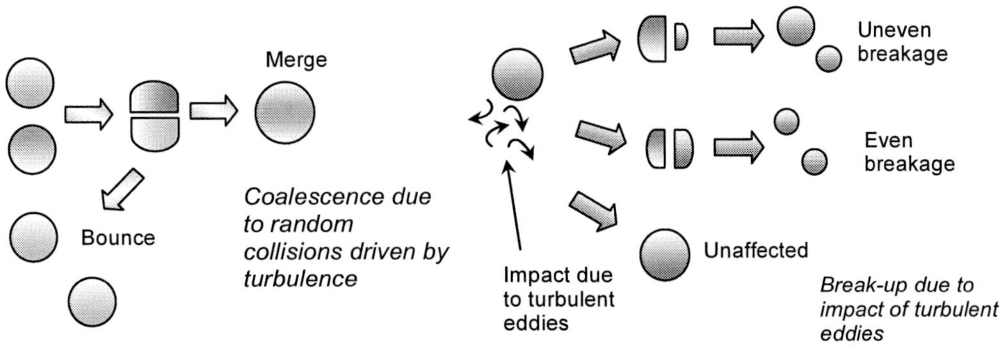

Figure 11. Bubbles undergoing coalescence and breakage.

$$\frac{\Omega(v_j : v_i)}{(1-\alpha_g)\, n_j} = C\left(\frac{\varepsilon}{d_j^2}\right)^{1/3} \int_{\xi_{\min}}^{1} \frac{(1+\xi)^2}{\xi^{11/3}} \exp\left(-\frac{12 c_f \sigma}{\beta \rho_l \varepsilon^{2/3} d_j^{5/3} \xi^{11/3}}\right) d\xi \tag{44}$$

where $\xi = \lambda/d_j$ is the size ratio between an eddy and a particle in the inertial sub-range and consequently $\xi_{min} = \lambda_{min}/d_j$; and C and β are determined respectively from fundamental consideration of drops or bubbles breakage in turbulent dispersion systems to be 0.923 and 2.0. The variable c_f denotes the increase coefficient of surface area: $c_f = \left[f_{BV}^{2/3} + (1 - f_{BV})^{2/3} - 1 \right]$ where f_{BV} is the breakage volume fraction.

For the merging of bubbles, the coalescence of two bubbles is assumed to occur in three stages. The first stage involves the bubbles colliding thereby trapping a small amount of liquid between them. This liquid film then drains until it reaches a critical thickness and the last stage features the rupturing of the liquid film subsequently causing the bubbles to coalesce. The collisions between bubbles may be caused by turbulence, buoyancy and laminar shear. Only random collision driven by turbulence is considered in the present model. The coalescence rate considering turbulent collision taken from Prince and Blanch [70] can be expressed as:

$$\chi_{ij} = \frac{\pi}{4}\left[d_i + d_j\right]^2 \left(u_{ti}^2 + u_{tj}^2\right)^{0.5} \exp\left(-\frac{t_{ij}}{\tau_{ij}}\right) \tag{45}$$

where τ_{ij} is the contact time for two bubbles given by $(d_{ij}/2)^{2/3}/\varepsilon^{1/3}$ and t_{ij} is the time required for two bubbles to coalesce having diameter d_i and d_j estimated to be $\{(d_{ij}/2)^3 \rho_l / 16\sigma\}^{0.5} \ln(h_0/h_f)$. The equivalent diameter d_{ij} is calculated as suggested by Chesters and Hoffman [71]: $d_{ij} = (2/d_i + 2/d_j)^{-1}$. According to Prince and Blanch [70], for air-water systems, experiments have determined h_0, initial film thickness and, h_f, critical film thickness at which rupture occurs to be 1×10^{-4} m and 1×10^{-8} m respectively. The turbulent velocity u_t in the inertial subrange of isotropic turbulence by Rotta [72] is given by:

$$u_t = 1.4 \varepsilon^{1/3} d^{1/3} \tag{46}$$

According to Fleischer et al. [73], the following population balance equation can be expressed in the form:

$$\frac{\partial n(V,\bar{x},t)}{\partial t} + \nabla \cdot \left(\bar{u}_g n(V,\bar{x},t)\right) = G(V,\bar{x},t) \tag{47}$$

where $n(V, \bar{x}, t)$ is the bubble number density distribution per unit mixture of which is a function of the spatial range $\bar{x}$ for a given time t and volume V. On the right hand side, the term $G(V, \bar{x}, t)$ contains the bubble source/sink rates per unit mixture volume due to the bubble interactions such as coalescence and break-up as described by equations (44) and (45) as well as phase change, which will be detailed below. For the case of nucleate boiling and condensation in a subcooled boiling flow, the phase change term includes the rate of change of bubble population with specific volumes. To effectively employed the number density transport equation in the context of CFD, Pohorecki et al. [74] suggested dividing equation (47) into N classes to classify the range of bubble sizes that may be present within the flow volume, *viz.*,

$$\frac{\partial n_i}{\partial t} + \nabla \cdot \left(\bar{u}_g n_i\right) = \left(\sum_j R_j\right)_i + \left(R_{ph}\right)_i \tag{48}$$

where $\left(\sum_j R_j\right)_i$ represents the net change in the number density distribution due to coalescence and break-up processes. This interaction term $\left(\sum_j R_j\right)_i$ $(= P_C + P_B - D_C - D_B)$ contains the source rates of P_C, P_B, D_C and D_B, which are respectively, the production rates due to coalescence and break-up and the death rate to coalescence and break-up of bubbles formulated as:

$$P_C = \frac{1}{2}\sum_{k=1}^{N}\sum_{l=1}^{N} \chi_{i,kl} n_i n_j \quad \chi_{i,kl} = \chi_{kl} \ \textit{if} \ v_k + v_l = v_i \quad \textit{else} \quad \chi_{i,kl} = 0 \ \textit{if} \ v_k + v_l \neq v_i$$

$$P_B = \sum_{j=i+1}^{N} \Omega\left(v_j : v_i\right) n_j \, ; \; D_C = \sum_{j=1}^{N} \chi_{ij} n_i n_j \, ; \; D_B = \Omega_i n_i \tag{49}$$

To account for the range of bubble classes n_i represented in the vapor phase, the continuity equation of vapor phase is altered to accommodate the additional source terms S_i due to break-up and coalescence mechanisms according to:

$$\frac{\partial \rho_g \alpha_g f_i}{\partial t} + \nabla \cdot \left(\rho_g \alpha_g \bar{u}_g f_i\right) = S_i - f_i \Gamma_{lg} \tag{50}$$

The term $f_i\Gamma_{lg}$ represents the mass transfer due to condensation redistributed for each of the discrete bubble classes. The gas void fraction along with the scalar fraction f_i are related to the number density of the discrete bubble ith class n_i (similarly to the jth class n_j) as $\alpha_g f_i = n_i v_i$. The size distribution of the dispersed phase is therefore defined by the scalar f_i.

The term $\left(R_{ph}\right)_i$ in equation (48) comprises of the essential formulation of the source/sink rates for the phase change processes associated with subcooled boiling flow. At the heated surface, bubbles form at activated cavities known as active nucleation sites. The bubble nucleation rate from these sites can be expressed as:

$$\phi_{WN} = \frac{N_a f \xi_H}{A_C} \tag{51}$$

where ξ_H and A_C are the heated perimeter and the cross-sectional area of the boiling channel respectively. Since the bubble nucleation process only occurs at the heated surface, this heated wall nucleation rate is not included in $\left(R_{ph}\right)_i$ but rather specified as a boundary condition to the continuity equation of vapor phase apportioned to the discrete bubble class n_i based on the bubble departure criteria on the heated surface. The bubble sink rate due to condensation in a control volume for each bubble class can be determined from:

$$\phi_{COND} = -\frac{n_i}{V_B} A_B \frac{dR_B}{dt} \tag{52}$$

The following holds for the bubble condensation velocity [68]:

$$\frac{dR_B}{dt} = \frac{h\left(T_{sat} - T_l\right)}{\rho_g h_{fg}} \tag{53}$$

Substituting equation (53) into (52) and given that the bubble surface area A_B and volume V_B based on the Sauter bubble diameter are respectively πD_s^2 and $\pi D_s^3/6$, equation (51) can be rearranged as:

$$\left(R_{ph}\right)_i = \phi_{COND} = -\frac{1}{\rho_g \alpha_g}\left(\frac{6\alpha_g}{D_s}\right)\left[\frac{h\left(T_{sat} - T_l\right)}{h_{fg}}\right] n_i = -\frac{1}{\rho_g \alpha_g}\left[\frac{h a_{if}\left(T_{sat} - T_l\right)}{h_{fg}}\right] n_i \tag{54}$$

Numerical simulations have been performed in [64] for the experimental setup of Yun et al. [62] and Lee et al. [63] to demonstrate the important effect of equation (54) in a subcooled liquid on the local bubble distribution. The range of wall heat fluxes and associated flow conditions considered for the model are stipulated in Figures 12 and 13 respectively. To account for the non-uniform bubble size distribution, bubbles ranging from 0 mm to 9.0 mm were equally divided into 15 classes. This resulted in 15 continuity equations for the vapor phase coupled with a single continuity equation of the liquid phase. Instead of considering 16 different complete phases, it was assumed for the sake of computational efficiency that each bubble class traveled at the same mean algebraic velocity to reduce the computational time. Only a single momentum equation was thus solved for the vapor phase. Prince and Blanch [70] have also indicated additional mechanisms that could be accounted for bubble coalescence such as buoyancy and laminar shear collisions. Indeed collisions caused by buoyancy could not be accounted as it has been assumed that all the bubbles from each class traveled at the same speed. Calculations have also shown that laminar shear collisions were negligible because of the low superficial gas velocities in the range of conditions considered for the subcooled boiling flow investigations.

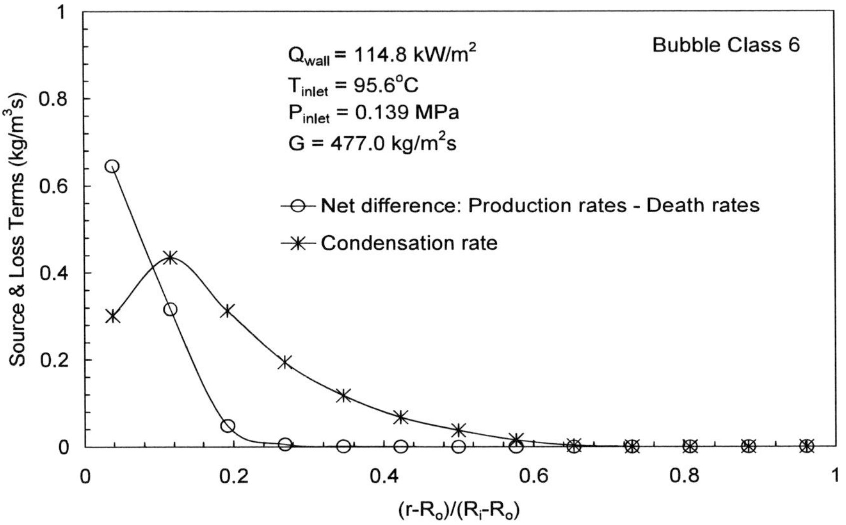

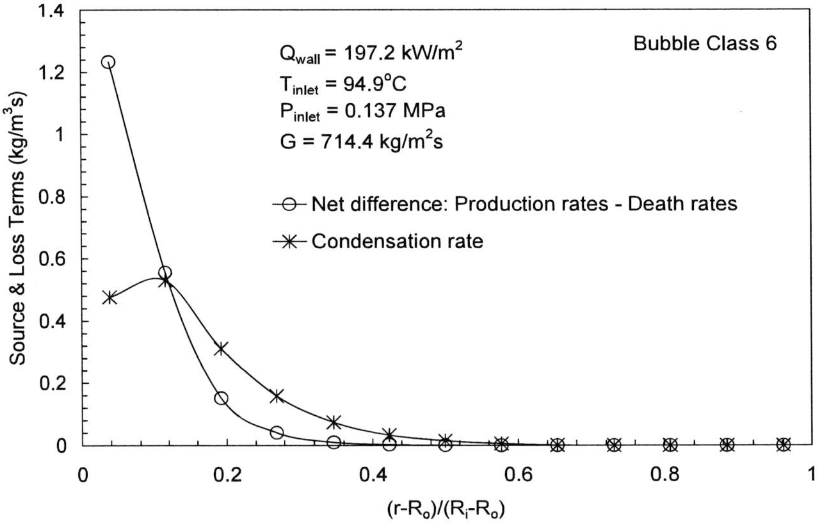

Figure 12. Radial distributions of the net rate $\sum_{j=1}^{4} R_j$ and condensation rate R_{ph} of population balance equation for bubble class 6.

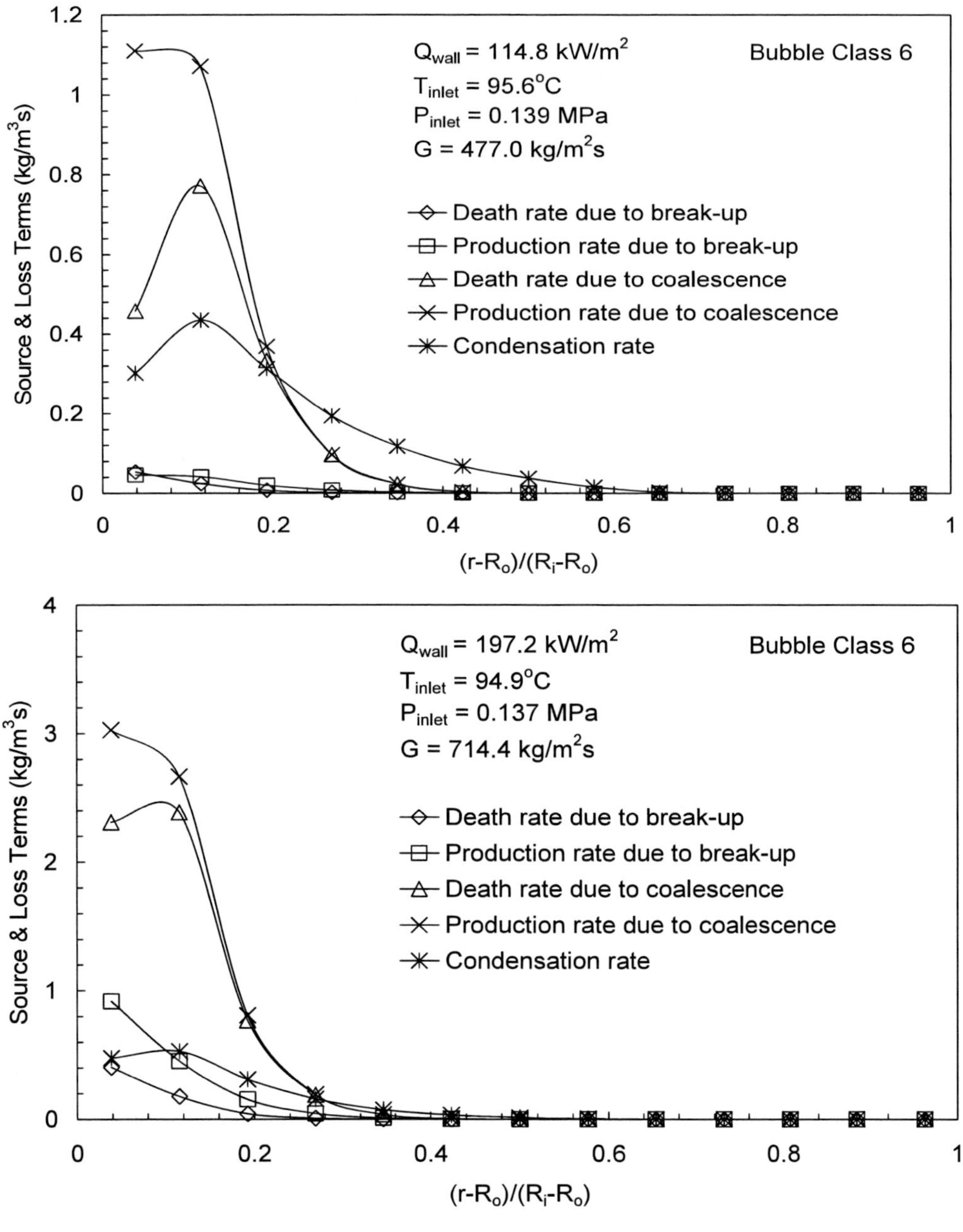

Figure 13. Radial contributions of the production rates of P_C and P_B and death rates of D_C and D_B as well as the condensation rate R_{ph}. of population balance equation for bubble class 6.

The distributions of the net rate $\sum_{j=1}^{4} R_j$ $(= P_C + P_B - D_C - D_B)$ and condensation rate R_{ph} of Equation (54) are presented in Figure 12. Near the heated wall, the net rate of $\sum_{j=1}^{4} R_j$ being higher than the condensation rate indicated a positive source contribution for the increase in bubble size. Nevertheless, further away from the heated surface, with the condensation rate significantly higher than the net rate and the phenomenon dominating between $(r-R_i)/(R_o-R_i) = 0.2$ and $(r-R_i)/(R_o-R_i) = 0.5$, the model predicted the gradual fragmentation of bubbles as they migrated towards the unheated wall. Figure 13 details the contributions of the production rates of P_C and P_B and death rates of D_C and D_B as well as the condensation rate R_{ph}. The dominant rate near the heated wall was the production rate due to coalescence. An important contribution through the death rate due to coalescence was also noted. The significance of the former indicated a source due to the merging of smaller bubbles forming bigger bubbles. Nevertheless, the latter indicated a source contribution due to the breakage of larger bubbles. The dominant coalescence rates found near the heated wall corresponded to the proper modeling of the physical bubble mechanisms that have been observed in experiments (see Figure 9). The condensation rate gradually diminishing away from the heated wall correctly modeled the condensation phenomenon as also observed through the experimental photographs and confirmed through Gopinath's [68] observations (see Figure 10).

IMPROVEMENTS TO WALL HEAT FLUX PARTITIONING MODEL

For vertical upward forced convective boiling flow as described by the experiment shown in Figure 9, it has been observed that bubbles have a tendency to slide before lifting off into the bulk subcooled liquid. In the case where the bubbles may merge with other nucleating bubbles whilst sliding, a smaller number of bubbles will be lifted off from the heated surface area than the actual number of active nucleation sites. As already indicated in Basu et al. [36,37] and reviewed in Warrier and Dhir [38] and Sateesh et al. [74], the transient conduction in such a case due to sliding bubbles becomes the dominant mode of heat transfer. It is therefore imperative that the wall heat partition model incorporates the heat transfer component due to these sliding bubbles.

In order to remove the application uncertainty of correlations empirically determined for the departing bubble diameter, a force balance model has been developed in Yeoh & Tu [75]. It has been demonstrated that the bubble growth and its maximum detached bubble size contributed significantly towards the void growth and heat transfer within the bulk liquid.

As observed through equations (29)-(33), numerous studies have attempted to tie the bubble departure frequency to some other parameter, namely the departing bubble diameter. Although the relationship between the bubble departure frequency and departing bubble diameter offers an attractive means of determining the frequency, application of these relationships, predominantly correlated from pool boiling data, remains contentious for forced convective boiling flow. For example, Cole's bubble frequency relationship (equation (29)) is frequently employed in many investigations of subcooled boiling flows. In order to eliminate the uncertainty of evaluating the bubble frequency, it is proposed that the fundamental theory based on the description of an ebullition cycle in nucleate boiling, which are: (1) waiting period t_w (transient conduction of heat to liquid); (2) growth period t_g: (a) bubble growth rate, (b) evaporation process, (c) agitation of liquid around the bubble and (d) termination of bubble whether by departure or collapse, is employed instead, *viz.*,

$$f = \frac{1}{t_g + t_w} \tag{55}$$

Part of the improvements made to the wall heat partition model to be further described in detail below incorporates the sliding period as well as the fundamental bubble frequency theory of equation (55). When combined with the force balance model, the departing bubble diameter is determined mechanistically. In general, for a given heat flux, a high frequency precedes small bubble diameter, while a low frequency usually results in large bubble diameter

Wall Heat Flux Partitioning Model Incorporating Bubble Sliding

Enhancement in heat transfer during forced convective boiling can be attributed by the presence of both sliding and stationary bubbles. There are essentially two mechanisms: (i) The latent heat transfer due to microlayer evaporation and (ii) Transient conduction as the disrupted thermal boundary layer reforms during the waiting period (i.e. incipience of the next bubble at the same nucleation site).

Transient conduction occurs in regions at the point of inception and in regions being swept by sliding bubbles. For a stationary bubble, the heat flux is Q_{tc} given by

$$Q_{tc} = 2\sqrt{\frac{k_l \rho_l C_{pl}}{\pi t_w}}(T_s - T_l) R_f N_a \left(K \frac{\pi D_d^2}{4}\right) t_w f + 2\sqrt{\frac{k_l \rho_l C_{pl}}{\pi t_w}}(T_s - T_l) R_f N_a \left(\frac{\pi D_d^2}{4}\right)(1 - t_w f) \tag{56}$$

where D_d is the bubble departure diameter, T_s is the temperature of the heater surface and T_l is the temperature of the liquid. Equation (56) indicates that some fraction of the nucleation sites will undergo transient conduction while the remaining will be in the growth period. For a sliding bubble, the heat flux Q_{tcsl} due to transient conduction that takes place during the sliding phase and the area occupied by the sliding bubble at any instant of time is given by

$$Q_{tcsl} = 2\sqrt{\frac{k_l \rho_l C_{pl}}{\pi t_w}}(T_s - T_l) R_f N_a l_s K D t_w f + 2\sqrt{\frac{k_l \rho_l C_{pl}}{\pi t_w}}(T_s - T_l) R_f N_a f t_{sl} \left(\frac{\pi D^2}{4}\right)(1 - t_w f) \tag{57}$$

where the average bubble diameter D is given by $D = (D_d + D_l)/2$ and D_l is the bubble lift-off diameter. Strictly speaking, the Kurul & Podowski's model is only applicable for subcooled boiling flows where bubbles are immediately released into the bulk subcooled liquid hence the absence of the bubble sliding phenomenon. This may be possibly true for pool boiling flows in a horizontal orientation.

The reduction factor R_f appearing in equations (56) and (57) depicts the ratio of the actual number of bubbles lifting off per unit area of the heater surface to the number of active nucleation sites per unit area, *viz.*, $R_f = 1/(l_s/s)$ where l_s is the sliding distance and s is the spacing between nucleation sites. In the present study, it shall be assumed that the nucleation sites are distributed in a square grid and that the bubbles slide only in the direction of the fluid flow [36]. The spacing between nucleation sites can thus be approximated as $s = 1/\sqrt{N_a}$. The factor R_f is obtained alongside with the sliding distance evaluated from the force balance model (to be described below). The significance of this factor provides the information whereby the bubble departing from its site of origin merges with other nucleating bubbles at adjacent sites. It is noted that for the case where the sliding distance l_s is less than the spacing s, $R_f = 1$.

Forced convection will always prevail at all times in areas of the heater surface that are not influenced by the stationary and sliding bubbles. The fraction of the heater area for stationary and sliding bubbles is given by

$$1 - A_q = 1 - R_f\left[N_a\left(K\frac{\pi D_d^2}{4}\right)t_w f + N_a\left(\frac{\pi D_d^2}{4}\right)(1 - t_w f) + N_a l_s K D t_w f + N_a f t_{sl}\left(\frac{\pi D^2}{4}\right)(1 - t_w f)\right] \tag{57}$$

Force Balance Model for Bubble Departure and Bubble Lift-off

The development of the force balance model concentrates on the various forces that influence the growth of a bubble during flow conditions in the directions parallel and normal to a vertical heating surface. These forces are formulated according to the studies performed by Klausner et al. [76] and Zeng et al. [77]. Figure 14 illustrates the forces acting on the bubble in the *x-direction* and *y-direction*; they are respectively,

$$\Sigma F_x = F_{sx} + F_{dux} + F_{sL} + F_h + F_{cp} \tag{58}$$

and

$$\Sigma F_y = F_{sy} + F_{duy} + F_{qs} + F_b \tag{59}$$

where F_s is the surface tension force, F_{du} is the unsteady drag due to asymmetrical growth of the bubble and the dynamic effect of the unsteady liquid such as the history force and the added mass force, F_{sL} is the shear lift force, F_h is the force due to the hydrodynamic pressure, F_{cp} is the contact pressure force accounting for the bubble being in contact with a solid rather than being surrounded by liquid, F_{qs} is the quasi steady-drag in the flow direction, and F_b is the buoyancy force. In addition, g indicates the gravitational acceleration; α, β and θ_i are the advancing, receding and inclination angles respectively; d_w is the surface/bubble contact diameter; and d is the vapor bubble diameter at the wall.

The forces acting in the *x-direction* can be estimated from:

$$F_{sx} = -d_w \sigma \frac{\pi}{\alpha - \beta}\left[\cos\beta - \cos\alpha\right];\ F_{dux} = -F_{du}\cos\theta_i;$$

$$F_{sL} = \frac{1}{2} C_L \rho_l \Delta U^2 \pi r^2;\ F_h = \frac{9}{4}\rho_l \Delta U^2 \frac{\pi d_w^2}{4};\ F_{cp} = \frac{\pi d_w^2}{4}\frac{2\sigma}{r_r}$$

In the *y-direction*, they are:

$$F_{sy} = -d_w \sigma \frac{\pi(\alpha - \beta)}{\pi^2 - (\alpha - \beta)^2}\left[\sin\alpha + \sin\beta\right];\ F_{duy} = -F_{du}\sin\theta_i;$$

$$F_{qs} = 6 C_D \mu_l \Delta U \pi r;\ F_b = \frac{4}{3}\pi r^3 (\rho_l - \rho_g) g$$

From the various forces described along the *x-direction* and *y-direction*, r is the bubble radius, ΔU is the relative velocity between the bubble centre of mass and liquid, C_D and C_L are the respective drag and shear lift coefficients and r_r is the curvature radius of the bubble at the reference point on the surface $x = 0$, which is $r_r \sim 5r$ [76].

The drag coefficient C_D and shear lift coefficient C_L appearing in the drag and lift forces are determined according to the relationships proposed by Klausner et al. [76], *viz.*,

$$C_D = \frac{2}{3} + \left[\left(\frac{12}{Re_b}\right)^n + 0.796^n\right]^{-1/n} \tag{60}$$

$$C_L = 3.877 G_s^{1/2}\left[\frac{1}{Re_b^2} + 0.014 G_s^2\right]^{1/4} \tag{61}$$

where n = 0.65 and $Re_b = \rho_l \Delta U d/\mu_l$ is the bubble Reynolds number. The dimensionless shear rate G_s is: $(dU/dx)(r/\Delta U)$. The gradient dU/dx can be determined through the universal velocity profile for turbulent flow:

$$\frac{U}{u_\tau} = 2.5\ln\left(9.8x^+\right) \tag{62}$$

where u_τ is the friction velocity and $x^+ = \rho_l u_\tau x/\mu_l$ is the non-dimensional normal distance from the heated wall. In equation (62), the velocity profile is assumed to be applicable for the time-averaged velocity distribution in the vicinity of the heated wall. Adjacent velocities, determined through the two-fluid model, are used to obtain the varying local friction velocities through equation (62). These friction velocities are subsequently used to evaluate the gradients dU/dx along the heated wall to determine the shear rate G_s.

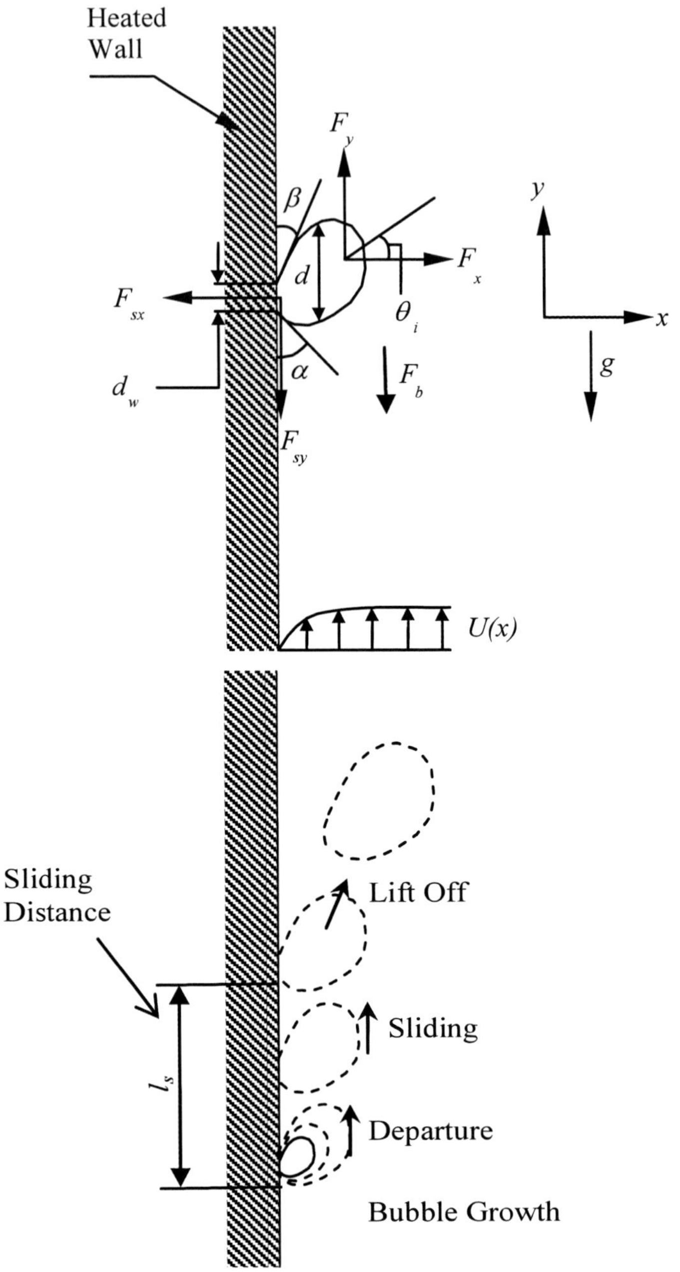

Figure 14. Schematic drawings illustrating the forces acting on a growing vapor bubble (above) and a bubble departing, sliding and lifting off from a vertical heated surface (below).

The growth force F_{du} is modeled by considering a hemispherical bubble expanding in an inviscid liquid, which is given by Zeng et al. [77] as

$$F_{du} = \rho_l \pi r^2 \left(\frac{3}{2} C_s \dot{r}^2 + r\ddot{r} \right) \tag{63}$$

where ($\dot{}$) indicates differentiation with respect to time. The constant C_s is taken to be $20/3$ according to [77]. In estimating the growth force, additional information on the bubble growth rate is required. As in [77], a diffusion controlled bubble growth solution by Zuber [78] is adopted:

$$r(t) = \frac{2b}{\sqrt{\pi}} Ja\sqrt{\eta t}\ ;\ Ja = \frac{\rho_l C_{pl} \Delta T_w}{\rho_g h_{fg}}\ ;\ \eta = \frac{\lambda_l}{\rho_l C_{pl}} \tag{64}$$

where Ja is the Jakob number, η is the liquid thermal diffusivity and b is an empirical constant that is intended to account for the asphericity of the bubble. For the range of heat fluxes investigated in this investigation, b is taken to be 0.21 based on a similar subcooled boiling study performed by Steiner et al. [35], which was experimentally verified through their in-house measurements with water as the working fluid.

While a vapor bubble remains attached to the heated wall, the sum of the parallel and normal forces must satisfy the following conditions: $\Sigma F_x = 0$ and $\Sigma F_y = 0$. For a sliding bubble case, the former establishes the bubble departure diameter (D_d) while the latter yields the bubble lift-off diameter (D_l). The growth period t_g appearing in equation (55) can be readily evaluated based on the availability of the bubble size at departure from its nucleation site through equation (64). The lift off period t_l can also be similarly calculated based on the bubble lift-off diameter. The difference between the bubble lift-off and bubble growth periods provides the period for the sliding bubble; the sliding distance l_s can subsequently be determined (see Figure 14). An estimation on this sliding distance can be determined according to the experimental correlation of Maity [79] as $l_s = (2/3) C_v t_{sl}^{3/2}$ where t_{sl} is the sliding time ($t_l - t_g$) and C_v is an acceleration coefficient correlated in terms of the tangential liquid velocity (u_l) adjacent to the heated surface: $C_v = 3.2u_l + 1$. This coefficient reflects the increase in bubble velocity with time after it begins to slide away from a nucleation site.

The bubble waiting time t_w is determined through the occurrence of transient conduction when a bubble slides or lifts off of which the boundary layer gets disrupted and cold liquid comes in contact with the heated wall. Assuming that the heat capacity of the heater wall $\rho_s C_{ps} \delta_s$ is very small, the conduction process can be modeled by considering one-dimensional transient heat conduction into a semi-infinite medium with the liquid at a temperature T_l and the heater surface at a temperature T_s. The wall heat flux can be approximated by

$$Q_w = \frac{k_l (T_s - T_l)}{\delta_l} \tag{65}$$

where δ_l ($= \sqrt{\pi \eta t}$) is the thickness of the thermal boundary layer. If the temperature profile inside this layer is taken to be linear according to Hsu & Graham [80], it can thus be expressed as

$$T_b = T_w - \frac{(T_s - T_l) x}{\delta_l} \tag{66}$$

where x is the normal distance from the wall. Based on the criterion of the incipience of boiling from a bubble site inside the thermal boundary layer, the bubble internal temperature for a nucleus site (cavity) with radius r_c is

$$T_b = T_{sat} + \frac{2\sigma T_{sat}}{C_2 r_c h_{fg} \rho_g} \quad \text{at } x = C_1 r_c \tag{67}$$

where $C_1 = (1+\cos\theta)/\sin\theta$ and $C_2 = 1/\sin\theta$. The angle θ refers here represents the bubble contact angle as described previously. By substituting equation (67) into equation (66), the waiting time t_w can be obtained as

$$t = t_w = \frac{1}{\pi\eta}\left[\frac{(T_s - T_l)C_1 r_c}{(T_w - T_{sat}) - 2\sigma T_{sat}/C_2 \rho_g h_{fg} r_c}\right]^2 \tag{68}$$

The cavity radius r_c can be determined by applying Hsu's criteria and tangency condition of equations (66) and (67), *viz.*,

$$t = \left[\frac{C_1 C_2 \rho_g h_{fg} r_c^2}{2\sigma T_{sat}}\right]^2 \frac{(T_s - T_l)^2}{\pi\eta} = \left[\frac{k_l}{Q_w}\right]^2 \frac{(T_s - T_l)^2}{\pi\eta} \tag{69}$$

From the above equation,

$$r_c = F\left[\frac{2\sigma T_{sat} k_l}{\rho_g h_{fg} Q_w}\right]^{1/2} \tag{70}$$

where

$$F = \left(\frac{1}{C_1 C_2}\right)^{1/2} = \left(\frac{\sin^2\theta}{1+\cos\theta}\right)^{1/2}$$

According to Basu et al. [79], the factor F indicates the degree of flooding of the available cavity size and the wettability of the surface. If the contact angle $\theta \to 0$, all the cavities will be flooded. Alternatively, as $\theta \to 90^{\circ}$, $F \to 1$, all the cavities will not be flooded (i.e. they contain traces of gas or vapor).

In reality, the surface/bubble contact diameter d_w evolves from the point of inception until the point of departure or lift-off. Here, a correlation based on the experimental data of Maity [79] as a function of the bubble contact angle θ is employed to determine the ratio of the bubble base diameter d_w to the bubble diameter at departure or lift-off, which is given as $C = 1 - \exp(-2\theta^{0.6})$. Experimental observations by Klausner et al. [76] and Bibeau and Salcudean [81] have indicated that the advancing angle α and receding angle β varied quite substantially during the sliding phase. Considering the complexity of the bubble departure and bubble lift-off, and the difficulty in obtaining the measurements, the advancing and receding angles can be reasonably evaluated through the bubble contact angle θ as $\alpha = \theta + \theta'$ and $\beta = \theta - \theta'$. Klausner et al. [76] have employed an angle θ' of 4.5° in their theoretical analysis while Bibeau and Salcudean [81] have reported a value of 2.5°. According to Winterton [82], the angle θ' has nonetheless been postulated to be as high as 10°. In the present study, an angle θ' of 5° is adopted. For the inclination angle θ_i, a value of 10° that gave the best fit to the data by Klausner et al. [76] is employed for the current force balance model.

Model Predictions

The improved wall heat flux partitioning model coupled with the force balance model were assessed by comparing the model predictions against subcooled flow boiling of local radial measurements by Yun et al. [62] and Lee et al. [63] and axial measurements by Zeitoun and Shoukri [58]. Based on their experimental observations, the bubble contact angles have been taken to be at 35° for the local cases and 45° for the axial cases respectively. Experimental conditions for the local and axial data that have been used for comparison with the simulated results are presented in Table 2. Note that the experimental conditions for the local casse chosen herein are identical to the investigations while applying the empirical correlations in Figure 7.

The conservation equations for mass, momentum and energy of each phase were discretised using the finite volume technique. A total number 15 bubble classes were prescribed for the dispersed phases, which are illustrated in Table 3 for the local and axial cases. This representd an additional set of 15 transport equations of which they were progressively solved and coupled with the flow equations during the simulations.

Numerical calculations were performed for one-quarter of the cross-section, utilizing the symmetry property of the annulus geometry for the local as well as the axial cases. A body-fitted conformal system was employed to generate the three-dimensional mesh within the annular channel resulting in a total of 13 (radial) × 30 (height) × 3 (circumference) control volumes for the local case while a total of 8 (radial) × 20 (height) × 3 (circumference) control volumes were used for the axial case. Since wall function was used in the present study to bridge the wall and the fully turbulent region away from heater surface, the normal distance between the wall and the first node in the bulk liquid should be such that the corresponding x^+ was greater than 30. Grid independence was examined. In the mean parameters considered, further grid refinement did not reveal significant changes to the two-phase flow parameters.

For the local cases of L1, L2 and L3, the measured and predicted radial profiles of the Sauter mean bubble diameter (D_s), vapor void fraction (α_g) and interfacial area concentration (a_{if}) located at the measuring plane 1.61 m downstream of the beginning of the heated section are shown in Figures 15-17. Note again that in all the figures, the dimensionless parameter $(r-R_i)/(R_o-R_i) = 1$ indicates the inner surface of the unheated flow channel wall while $(r-R_i)/(R_o-R_i) = 0$ indicates the surface of the heating rod in the channel.

Table 2. Experimental conditions for local (L1, L2, L3) and axial (A1, A2, A3) cases

	P_{inlet} (MPa)	Q_w (kW/m^2)	G (kg/m^2 s)	T_{inlet} (°C)	$T_{sub,inlet}$ (°C)
L1	*0.143*	*152.9*	*474.0*	*96.9*	*13.4*
L2	*0.137*	*197.2*	*714.4*	*94.9*	*13.8*
L2	*0.143*	*251.5*	*1059.2*	*92.1*	*17.9*
A1	*0.137*	*286.7*	*156.2*	*91.9*	*14.9*
A2	*0.150*	*508.0*	*264.3*	*94.6*	*16.6*
A3	*0.150*	*705.0*	*411.7*	*88.9*	*22.5*

Table 3. Evaluated diameters of each discrete bubble classes for the local and axial cases

Bubble class	d_{local} (mm)	d_{axial} (mm)
1	0.45	0.29
2	0.94	0.61
3	1.47	0.95
4	2.02	1.31
5	2.58	1.67
6	3.14	2.03
7	3.71	2.40
8	4.27	2.76
9	4.83	3.13
10	5.40	3.49
11	5.96	3.86
12	6.53	4.23
13	7.10	4.59
14	7.66	4.96
15	8.23	5.32

Experimental photographs (see Figure 9) clearly showed large bubble sizes being present away from the heated wall but not at the heated wall, which this trend has been correctly modeled and predicted by the population balance model. The presence of larger bubbles away from the heated wall confirmed the occurrence of bubble coalescence. In all the three cases, good agreement was achieved between the predicted and measured profiles with the maximum predicted Sauter bubble diameters for L1, L2 and L3 were obtained at about 4.9 mm, 4.7 mm and 3.9 mm while the experimental maximum bubble sizes were measured at about 4.7 mm, 4.9 mm and 4.1 mm. Further away from the heated wall, the reduction of the bubble sizes indicated the bubbles condensing due to the subcooled bulk liquid. It should be noted that the local determination of the Sauter bubble diameter by the population balance model required the appropriate determination of the bubble lift-off diameters through the force balance model. At the measuring location, the bubble departure diameters were predicted to be about 0.56 mm, 0.58 mm and 0.59 mm while the bubble lift-off diameters were approximately attained as 1.4 mm 1.3 mm and 1.2 mm respectively. From these bubble sizes, the ratio of the bubble lift-off diameter to the bubble departure diameter achieved an approximate value of two, which also showed good agreement against experimental observations of Basu et al. [36]. For the bubble frequency, the predicted growth and waiting times for L1 were found to yield about 5.3 ms and 0.74 ms at wall superheat and subcooling temperatures of 18.1°C and 4.1°C, L2, about 5.1 ms and 0.63 ms at wall superheat and subcooling temperatures of 18.5°C and 4.5°C, and L3, about 4.9 ms and 0.67 ms at wall superheat and subcooling temperatures of 19.6°C and 8.8°C. These predicted growth and waiting times were of a similar order according to the values correlated by Basu et al. [36] though they have exerted that the application of their correlated relationships should be used with care since they have been formulated through a limited database. At the point of bubble lifting off into the bulk liquid, the shear lift force was seen as the dominant force in overcoming the surface tension forces acting on the vapor bubble. The calculated shear lift and surface tension forces were respectively:

L1: $F_{sx} \sim 5.06 \times 10^{-5}$ N and $F_{sL} \sim 1.78 \times 10^{-4}$ N;
L2: $F_{sx} \sim 5.06 \times 10^{-5}$ N and $F_{sL} \sim 1.76 \times 10^{-4}$ N; and
L3: $F_{sx} \sim 5.06 \times 10^{-5}$ N and $F_{sL} \sim 1.73 \times 10^{-4}$ N.

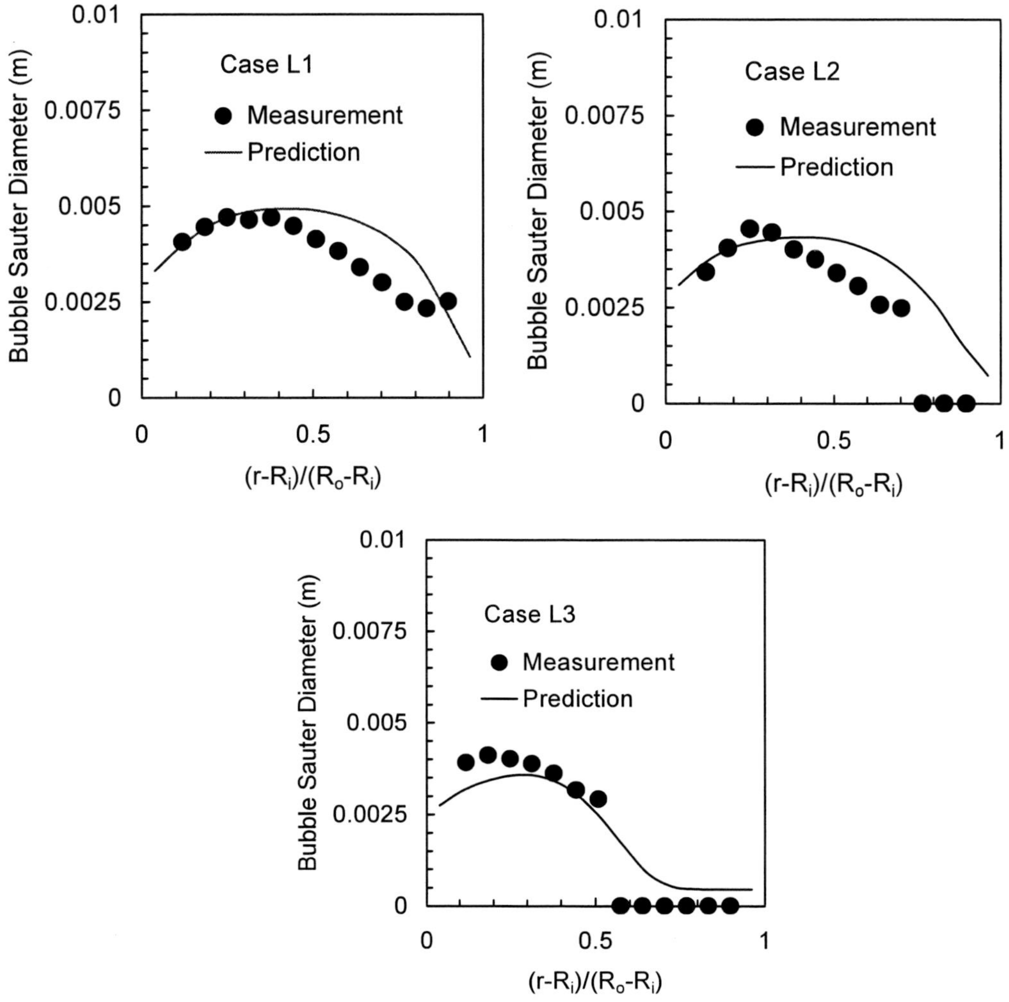

Figure 15. Local Sauter bubble diameter profiles at the measuring plane.

Peak local void fractions observed near the heated surface in Figure 16 represnted a typical chracteristic commonly found in subcooled boiling flow. This high local void fraction found there was explicitly due to the large number of bubbles generated from the active nucleation sites on the heated surface. Here, large amount of bubbles were generated from these active nucleation sites when the temperature on the heated surface exceeded the saturation temperature. A proper prediction of the active nucleation site densities was therefore crucial. As these bubbles reached a critical size, they detached from the heated surface and migrated laterally toward the subcooled liquid core under the competing process of bubble coalescence and condensation as aforementioned.

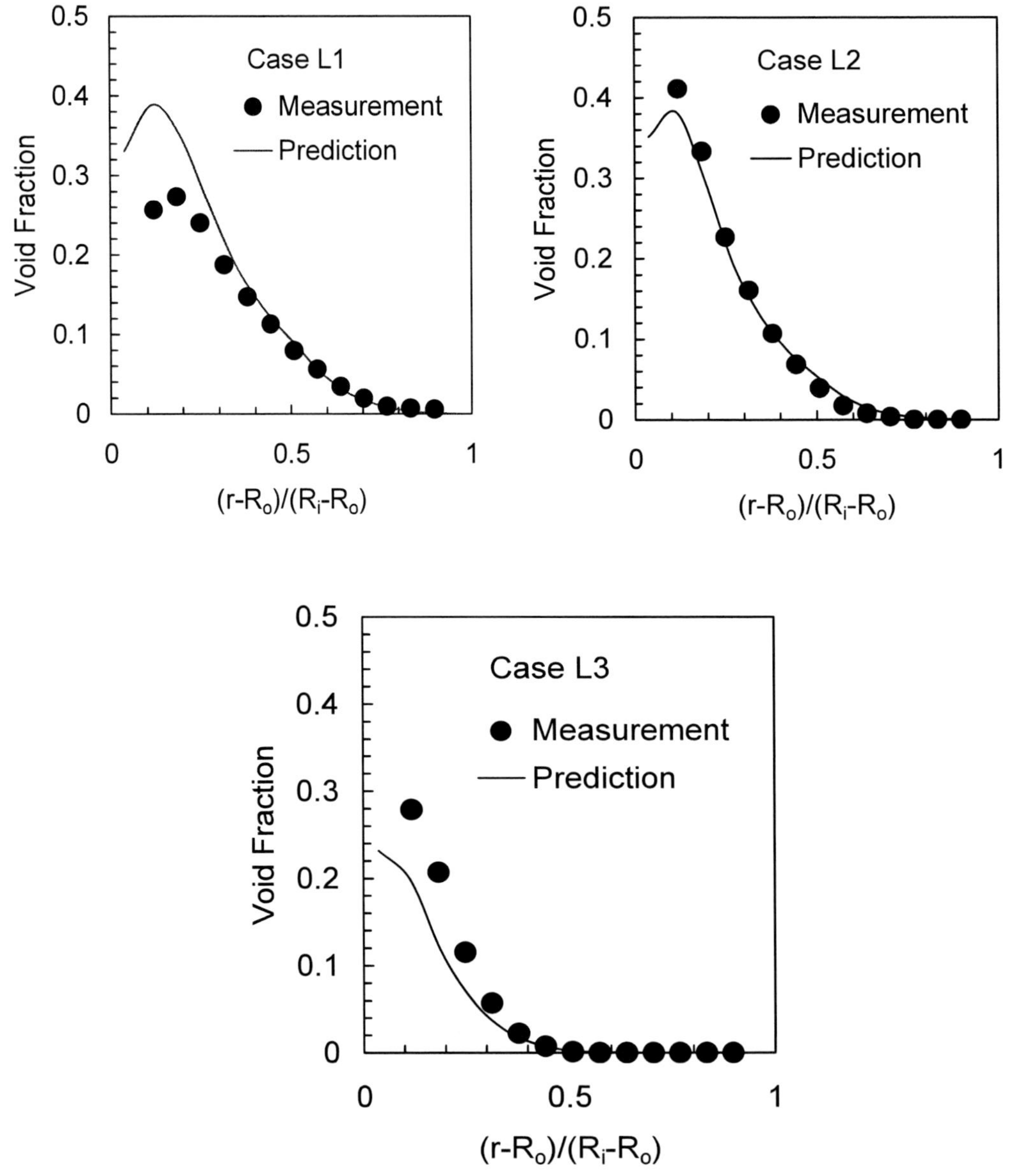

Figure 16. Local void fraction profiles at the measuring plane.

Figure 17 describes the local interfacial area concentration radial distribution. Good agreement was achieved between the measured and predicted interfacial area concentration for all the cases except for case L1 where the interfacial area concentration was significantly over predicted near the heated wall. This could be inferred by the high void fraction predicted near the heated wall as seen in Figure 16. The trend of the interfacial area concentration distributions for all the three cases were correctly modeled and predicted by the two-fluid and population balance models.

The radial profiles of the axial component of the local vapor velocity are shown in Figure 18 while Figure 19 presents the radial profiles of the local liquid velocity for experimental conditions L1, L2 and L3.

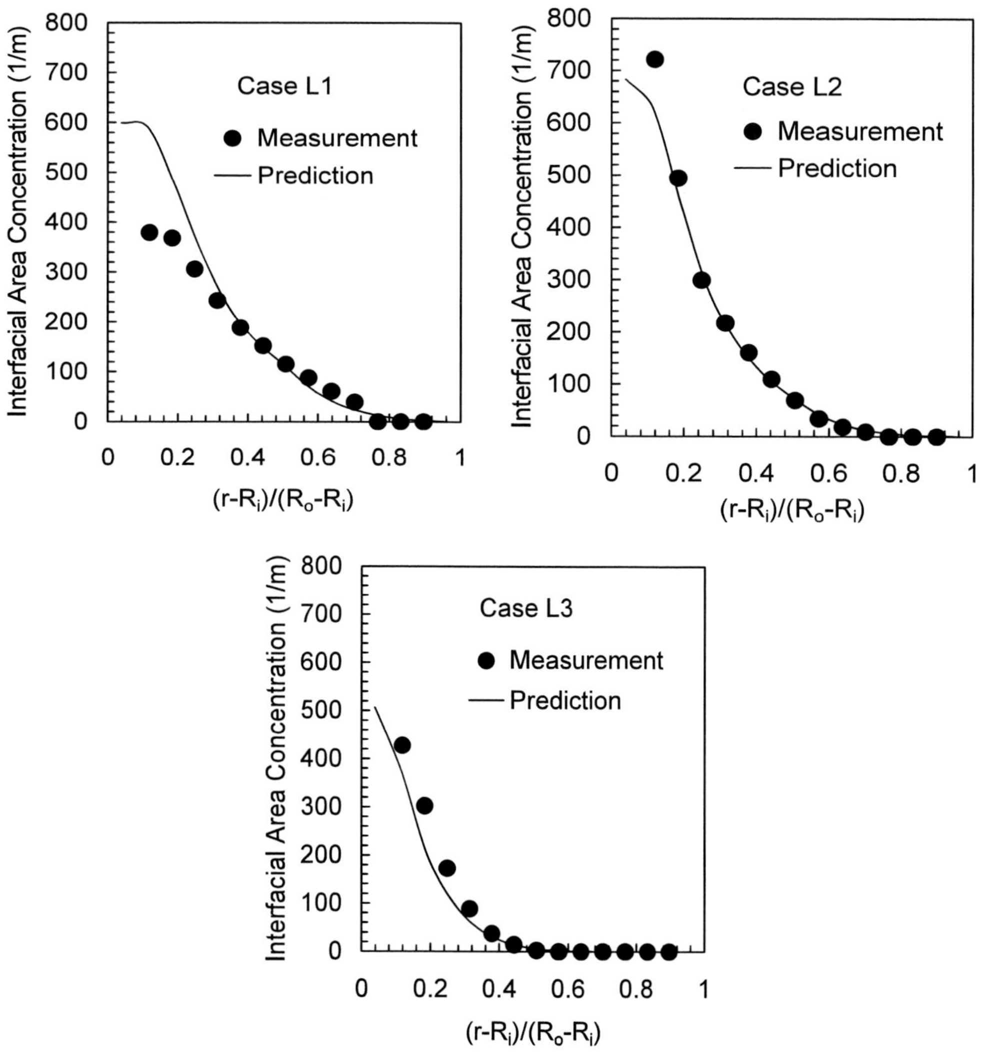

Figure 17. Local interfacial area concentration profiles at the measuring plane.

The vapor velocity being greater than the liquid velocity was due to buoyancy force caused by density difference. As observed in the experiment, the vapor velocity was higher at the centre than the velocities near the heating rod. This was probably due to the buoyancy effect being enhanced by the migration of the large bubbles there, which was again confirmed by high-speed photography in Lee et al. [63]. Vapor velocity predicted by the model showed that higher velocity values approaching the heated boundary. This was mainly due to the assumption where each bubble class has been considered to be traveling at the same mean algebraic velocity due to a single momentum equation being solved for the vapor phase. The philosophy behind adopting this approach was to hasten the computational time and to reduce the computational resources. The discrepancies between the predicted and measured velocities found near the heated wall clearly demonstrated the inadequacy of the adopted approach.

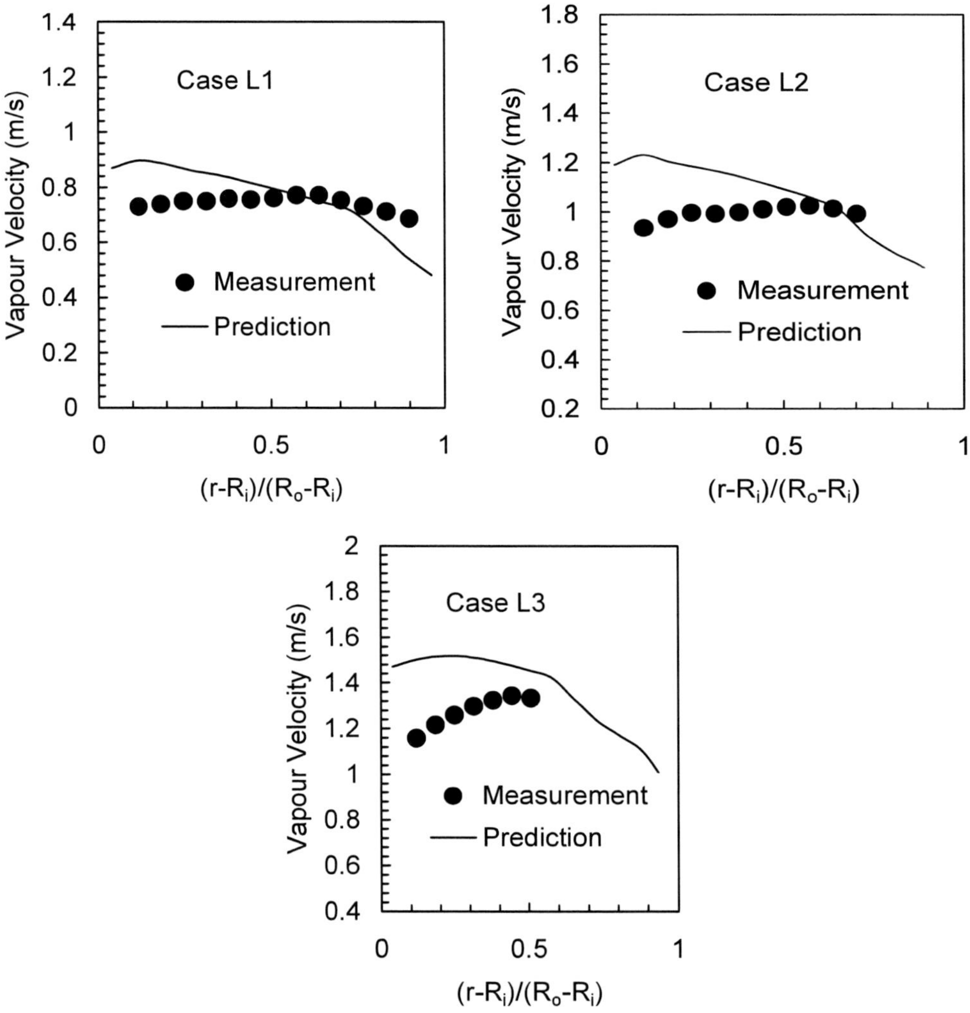

Figure 18. Local vapor velocity profiles at the measuring plane.

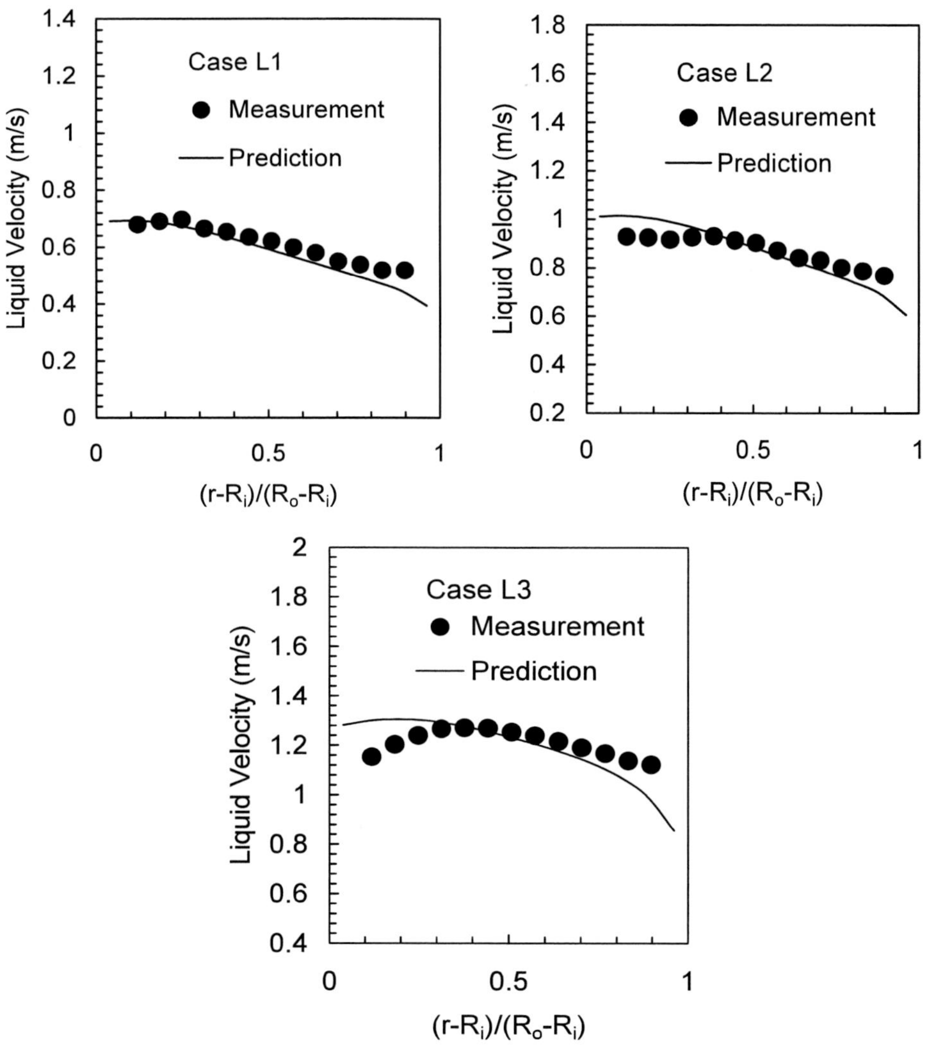

Figure 19. Local liquid velocity profiles at the measuring plane.

Within the channel space, different size bubbles can be expected to travel with different speeds. As an initial step towards resolving the problem, an algebraic slip model could be proposed to account for bubble separation. The terminal velocities for each of the bubbles can be considered through applying an algebraic relationship suggested by Clift et al. [83], which could be then used to evaluate the individual bubble slip velocities. Alternatively, the consideration of additional momentum equations to cater for each of the 15 bubble classes would increase the computational resources tremendously and deem impractical. Ongoing investigations are currently undertaken to test a pertinent choice of two or three dominant groups of bubbles transformed into the Eulerian phases to sufficiently accommodate the hydrodynamics of wide bubble size distributed bubbly flows such as the recent development of the inhomogeneous model of the population balance approach based on the multiple size group (MUSIG) in ANSYS-CFX11. Nevertheless, good agreement was achieved for the liquid velocities between the predictions and experimental values at the measuring plane in the liquid phase. These velocities showed a closer resemblance to the measurements than the predicted profiles of the vapor velocity.

For the axial Cases A1, A2 and A3, the measured and predicted profiles of Sauter bubble diameter normalized by the length scale $\sqrt{\sigma/g(\rho_l-\rho_g)}$, interfacial area concentration and the void fraction along the heated section are presented in Figures 20-22.

The increasing bubble sizes along the axial length clearly showed the bubble coalescence intensifying as the bubbles traveled downstream. Predicted Sauter bubble diameter profiles were seen to agree well with the experimental data as shown in Figure 19. At a point before the channel exit, the predicted growth and waiting times for A1 were ascertained to be about 4.6 ms and 0.5 ms at wall superheat and subcooling temperatures of 18.9°C and 4.3°C; A2, about 4.8 ms and 0.23 ms at wall superheat and subcooling temperatures of 22.8°C and 2.3°C; and A3, about 4.8 ms and 0.24 ms at a wall superheat and subcooling temperatures of 23.0°C and 7.0°C. As expected, the waiting times especially for A2 and A3 were much lower than A1 due to the higher wall heat fluxes.

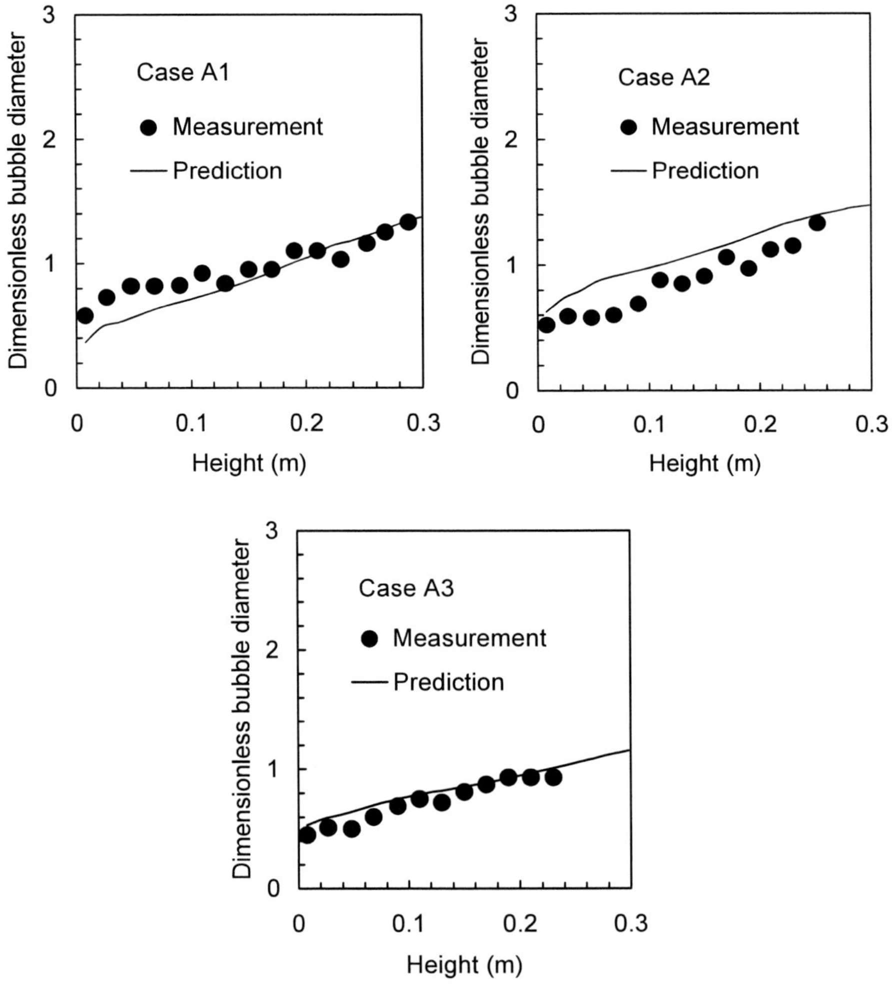

Figure 20. Axial dimensionless Sauter bubble diameter void fraction profiles.

Figure 21 describes the axial interfacial area concentration distribution while Figure 22 demonstrates the axial void fraction evolution along the heated section. Owing to the good agreement achieved between the predictions and measurements of the void fraction, the interfacial area concentration were also found to be in good agreement with the measured values, for the given axial bubble size evolution. The predicted void fraction profiles clearly represented the subcooled boiling characteristic as depicted in Figure 1: low void fraction region followed by a second region, in which the void fraction increased significantly. The predicted void fraction profiles also correctly established a plateau at the initial stages of boiling. The sharp increase of the void fraction profiles approaching the channel exit was due to the increase of bubble sizes decreasing the interfacial area concentration thereby causing a reduction in the condensation rate per unit volume of the channel. This was also associated with the reduction in the relative importance of condensation at the bubble interface due to the decrease of liquid subcooling, which encouraged more bubbles to merge.

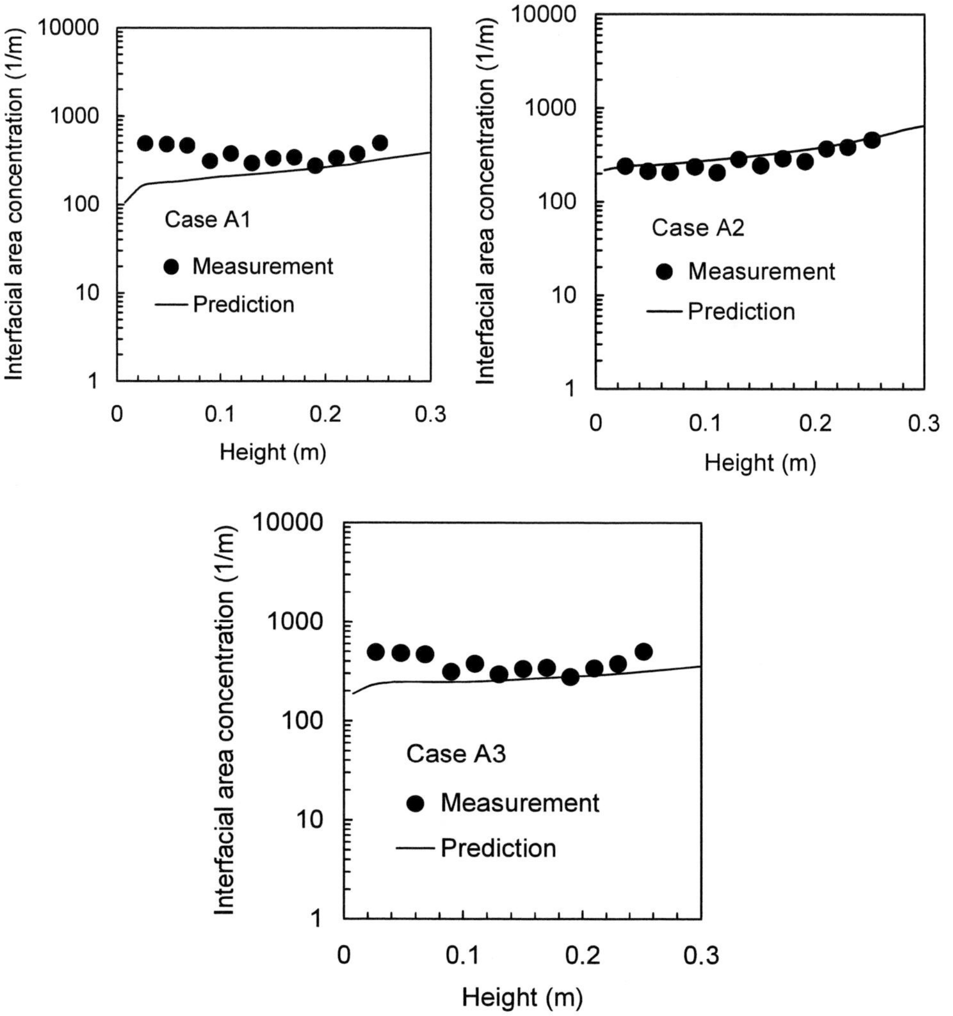

Figure 21. Axial interfacial area concentration profiles.

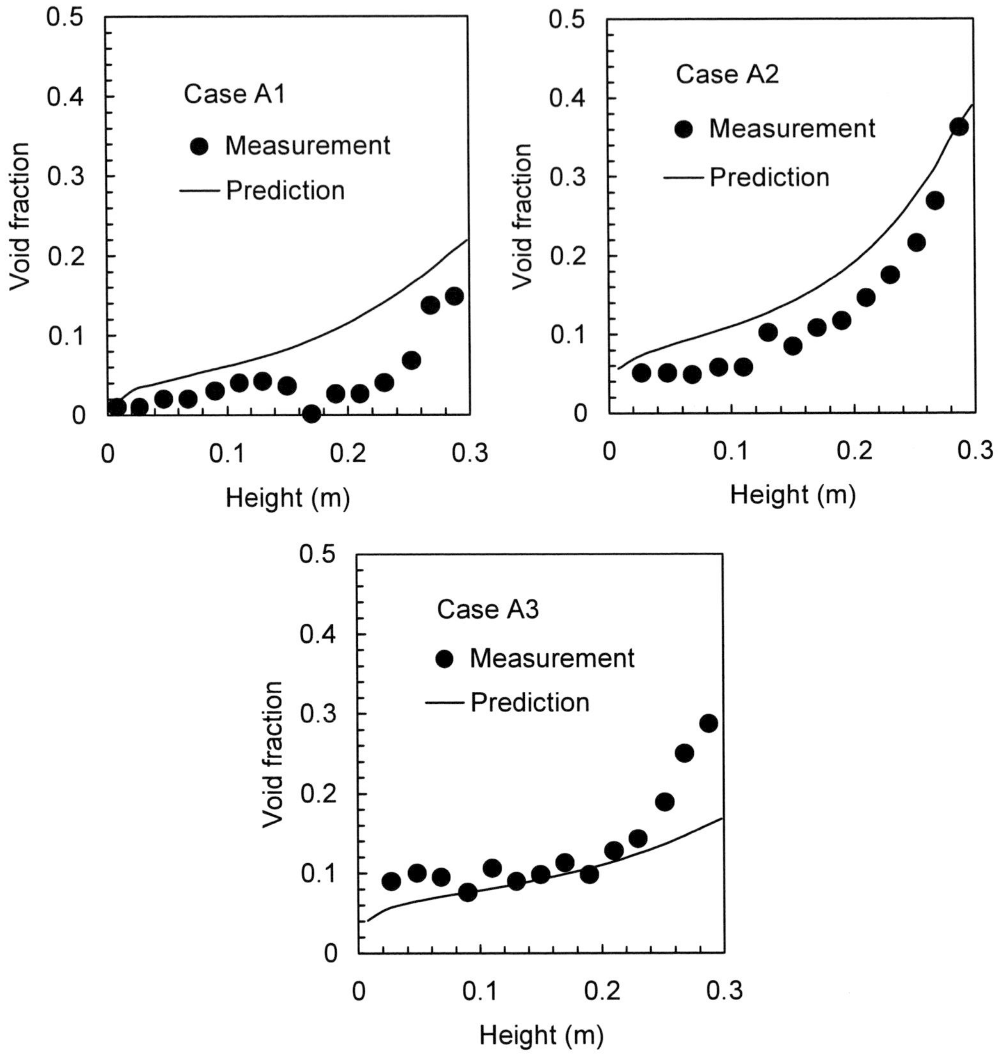

Figure 22. Axial void fraction profiles.

Conclusion

A review of both empirical correlations and mechanistic models to handle subcooled boiling flow has been presented. Empirical correlations which are usually based on curve fitting experimental datasets are unable to account for the physical mechanisms that are unherently present within the complex boiling process. Significant differences between the predictions and experimental data can thus occur when the conditions for which they were developed are not duplicated. This represents the major drawback of the application of such models.

In this article, a two-fluid model coupled with population balance approach is presented for handling low-pressure subcooled boiling flow. The increase in complexity of modeling such flow derives from the additional consideration of the gas or liquid undergoing a phase transformation. Formulation of an improved wall heat flux partitioning model that account for sliding bubbles alongside with the fundamental consideration of bubble frequency theory during low-pressure subcooled flow boiling is also presented. This model when coupled with a force balance model to predict forces acting on a vapor bubble growing under subcooled boiling flow demonstrates the capability to accommodate more complex analyses of bubble growth, bubble departure and bubble lift-off over a wide range of wall heat fluxes and flow conditions. When assessed against local and axial measurements, this present model reveals remarkable agreement for the local and axial profiles of the void fraction, Sauter bubble diameter and interfacial area concentration for different experimental conditions. For the local case, good agreement is also achieved by the present model with regards to the radial distributions of the liquid velocity profiles. Significant weakness is however found for the vapor velocity distribution, which requires additional modeling. For the axial case, a plateau occurring at the initial boiling stages, typical of subcooled flow boiling at low pressures, followed by the significant rise of the void fraction downstream of the annular channel is predicted by the present model.

References

[1] Bergles, A. E., & Rohsenow, W. M. (1964). The determination of forced convection, surface boiling heat transfer, *ASME J. Heat Transfer*, *86*, 365-372.

[2] Mikic, B. B., & Rohsenow, W. M. (1969). A new correlation of pool boiling data including the effect of heating surface characteristics, *ASME J. Heat Transfer*, *91*, 245-256.

[3] Chen, J. C. (1966). A correlation for boiling heat transfer to saturated fluids in convective flows, *Ind. Eng. Chem. Process Des. Dev.*, *5*, 322-329.

[4] Kandlikar, S. G. (1998). Heat transfer characteristics in partial boiling, fully developed boiling, and significant void flow regions of subcooled flow boiling, *ASME J. Heat Transfer*, *120*, 395-401.

[5] McAdams, W. H., Kennel, W. E., Minden, C. S., Carl, R., Picornell, P. M., & Dew, J. E. (1949). Heat transfer at high rates to water with surface boiling, *Ind. Eng. Chem.*, *41*, 1945-1953.

[6] Thom, J. R. S., Walker, W. M., Fallon, T. A., & Reising, G. F. S. (1965). Boiling in subcooled water during flow up heated tubes or annuli, presented at the Symposium on Boiling Heat Transfer in Steam Generating Units and Heat Exchangers, Institute of Mechanical Engineers, London.

[7] Griffith, P., Clark, J. A., & Rohsenow, W. M. (1958). Void volumes in subcooled boiling, *ASME Paper 58-HT-19.*, U.S. National Heat Transfer Conference, Chicago.

[8] Zuber, N., & Findlay, J. (1965). Average volumetric concentration in two-phase flow systems, *ASME J. Heat Transfer*, *87*, 453-462.

[9] Lahey, R. T. Jr. (1978). A mechanistic subcooled boiling model, Proceedings of the 6th International Heat Transfer Conference, Toronto, Canada, Hemisphere Publishing Corporation, Washington, *1*, 293-297.

[10] Chatoorgoon, V., Dimmick, G. R., Carver, M. B., Selander, W. N., & Shoukri, M. (1992). Application of generation and condensation models to predict subcooled boiling void at low pressure, *Nuc. Technol.*, *98*, 366-378.

[11] Zeitoun, O. (1994). Subcooled flow boiling and condensation, Ph.D. Thesis, McMaster University, Hamilton, Ontario, Canada.

[12] Krishna, R., Urseanu, M. I., van Baten, J. M., & Ellenberger J. (1999). Influence of scale on the hydrodynamics of bubble columns operating in the churn-turbulent regime: experiments vs. Eulerian simulations, *Chem. Eng. Sci.*, *54*, 4903-4911.

[13] Shimizu, K., Takada, S., Minekawa, K., & Kawase, Y. (2000). Phenomenological model for bubble column reactors: prediction of gas hold-up and volumetric mass transfer coefficients, *Chem. Eng. J.*, *78*, 21-28.

[14] Pohorecki, R., Moniuk, W., Bielski, P., & Zdrojkwoski, A. (2001). Modeling of the coalescence/redispersion processes in bubble columns, *Chem. Eng. Sci.*, *56*, 6157-6164.

[15] Olmos, E., Gentric, C., Vial Ch., Wild, G., & Midoux, N. (2001). Numerical simulation of multiphase flow in bubble column reactors. Influence of bubble coalescence and break-up, *Chem. Eng. Sci.*, *56*, 6359-6365.

[16] Ramkrishna, D., & Mahoney, A. W. (2002). Population balance modeling. Promise for the future, *Chem. Eng. Sci.*, *57*, 595-606.

[17] Ranz, W. E., & Marshall, W. R. (1952). Evaporation from droplets: parts I and II, *Chem. Eng. Prog.*, *48*, 141-148.

[18] Ishii, M., & Zuber, N. (1979). Drag coefficient and relative velocity in bubbly, droplet or particulate flows, *AIChE J.*, *25*, 843-855.

[19] Drew, D. A., & Lahey, R. T. Jr. (1979). Application of general constitutive principles to the derivation of multi-dimensional two-phase flow equation, *Int. J. Multiphase Flow*, *5*, 243-264.

[20] Lopez de Bertodano, M. (1992). Tubulent bubbly two-phase flow in a triangular duct, Ph.D. Thesis, Rensselaer Polytechnic Institute, Troy, New York.

[21] Antal, S. P, Lahey, R. T. Jr., & Flaherty, J. E. (1991). Analysis of phase distribution and turbulence in dispersed particle/liquid flows, *Chem. Eng. Comm.*, *174*, 85-113.

[22] Anglart, H., & Nylund, O. (1996). CFD application to prediction of void distribution in two-phase bubbly flows in rod bundles, *Nuc. Sci. Eng.*, *163*, 81-98.

[23] Lahey, R. T. Jr., & Drew, D. A. (2001). The analysis of two-phase flow and heat transfer using multidimensional, four field, two-fluid model, *Nuc. Eng. Des.*, *204*, 29-44.

[24] Takagi, S., & Matsumoto, Y. (1998). Numerical study on the forces acting on a bubble and particle, Proceedings of the 3rd International Conference on Mulitphase Flow, Lyon, France.

[25] Tomiyama, A. (1998). Struggle with computational bubble dynamics, Proceedings of the 3rd International Conference on Mulitphase Flow, Lyon, France.

[26] Wang, S. K., Lee, S. J., Lahey, R. T. Jr., & Jones, O. C. (1987). 3-D turbulence structure and phase distribution measurements in bubbly two-phase flows, *Int. J. Multiphase Flow*, *13*, 327-343.

[27] Kurul, N., & Podowski, M. Z. (1990). Multi-dimensional effects in forced convection sub-cooled boiling. Proceedings of the 9th Heat Transfer Conference, Jerusalem, Israel, Hemisphere Publishing Corporation, *2*, 21-26.

[28] Menter, F. R. (1994). Two-equation eddy viscosity turbulence models for engineering applications, *AIAA J.*, *32*, 1598-1605.

[29] Cheung, S. C. P., Yeoh, G. H., & Tu, J. Y. (2007). On the modeling of population balance in isothermal vertical bubbly flows – Average bubble number density approach, *Chem. Eng. Processing*, *46*, 742-756.

[30] Frank, T., Shi, J., & Burns, F. A. D. (2004). Validation of Eulerian multiphase flow models for nuclear safety application, Proceedings of the 3rd Symposium on Two-Phase Modeling and Experimentation, Pisa, Italy.

[31] Sato, Y., Sadatomi, M., & Sekoguchi, K. (1981). Momentum and heat transfer in two-phase bubbly flow-I. *Int. J. Multiphase Flow*, *7*, 167-178.

[32] Del Valle, V. H. M., & Kenning, D. B. R. (1985). Subcooled flow boiling at high heat flux, *Int. J. Heat Mass Transfer*, *28*, 1907-1920.

[33] Graham, R. W., & Hendricks, R. C. (1967). Assessment of convection, conduction and evaporation in nucleate boiling, NASA TND-3943.

[34] Judd, R. L., & Hwang, K. S. (1976). A comprehensive model for nucleate pool boiling heat transfer including microlayer evaporation, ASME J. Heat Transfer, 98, 623-629.

[35] Steiner, H., Kobor, A., & Gebhard, L. (2005). A wall heat transfer model for subcooled boiling flow, *Int. J. Heat Mass Transfer*, *48*, 4161-4173.

[36] Basu, N., Warrier, G. R., & Dhir, V. K. (2005). Wall heat flux partitioning during subcooled flow boiling: Part I – Model development, *ASME J. Heat Transfer*, *127*, 131-140.

[37] Basu, N., Warrier, G. R., & Dhir, V. K. (2005). Wall heat flux partitioning during subcooled flow boiling: Part II – Model validation, *ASME J. Heat Transfer*, *127*, 141-148.

[38] Warrier, G. R., & Dhir, V. K. (2006). Heat transfer and wall heat flux partitioning during subcooled flow nucleate boiling, *ASME J. Heat Transfer*, *128*, 1243-1256.

[39] Bowring, R. W. (1962). Physical model based on bubble detachment and calculation of steam voidage in the subcooled region of a heated channel. *Report HPR-10,* Institute for Atomenergi, Halden, Norway.

[40] Kenning, D. B. R., & Del Valle, V. H. M (1981). Full developed nucleate boiling: overlap of areas of influence and interference between bubble sites, *Int. J. Heat Mass Transfer*, *24*, 1025-1032.

[41] Tu, J. Y., & Yeoh, G. H. (2002). On numerical modelling of low-pressure subcooled boiling flows. *Int. J. Heat Mass Transfer*, 45, 1197-1209.

[42] Yeoh, G. H., & Tu, J. Y. (2004). Population balance modeling for bubbly flows with heat and mass transfer, *Chem. Eng. Sci.*, *59*, 3125-3139.

[43] Yeoh, G. H., & Tu, J. Y. (2005). Thermal hydraulic modeling of bubbly flows with heat and mass transfer, *AIChE J.*, *51*, 8-27.

[44] Tolubinsky, V. I. and Kostanchuk, D. M. (1970). Vapor bubbles growth rate and heat transfer intensity at subcooled water boiling, Fourth International Heat Transfer Conference, 5, Paper No. B-2.8, Paris, France.

[45] Unal, H. C. (1976). Maximum bubble diameter, maximum bubble growth time and bubble growth rate, *Int. J. Heat Mass Transfer*, *19*, 643-649.

[46] Fritz, W. (1935). Berechung des Maximalvolumes von Dampfblasen, *Phys. Z.*, *36*, 379.

[47] Kocamustafaogullari, G., & Ishii, M. (1995). Foundation of the interfacial area transport equation and its closure relations, *Int. J. Heat Mass Transfer*, *38*, 481-493.

[48] Cole, R. (1960). A photographic study of pool boiling in the region of the critical heat flux, *AIChE J.*, *6*, 533-542.

[49] Ivey, H. J. (1967). Relationships between bubble frequency, departure diameter and rise velocity in nucleate boiling, *Int. J. Heat Mass Transfer*, *10*, 1023-1040.

[50] Stephan, K. (1992). Heat transfer in condensation and boiling, Springer-Verlag, NewYork.

[51] Peebels, F. N. and Garber, H. J. (1953). Studies on the motion of gas bubbles in liquids, *Chem. Eng. Prog.*, *49*, 88-97.

[52] Jakob, M. (1949). Heat Transfer, 1, Chapter 29, Wiley and Sons, New York.

[53] Lemmert, M, & Chwala, J. M. (1977). Influence of flow velocity on surface boiling heat transfer coefficient. Academic Press and Hemisphere, New York and Washington.

[54] Končar, B., Kljenak, I., & Mavko, R. (2004). Modeling of local two-phase flow parameters in upwards subcooled flow boiling at low pressure, *Int. J. Heat Mass Transfer*, *47*, 1499-1513.

[55] Basu, N., Warrier, G. R., & Dhir, V. K. (2002). Onset of nucleate boiling and active nucleation site density during subcooled flow boiling, ASME J. Heat Transfer, 124, 717-728.

[56] Hibiki, T., & Ishii, M. (2003). Active nucleation site density in boiling systems, *Int. J. Heat Mass Transfer*, *46*, 2587-2601.

[57] Nylund, O., Becker, K. M., Eklund, R., Gelius, O., Haga, I., Hansson, P. T., Hernbord, G., & Åkerhielm, F. (1967). Measurements of hydrodynamics characteristics, instability thresholds, and burnout limits for 6-rod clusters in natural and forced circulation, FRIGG_1 Report, ASEA and AB Atomenergi, Sweden.

[58] Zeitoun, O., & Shoukri, M. (1996). Bubble behavior and mean diameter in subcooled flow boiling, *ASME J. Heat Transfer*, *118*, 110-116.

[59] Zeitoun, O., & Shoukri, M. (1996). Axial void fraction profile in low pressure subcooled flow boiling, *Int. J. Heat Mass Transfer*, *40*, 867-879.

[60] Donveski, B., & Shoukri, M. (1989). Experimental study of subcooled flow boiling and condensation in an annular channel, Thermofluids Report No. ME/89/TF/R1, Department of Mechanical Engineering, McMaster University, Hamilton, Ontario, Canada.

[61] Dimmick, G. R., & Selander, W. N. (1990). A dynamic model for predicting subcooled void: experimental results and model development, EUROTHERM Seminar #16, Pisa, Italy.

[62] Lee, T. H., Park. G.-C, & Lee, D. J. (2002). Local flow characteristics of subcooled boiling flow of water in a vertical annulus, *Int. J. Multiphase Flow*, *28*, 1351-1368.

[63] Yun, B. J., Park, G.-C., Song, C. H., & Chung, M. K. (1997). Measurements of local two-phase flow parameters in a boiling flow channel, Proceedings of the OECD/CSNI Specialist Meeting on Advanced Instrumentation and Measurement Techniques.

[64] Tu, J. Y., Yeoh, G. H., Park, G.-C., & Kim, M.-O. (2005). On population balance approach for subcooled boiling flow prediction, *ASME J. Heat Transfer*, *127*, 253-264.

[65] Krepper, E., Končar, B., & Egorov, Y. (2007). CFD modelling of subcooled boiling – concept, validation and application to fuel assembly design, *Nuc. Eng. Des.*, *237*, 716-731.

[66] Bonjour, J., & Lallemand, M. (2001). Two-phase flow structure near a heated vertical wall during nucleate pool boiling, *Int. J. Multiphase Flow*, *27*, 1789-1802.

[67] Prodanovic, V., Fraser, D., & Salcudean, M. (2002). Bubble behaviour in subcooled flow boiling of water at low pressures and low flow rates, *Int. J. Multiphase Flow*, *28*, 1-19.

[68] Gopinath, R., Basu, N. & Dhir, V. K. (2002) Interfacial heat transfer during subcooled flow boiling, *Int. J. Heat Mass Transfer*, *45*, 3947-3959.

[69] Luo, H., & Svendsen, H. (1996). Theoretical model for drop and bubble break-up in turbulent dispersions, *AIChE J.*, *42*, 1225-1233.

[70] Prince, M. J, & Blanch, H. W. (1990). Bubble coalescence and break-up in air-sparged bubble column, *AIChE J.*, *36*, 1485-1499.

[71] Chesters, A. K., & Hoffman, G. (1982). Bubble coalescence in pure liquids, *Appl. Sci. Res.*, *38*, 353-361.

[72] Rotta, J. C. (1972). Turbulente Stromungen. Stuttgart: B. G. Teubner, 1972.

[73] Fleischer, C., Becker, S., & Eigenberger, G. (1996). Detailed modeling of the chemisorption of CO_2 into NaOH in a bubble column, *Chem. Eng. Sci.*, *51*, 1715-1724.

[74] Sateesh, G., Sarit, K. D., & Balakrishnan, A. R. (2005). Analysis of pool boiling heat transfer: effect of bubbles sliding on the heating surface, *Int. J. Heat Mass Transfer*, *48*, 1543-1553.

[75] Yeoh, G. H., & Tu, J. Y. (2005). A unified model considering force balances for departing vapour bubbles and population balance in subcooled boiling, *Nuc. Des. Eng.*, *235*, 1251-1265.

[76] Klausner, J. F., Mei, R., Bernhard, D. M., & Zeng, L. Z. (1993). Vapor bubble departure in forced convection boiling, *Int. J. Heat Mass Transfer*, *36*, 651-662.

[77] Zeng, L. Z., Klausner, J. F., Bernhard, J. F., & Mei, R. (1993). A unified model for the prediction of bubble detachment diameters in boiling systems – II. Flow boiling, *Int. J. Heat Mass Transfer*, *36*, 2271-2279.

[78] Zuber, N. (1961). The dynamics of vapor bubbles in nonuniform temperature fields, *Int. J. Heat Mass Transfer*, *2*, 83-98.

[79] Maity, S. (2000). Effect of velocity and gravity on bubble dynamics, M.S. Thesis, University of California, Los Angeles, U.S.A.

[80] Hsu, Y. Y., & Graham, R. W. (1976). Transport process in boiling and two-phase systems, Hemisphere Publishing Company, Washington.

[81] Bibeau, E. L., & Salcudean, A. (1994). A study of bubble ebullition in forced convective subcooled nucleate boiling at low pressure, *Int. J. Heat Mass Transfer*, *37*, 2245-2259.

[82] Winterton, R. H. S. (1984). Flow boiling: prediction of bubble departure, *Int. J. Heat Mass Transfer*, *27*, 1422-1424.

[83] Clift, R., Grace, J. R., & Weber, M. E. (1978). Bubbles, drops and particles, Academic Press, New York.

In: Applied Mathematical Modeling
Editor: Eetu N. Virtanen, pp. 145-184

ISBN: 978-1-60021-975-7

Chapter 3

A SECOND-ORDER SCHEME OF PRECISE TIME-STEP INTEGRATION METHOD FOR DYNAMIC ANALYSIS WITH RESPECT TO LONG-TERM INTEGRATION AND TRANSIENT RESPONSES

***Qing-Hua Qin*[1] *and Hang Ma*[2]**
[1]Department of Engineering, Australian National University,
Canberra, ACT 0200, Australia
[2]Department of Mechanics, College of Sciences, Shanghai University,
Shanghai 200444, China

ABSTRACT

In this chapter, a second-order scheme of precise time-step integration (PTI) method is introduced for dynamic analysis with respect to long-term integration and transient responses while spatial discretization is realized with the differential quadrature method. Rather than transforming into first-order equations, a recursive scheme is presented in detail for direct solution of the homogeneous part of second-order differential and algebraic equations. The sine and cosine matrices involved in the scheme are calculated using the so-called 2^N algorithm, and the corresponding particular solution is also presented where the excitation vector is approximated by the truncated Taylor series. The performance and numerical behaviors of the second-order scheme of the PTI method are tested by a series of numerical examples in comparison with the first-order scheme or with the traditional time-marching Newmark- β method as the reference. The issue of spurious high-frequency responses resulting from spatial discretization for shock-excited structural dynamic analysis is also studied in the framework of the second-order PTI method. The effects of spatial discretization, numerical damping and time step on solution accuracy are explored by analyzing longitudinal vibrations of a shock-excited rod with rectangular, half-triangular and Heaviside step impact.

The chapter is organized as follows: In Section 1, previous research into time-marching algorithms is briefly reviewed. Following a brief introductory overview of the first-order scheme, the second-order PTI method is introduced in detail in Section 2 together with an analysis of the accuracy that can be achieved, the treatment of particular solutions, and solution behavior with the second-order scheme. Numerical examples for smooth responses of various wave equations and wave propagations are presented using the second-order PTI method in Section 3. Numerical examples of impact responses are presented and studied extensively in Section 4 with respect to the effect of spatial discretization, modal types that the discretization can afford, and numerical damping, etc., with some concluding remarks presented in Section 5.

Numerical tests show that for the analysis of smooth and shock-excited responses, both efficiency and accuracy can be improved considerably with the second-order PTI method. For the analysis of shock-excited responses, spatial discretization can provide a reasonable number of model types for any given error tolerances. An appropriate spatial discretization scheme for a given structure is usually required in order to obtain an accurate and meaningful numerical solution, especially for describing the fine detail of traction responses with sharp changes. Under the framework of the PTI method, the numerical damping that is often required in traditional integration schemes is found to be unnecessary, and there is no restriction on the size of time steps, because the PTI method can usually produce results with machine-like precision and is an unconditionally stable explicit method.

Keywords: precise time-step integration method, second-order scheme, initial problem, transient dynamic analysis, differential quadrature method, spatial discretization

1. INTRODUCTION

In the middle of 1990s, Zhong and Williams proposed a precise time-step integration (PTI) method [1-2] for the analysis of structural dynamics, which is of great engineering importance. In the PTI method, after transforming the second-order equation into a first-order equation, the solution of the homogeneous part is expressed by matrix exponentials, which can be computed precisely using the so-called 2^N algorithm. Shortly afterwards, Lin et al. extended this method to dealing with sinusoidal loading and more general loading forms using Fourier series [3]. Shen et al. improved the efficiency of the method by a parallel computing method [4]. Wang et al. transformed the governing equations into an equivalent homogeneous form by expanding the dimensions of the problems to avoid computing the particular solutions separately [5]. Here the term ‘precise integration’ is specially referred as to a class of time integration procedure, which uses a recurrence formula to reduce computing effort and simplifies the exponential matrix method [1-2]. It is a numerical rather than an analytical method, with which precision can be achieved for solution of the homogeneous part on ordinary computers. The PTI method is an unconditionally stable explicit method, which has been described in several research publications, exhibiting high precision in comparison with traditional methods. Since the original work of Zhong and Williams, the PTI method has been applied to solve a number of problems such as the dynamic structural responses of a continuous beam under moving loads [6] or with holonomic constraints [7] or under evolutionary random excitations [8-10], time-dependent partial differential equations [11], asymmetric Riccati differential equations [12-13] and dynamic elastic-plastic hardening and softening problems [14].

It is well known that analysis of the dynamic responses of structures plays an important role in practical engineering. These responses are controlled by second-order differential equations, which are traditionally analyzed by means of direct integration or step-by-step schemes, among which are implicit methods such as the Newmark, Wilson-θ or Houbolt schemes, and explicit methods such as the central difference scheme. Compared with traditional methods, the accuracy of the PTI method has been greatly improved, up to machine-like precision. However, the PTI method can be further improved by increasing its computational efficiency. This is possible because it is based on a system of first-order differential equations. The second-order differential equations under consideration must thus be transformed into first-order equations before numerical analysis can be performed. This would make the system matrix relatively larger and the scheme less efficient. In Section 2, following a brief introduction of the first-order scheme, the PTI method is extended to direct solution of second-order algebraic and differential equations, whereby efficiency and accuracy can be further improved [15]. This is considered significant especially for dynamic analysis with an emphasis on long-term dynamic evolution. The sine and cosine matrices involved in the second-order scheme are calculated using the so-called 2^N algorithm. The corresponding particular solution is presented incorporated with the second-order scheme, where the excitation vector is approximated by the truncated Taylor series. Then the solution features of the PTI method are discussed with the aid of mode decomposition to show how damping effects can be introduced into the algorithm. In Section 3, using the differential quadrature method [16] for spatial discretization, numerical examples for smooth responses of various wave equations and wave propagations are presented using the second-order PTI method. For transient responses under impact loading, spurious high-frequency responses resulting from spatial discretization in time-step algorithms for structural dynamic analysis have long been an issue of concern in the framework of traditional finite difference methods [17-19]. Such algorithms should be not only numerically dissipative in a controllable manner, but also unconditionally stable so that the time-step size can be governed solely by the accuracy requirement. This issue is considered in the framework of the second-order PTI method. In Section 4, taking the Newmark-β method as a reference, the performance and numerical behaviors for transient responses under impact loading are studied extensively [20] with respect to the effect of spatial discretization, the modal types that discretization can afford and the numerical damping, etc. Section 5 presents concluding remarks. Numerical tests show that both the efficiency and accuracy of the solution can be improved considerably with the proposed second-order algorithm.

2. The Precise Time-step Integration Method

In dynamic analysis of a structure, the equation of motion of a structural model can be written as the following second-order ordinary differential equations in conjunction with the initial conditions written in matrix form:

$$\mathbf{M}\ddot{\mathbf{u}} + \mathbf{B}\dot{\mathbf{u}} + \mathbf{K}\mathbf{u} = \mathbf{f}(t) \tag{1}$$

$$\mathbf{u}(t_0) = \mathbf{u}_0,\ \dot{\mathbf{u}}(t_0) = \dot{\mathbf{u}}_0 \tag{2}$$

where $\mathbf{M}$, $\mathbf{B}$ and $\mathbf{K}$ represent time-invariant mass, physical damping and stiffness matrices of order n of the structure, respectively, with $\mathbf{M}$ being assumed positively definite. $\mathbf{f}(t)$ represents the excitation vector. The superscript dot indicates differentiation of state variable with respect to time. The matrices $\mathbf{M}$, $\mathbf{B}$ and $\mathbf{K}$ can be obtained by means of various spatial discretization methods such as finite element method, boundary element or differential quadrature method.

In this section, both the first-order and the second-order schemes for the PTI method are described for solving equation (1), together with the 2^N algorithm for computing numerically the sine and cosine matrix functions related to the second-order scheme. The homogeneous solution of equation (1) is discussed first. Then the treatment for the particular solution with the aid of Taylor expansion is detailed. In order to show the solution features of the PTI method and how damping effects can be introduced into the algorithm, the mode decomposition method is used to study the correlation between high-frequency modes and spatial discretization within the framework of the PTI method.

2.1. The First-order Scheme

In the first-order PTI method, following Zhong and Zhong and Williams [1,2], in which they introduced a new variable $\mathbf{v} = \left\{\mathbf{u}^T, (\mathbf{M}\dot{\mathbf{u}} + \mathbf{B}\mathbf{u}/2)^T\right\}^T$, equation (1) is first transformed into the form of first-order differential equations as follows:

$$\dot{\mathbf{v}} = \mathbf{H}\mathbf{v} + \mathbf{r} \tag{3}$$

where

$$\mathbf{r} = \begin{Bmatrix} \mathbf{0} \\ \mathbf{f} \end{Bmatrix},\quad \mathbf{H} = \begin{bmatrix} -\mathbf{M}^{-1}\mathbf{B}/2 & \mathbf{M}^{-1} \\ \mathbf{B}\mathbf{M}^{-1}\mathbf{B}/4 - \mathbf{K} & -\mathbf{B}\mathbf{M}^{-1}/2 \end{bmatrix} \tag{4}$$

The general solutions of equation (1) for the first-order PTI method can be obtained analytically in the form [1,2]

$$\mathbf{v}(t) = \mathbf{v}_h(t) + \mathbf{v}_p(t) = \exp[\mathbf{H}(t - t_0)]\,\mathbf{v}(t_0) + \int_{t_0}^{t} \exp\{\mathbf{H}(t - \eta)\}\mathbf{r}(\eta)d\eta \tag{5}$$

where $\mathbf{v}(t)$ and $\mathbf{v}(t_0)$ represent the values of the solution vector at time t and initial time t_0, respectively. Letting Δt be the length of a constant time step and denoting $\tau = t - \eta$, the explicit recursive formula for the general solutions of equation (1) at the $(j+1)$th time step, i.e. $t = t_0 + (j+1)\Delta t$, can then be written as

$$\mathbf{v}_{j+1} = \exp(\mathbf{H}\Delta t)\,\mathbf{v}_j + \int_0^{\Delta t} \exp(\mathbf{H}\tau)\mathbf{r}(t_j + \Delta t - \tau)d\tau,\ (0 \le \tau \le \Delta t) \tag{6}$$

The matrix $\mathbf{H}$ in equation (6) is a constant matrix and $\exp(\mathbf{H}\Delta t)$ is known as the amplification matrix, or the system matrix. The key point of the PTI method is the calculation of $\exp(\mathbf{H}\Delta t)$ as accurately as possible [1-2] because the accuracy of equation (6) depends on how accurately the exponential matrix function $\exp(\mathbf{H}\Delta t)$ can be evaluated. Due to its wide application in engineering problems, calculation of the matrix exponent has been discussed in the literature [22]. Zhong and Williams [1-2] proposed an accurate and efficient method for the calculation of $\exp(\mathbf{H}\Delta t)$ called the 2^N algorithm, which has the form

$$\exp(\mathbf{H}\Delta t) = [\exp(\mathbf{H}\tau)]^m \tag{7}$$

where $\tau = \Delta t / m$, $m = 2^N$ and N is a positive integer. It has been argued [2] that precise integration can achieve high accuracy because τ is, in general, very small, even for a small value of N. For example, if $N = 20$ then $\tau = \Delta t / 1048576$. In view of structural dynamics, the value of τ is much smaller than the highest modal period of any conventional idealized structure. However, the first order scheme will make the system matrix relatively large. For example, if the size of $\mathbf{M}$, $\mathbf{B}$ and $\mathbf{K}$ in equation (1) is $n \times n$, then the size of $\mathbf{H}$ becomes $2n \times 2n$. Therefore, the efficiency of the method can be further improved, especially in the case of long-term integration, by introducing the second-order scheme described in the following subsection.

2.2. The Second-order Scheme

In the second-order PTI method [15], the differential equation in matrix form can also be solved analytically. In this subsection, the homogeneous solution of equation (1) is discussed by firstly assuming $\mathbf{f}(t) = \mathbf{0}$ as follows:

$$\mathbf{M}\ddot{\mathbf{u}}_h + \mathbf{B}\dot{\mathbf{u}}_h + \mathbf{K}\mathbf{u}_h = \mathbf{0} \tag{8}$$

with the initial conditions

$$\mathbf{u}_h(t_0) = \mathbf{u}_0,\ \dot{\mathbf{u}}_h(t_0) = \dot{\mathbf{u}}_0 \tag{9}$$

Suppose that the homogeneous solution has the form $\mathbf{u}_h = \exp(\mathbf{A}t)\mathbf{a}$ and insert it into equation (8), where $\mathbf{a}$ represents an arbitrary vector. We have the characteristic matrix equation as

$$\mathbf{MA}^2 + \mathbf{BA} + \mathbf{K} = \mathbf{0} \tag{10}$$

Solving equation (10), we obtain

$$\mathbf{A} = -\mathbf{D} \pm i\mathbf{\Omega} \tag{11}$$

where $i = \sqrt{-1}$. The matrices $\mathbf{D}$ and $\mathbf{\Omega}$ are written as follows, respectively:

$$\mathbf{D} = \frac{1}{2}\mathbf{M}^{-1}\mathbf{B} \tag{12}$$

$$\mathbf{\Omega} = \sqrt{\mathbf{J}} = \sqrt{-\mathbf{M}^{-2}\mathbf{B}/4 + \mathbf{M}^{-1}\mathbf{K}} \tag{13}$$

The matrix $\mathbf{\Omega}$ is the square root of the matrix $\mathbf{J}$, which can be computed by the technique of matrix decomposition

$$\mathbf{J} = \mathbf{Q\Lambda Q}^{-1} = -\mathbf{Q}\left(i\sqrt{\Lambda}\right)\mathbf{Q}^{-1}\,\mathbf{Q}\left(i\sqrt{\Lambda}\right)\mathbf{Q}^{-1} = -(i\mathbf{\Omega})^2 = \mathbf{\Omega}^2,\ (\lambda_k < 0) \tag{14a}$$

$$\mathbf{J} = \mathbf{Q\Lambda Q}^{-1} = \mathbf{Q}\sqrt{\Lambda}\mathbf{Q}^{-1}\,\mathbf{Q}\sqrt{\Lambda}\mathbf{Q}^{-1} = \mathbf{\Omega}^2,\ (\lambda_k > 0) \tag{14b}$$

That is

$$\mathbf{\Omega} = \mathbf{Q}\sqrt{\Lambda}\mathbf{Q}^{-1} \tag{15}$$

where $\mathbf{\Lambda}$ stands for the diagonal matrix which is composed of the eigenvalues of $\mathbf{J}$ as follows:

$$\mathbf{\Lambda} = diag\left(|\lambda_1|,\cdots,|\lambda_n|\right) \tag{16}$$

and the k th column in $\mathbf{Q}$ represents the eigenvector corresponding to the eigenvalue λ_k. The eigenvalues and eigenvectors of the matrix $\mathbf{J}$ can be computed numerically using the subroutines described in [21]. Here an assumption has to be made that all the eigenvalues λ_k ($k = 1,\cdots,n$) are distinct, real and non-zero ($\lambda_k \neq 0$). With the theory of ordinary differential equations [23], the homogeneous solution of equation (10) can be expressed as

$$\mathbf{u}_h = \exp[-\mathbf{D}(t-t_0)]\,[\cos\mathbf{\Omega}(t-t_0)\mathbf{c}_1 + \sin\mathbf{\Omega}(t-t_0)\mathbf{c}_2] \tag{17}$$

where $\mathbf{c}_1$ and $\mathbf{c}_2$ represent constant vectors to be determined by initial conditions (9). Then we have

$$\mathbf{u}_h = \exp[-\mathbf{D}(t-t_0)]\,\{[\cos\mathbf{\Omega}(t-t_0)+\mathbf{\Omega}^{-1}\mathbf{D}\sin\mathbf{\Omega}(t-t_0)]\mathbf{u}_0 + \mathbf{\Omega}^{-1}\sin\mathbf{\Omega}(t-t_0)\dot{\mathbf{u}}_0\} \tag{18}$$

$$\begin{aligned}\dot{\mathbf{u}}_h = -\exp\left[-\mathbf{D}(t-t_0)\right]\Big\{&\left(\mathbf{\Omega}+\mathbf{\Omega}^{-1}\mathbf{D}^2\right)\sin\mathbf{\Omega}(t-t_0)\,\mathbf{u}_0 \\ &+\left[\mathbf{\Omega}^{-1}\mathbf{D}\sin\mathbf{\Omega}(t-t_0) - \cos\mathbf{\Omega}(t-t_0)\right]\dot{\mathbf{u}}_0\Big\}\end{aligned} \tag{19}$$

Denoting $\mathbf{C}=\cos(\mathbf{\Omega}\,\Delta t)$, $\mathbf{S}=\sin(\mathbf{\Omega}\,\Delta t)$ and $\mathbf{E}=\exp(-\mathbf{D}\Delta t)$ and letting Δt be the length of a constant time step, the explicit recursive formula for the homogeneous solution at the $(k+1)$th time step, i.e. $t=t_0+(k+1)\Delta t$, can be written as

$$\mathbf{u}_{h,k+1} = \mathbf{E}\left[\left(\mathbf{C}+\mathbf{\Omega}^{-1}\mathbf{D}\mathbf{S}\right)\mathbf{u}_{h,k} + \mathbf{\Omega}^{-1}\mathbf{S}\dot{\mathbf{u}}_{h,k}\right] \tag{20}$$

$$\dot{\mathbf{u}}_{h,k+1} = -\mathbf{E}\left[\left(\mathbf{\Omega}+\mathbf{\Omega}^{-1}\mathbf{D}^2\right)\mathbf{S}\mathbf{u}_{h,k} + \left(\mathbf{\Omega}^{-1}\mathbf{D}\mathbf{S}-\mathbf{C}\right)\dot{\mathbf{u}}_{h,k}\right] \tag{21}$$

2.3. The 2^N Algorithm for Matrix Cosine and Sine Functions

The algorithm for calculating the exponential matrix $\mathbf{E}$ precisely was proposed by Zhong and Zhong and Williams [1,2]. Here we present the algorithm for calculating the matrix cosine and sine functions $\mathbf{C}$ and $\mathbf{S}$ precisely. This is obviously the key point of the scheme. We start from the following exponential relation by letting $\tau = \Delta t / m$ and $m = 2^N$

$$\exp(i\mathbf{\Omega}\Delta t) = \{\exp(i\mathbf{\Omega}\,\tau)\}^m \tag{22}$$

Expand the right-hand side of equation (22) to the form of Taylor series [22,23]

$$\exp(i\mathbf{\Omega}\,\tau) = \sum_{k=0}^{\infty}\frac{1}{k!}(i\mathbf{\Omega}\,\tau)^k \tag{23}$$

As mentioned in the previous subsection, τ is, in general, extremely small, and so the series can be truncated to retain limited terms as follows:

$$\exp(i\mathbf{\Omega}\,\tau) \approx \mathbf{I} + \mathbf{T}_0 + i\mathbf{F}_0 \tag{24}$$

where $\mathbf{I}$ refers the identity matrix of order n and

$$\mathbf{T}_0 = -\frac{1}{2}(\mathbf{\Omega}\tau)^2 + \frac{1}{24}(\mathbf{\Omega}\tau)^4 - \frac{1}{720}(\mathbf{\Omega}\tau)^6 \tag{25}$$

$$\mathbf{F}_0 = \mathbf{\Omega}\tau - \frac{1}{6}(\mathbf{\Omega}\tau)^3 + \frac{1}{120}(\mathbf{\Omega}\tau)^5 \tag{26}$$

With the following relations

$$\begin{aligned}&(\mathbf{I} + \mathbf{T}_k + i\mathbf{F}_k)^2 = \mathbf{I} + \left(\mathbf{T}_k^2 - \mathbf{F}_k^2 + 2\mathbf{T}_k\right) + 2i\mathbf{F}_k(\mathbf{I} + \mathbf{T}_k) = \mathbf{I} + \mathbf{T}_{k+1} + i\mathbf{F}_{k+1} \\ &(k = 0,1,\cdots,N)\end{aligned} \tag{27}$$

we can compute the matrices $\mathbf{T}_N$ and $\mathbf{F}_N$ recursively. In FORTRAN language, the loop statement for the computation looks like this:

do $k = 0, N-1$; $\mathbf{T}_{k+1} = \mathbf{T}_k^2 - \mathbf{F}_k^2 + 2\mathbf{T}_k$; $\mathbf{F}_{k+1} = 2\mathbf{F}_k(\mathbf{I} + \mathbf{T}_k)$; end do

Invoking the Euler formula $\exp(i\mathbf{\Omega}\Delta t) = \cos(\mathbf{\Omega}\Delta t) + i\sin(\mathbf{\Omega}\Delta t)$, we can write

$$\{\exp(i\mathbf{\Omega}\tau)\}^m = \mathbf{C} + i\mathbf{S} \approx (\mathbf{I} + \mathbf{T}_0 + i\mathbf{F}_0)^m = \mathbf{I} + \mathbf{T}_N + i\mathbf{F}_N \tag{28}$$

That is, $\mathbf{C} = \cos(\mathbf{\Omega}\Delta t) \approx \mathbf{I} + \mathbf{T}_N$ and $\mathbf{S} = \sin(\mathbf{\Omega}\Delta t) \approx \mathbf{F}_N$. The truncated error of the expansion equation (24) can be estimated as $(\mathbf{\Omega}\tau)^7 / 7!$. By substituting equation (15) into (23) and then taking the first seven terms, we have

$$\exp(i\mathbf{\Omega}\tau) \approx \mathbf{Q}\left[\sum_{k=0}^{6} \frac{1}{k!}\left(i\sqrt{\mathbf{\Lambda}}\tau\right)^k\right]\mathbf{Q}^{-1} \tag{29}$$

Thus the truncated error from the kth eigensolution corresponding to λ_k would be $\left(\tau\sqrt{|\lambda_k|}\right)^7 / 7!$. Suppose ε is the allowed truncation error, then we have $\Delta t\sqrt{|\lambda_k|} < 2^N (7!\varepsilon)^{1/7}$. If we neglect the inherent damping, the eigenvalue $|\lambda_k|$ would in fact be the kth angular frequency of the structure, that is, $|\lambda_k| = \omega_k$. It follows, by substituting $\omega_k = 2\pi / T_k$, where T_k is the kth natural period of the structure, that

$$\frac{\Delta t}{\sqrt{T_k}} < 2^N (7!\varepsilon)^{1/7} / \sqrt{2\pi} \tag{30}$$

Since 2^N is, in general, a very large number, the time step size Δt can be assumed to be relatively large in numerical computation, even for an extremely small value of ε. If N is a moderate number that is not too small, the accuracy of the algorithm would not be dominated by the step size Δt in the sense of numerical computation. For example, suppose $\varepsilon = 10^{-17}$, which reaches or exceeds computer precision, and $N = 20$, then we have from (30) that $\Delta t / \sqrt{T_k} < 2100$. This means that there would be no significant truncation error induced by using equation (24), even if the time step size is chosen to be 2100 times the square root of the kth natural period of the structure. In practice, taking into consideration the effect of higher modes, the contribution of high modes to the solution would be damped out because of the inherent damping effect of the structure. It follows that if $N = 20$ the precise integration stated above will provide essentially the exact homogeneous solution. In other words, the matrices $\mathbf{C}$ and $\mathbf{S}$ thus computed reflect the characteristics of the structure, including those of its higher mode. As a result of this, the accuracy of the general solution of equation (1) would be dominantly controlled by the accuracy of the particular solution.

2.4. The Particular Solution

If $\mathbf{f}(t) \neq \mathbf{0}$, the general solution of equation (1) consists of the homogeneous and the inhomogeneous or the particular solution which satisfies

$$\mathbf{M}\ddot{\mathbf{u}}_p + \mathbf{B}\dot{\mathbf{u}}_p + \mathbf{K}\mathbf{u}_p = \mathbf{f}(t) \tag{31}$$

The particular solution incorporated with the second-order PTI method can be derived [23] with the aid of the homogeneous solution (17). Suppose

$$\mathbf{u}_p = \mathbf{U}_1\mathbf{k}_1 + \mathbf{U}_2\mathbf{k}_2 \tag{32}$$

where $\mathbf{k}_1$ and $\mathbf{k}_2$ are vectors to be determined and

$$\mathbf{U}_1 = \exp[-\mathbf{D}(t - t_0)] \cos[\mathbf{\Omega}(t - t_0)] \tag{33a}$$

$$\mathbf{U}_2 = \exp[-\mathbf{D}(t - t_0)] \sin[\mathbf{\Omega}(t - t_0)] \tag{33b}$$

By solving the following matrix differential equations:

$$\begin{bmatrix} \mathbf{U}_1 & \mathbf{U}_2 \\ \dot{\mathbf{U}}_1 & \dot{\mathbf{U}}_2 \end{bmatrix} \begin{Bmatrix} \dot{\mathbf{k}}_1 \\ \dot{\mathbf{k}}_2 \end{Bmatrix} = \begin{Bmatrix} \mathbf{0} \\ \mathbf{M}^{-1}\mathbf{f} \end{Bmatrix} \tag{34}$$

we have

$$\mathbf{k}_1 = \int_{t_0}^{t} |\mathbf{\Delta}|^{-1} \begin{bmatrix} \mathbf{0} & \mathbf{U}_2 \\ \mathbf{M}^{-1}\mathbf{f} & \dot{\mathbf{U}}_2 \end{bmatrix} \mathbf{f}(\eta) d\eta, \ \mathbf{k}_2 = \int_{t_0}^{t} |\mathbf{\Delta}|^{-1} \begin{bmatrix} \mathbf{U}_1 & \mathbf{0} \\ \dot{\mathbf{U}}_2 & \mathbf{M}^{-1}\mathbf{f} \end{bmatrix} \mathbf{f}(\eta) d\eta \tag{35}$$

where

$$\mathbf{\Delta} = \begin{bmatrix} \mathbf{U}_1 & \mathbf{U}_2 \\ \dot{\mathbf{U}}_1 & \dot{\mathbf{U}}_2 \end{bmatrix} = \mathbf{\Omega} \exp[-2\mathbf{D}(t - t_0)] \tag{36}$$

Combining equations (32)-(36), yields

$$\mathbf{u}_p = \mathbf{M}^{-1}\mathbf{\Omega}^{-1} \int_{t_0}^{t} \exp[-\mathbf{D}(t-\eta)] \sin \mathbf{\Omega}(t-\eta) \mathbf{f}(\eta) d\eta \tag{37}$$

$$\dot{\mathbf{u}}_p = \mathbf{M}^{-1} \int_{t_0}^{t} \exp[-\mathbf{D}(t-\eta)] [\cos \mathbf{\Omega}(t-\eta) \ - \mathbf{\Omega}^{-1}\mathbf{D} \sin \mathbf{\Omega}(t-\eta)] \mathbf{f}(\eta) d\eta \tag{38}$$

By substituting the integral variable with $\tau = t - \eta$, the particular solution at the $(k+1)$th time step, i.e. $t = t_0 + (k+1)\Delta t$, can be written as

$$\mathbf{u}_{p,k+1} = \mathbf{M}^{-1}\mathbf{\Omega}^{-1} \int_{0}^{\Delta t} \exp(-\mathbf{D}\tau) \sin(\mathbf{\Omega}\tau) \mathbf{f}(t_k + \Delta t - \tau) d\tau \tag{39}$$

$$\dot{\mathbf{u}}_{p,k+1} = \mathbf{M}^{-1} \int_{0}^{\Delta t} \exp(-\mathbf{D}\tau) [\cos(\mathbf{\Omega}\tau) \ - \mathbf{\Omega}^{-1} \sin(\mathbf{\Omega}\tau)] \mathbf{f}(t_k + \Delta t - \tau) d\tau \tag{40}$$

In the first-order scheme, methods have been proposed [2-6] for the numerical treatment of particular solutions according to the properties of $\mathbf{f}(t)$. In the second-order scheme described in this chapter, however, suppose that $\mathbf{f}(t)$ are smooth functions within one time step. We can compute the particular solution with the aid of Taylor expansion as

$$\mathbf{f}(t_k + \Delta t - \tau) = \sum_{m=0}^{\infty} \frac{\mathbf{c}_m^k}{m!} (\Delta t - \tau)^m \tag{41}$$

where

$$\mathbf{c}_m^k = \mathbf{f}^{(m)}(t_k) \tag{42}$$

Define the integrals including the term $(\Delta t - \tau)^m$ in the Taylor series as

$$\mathbf{I}_S^m = \int_0^{\Delta t} \exp(-\mathbf{D}\tau)\sin(\boldsymbol{\Omega}\tau)(\Delta t - \tau)^m \, d\tau \tag{43}$$

$$\mathbf{I}_C^m = \int_0^{\Delta t} \exp(-\mathbf{D}\tau)\cos(\boldsymbol{\Omega}\tau)(\Delta t - \tau)^m \, d\tau \tag{44}$$

We can deduce the recurrence formula for calculation of the above integrals as follows:

$$\mathbf{I}_S^0 = \left(\mathbf{D}^2 + \boldsymbol{\Omega}^2\right)^{-1} (\boldsymbol{\Omega} - \mathbf{DES} - \boldsymbol{\Omega}\mathbf{EC}) \tag{45}$$

$$\mathbf{I}_C^0 = \left(\mathbf{D}^2 + \boldsymbol{\Omega}^2\right)^{-1} (\mathbf{D} + \boldsymbol{\Omega}\mathbf{ES} - \mathbf{DEC}) \tag{46}$$

$$\mathbf{I}_S^m = m\left(\mathbf{D}^2 + \boldsymbol{\Omega}^2\right)^{-1} \left(\frac{1}{m}\boldsymbol{\Omega}\Delta t^m - \mathbf{D}\mathbf{I}_S^{m-1} - \boldsymbol{\Omega}\mathbf{I}_C^{m-1}\right) \tag{47}$$

$$\mathbf{I}_C^m = m\left(\mathbf{D}^2 + \boldsymbol{\Omega}^2\right)^{-1} \left(\frac{1}{m}\mathbf{D}\Delta t^m + \boldsymbol{\Omega}\mathbf{I}_S^{m-1} - \mathbf{D}\mathbf{I}_C^{m-1}\right), \; (m = 1, 2, \cdots) \tag{48}$$

By combining the homogeneous solutions (20)-(21) and the particular solutions (39)-(40), we have the general solutions of equation (1) using the second-order scheme as follows:

$$\mathbf{u}_{k+1} = \mathbf{E}\left[\mathbf{C} + \boldsymbol{\Omega}^{-1}\mathbf{S}(\mathbf{u}_k + \dot{\mathbf{u}}_k)\right] + \boldsymbol{\Omega}^{-1}\mathbf{M}^{-1} \int_0^{\Delta t} \exp(-\mathbf{D}\tau)\sin(\boldsymbol{\Omega}\tau)\mathbf{f}(t_k + \Delta t - \tau)d\tau \tag{49}$$

$$\dot{\mathbf{u}}_{k+1} = -\mathbf{E}\left[\left(\boldsymbol{\Omega} + \boldsymbol{\Omega}^{-1}\mathbf{D}^2\right)\mathbf{S}\mathbf{u}_k + \left(\boldsymbol{\Omega}^{-1}\mathbf{DS} - \mathbf{C}\right)\dot{\mathbf{u}}_k\right] + \mathbf{M}^{-1} \int_0^{\Delta t} \exp(-\mathbf{D}\tau)\left[\cos(\boldsymbol{\Omega}\tau) - \boldsymbol{\Omega}^{-1}\mathbf{D}\sin(\boldsymbol{\Omega}\tau)\right]\mathbf{f}(t_k + \Delta t - \tau)d\tau \tag{50}$$

In order to show the performance of the second-order PTI method, numerical examples for smooth responses of various wave equations and wave propagations are presented in Section 3.

2.5. Solution Features of the PTI Method

In the course of the computation of sine and cosine matrices involved in the second-order PTI method, the 2^N algorithm is usually used. In this subsection, however, the mode decomposition method is used to show the solution features of the PTI method and to study how damping effects can be introduced into the algorithm more directly. To this end, the matrix $\mathbf{\Omega}$ in equation (13) is decomposed into the form in which the eigenvectors must be used:

$$\mathbf{\Omega} = \mathbf{Q} diag\left(\omega_1, \cdots, \omega_n\right) \mathbf{Q}^{-1} \tag{51}$$

where ω_k ($k = 1, \cdots, n$) stands for the eigenvalues or the natural circular eigen-frequencies of the discrete model. Suppose that they are distinct from each other and have been arranged in such a way that $\omega_1 < \cdots < \omega_n$. The values of ω_k can reflect the characteristics of the structure concerned, although they might be affected by the manner of spatial discretization. The k th column in the matrix $\mathbf{Q}$ represents the eigenvector corresponding to the eigenvalue ω_k. The eigenvalues and eigenvectors of $\mathbf{\Omega}$ can be computed numerically using the subroutines described in [21]. Suppose that the system-damping matrix is diagnosable, that is

$$\mathbf{D} = \mathbf{Q} diag\left(d_1, \cdots, d_n\right) \mathbf{Q}^{-1} \tag{52}$$

where $d_k = \xi_k \omega_k^0$ ($k = 1, \cdots, n$) stands for the physical damping related to the corresponding eigen-frequency. ξ_k and ω_k^0 represent the coefficients of damping ratio and the intrinsic circular eigen-frequencies, respectively. To study the solution features of the PTI, rewrite equations (20)-(21) in the form

$$\begin{Bmatrix} \mathbf{u}_{h,j+1} \\ \dot{\mathbf{u}}_{h,j+1} \end{Bmatrix} = \mathbf{T} \begin{Bmatrix} \mathbf{u}_{h,j} \\ \dot{\mathbf{u}}_{h,j} \end{Bmatrix} \tag{53}$$

where $\mathbf{T}$ is the amplification matrix or the system matrix of order $2n$ as follows:

$$\mathbf{T} = \begin{bmatrix} \mathbf{E} & \mathbf{0} \\ \mathbf{0} & \mathbf{E} \end{bmatrix} \begin{bmatrix} \mathbf{C} + \mathbf{\Omega}^{-1}\mathbf{D}\mathbf{S} & \mathbf{\Omega}^{-1}\mathbf{S} \\ -\left(\mathbf{\Omega} + \mathbf{\Omega}^{-1}\mathbf{D}^2\right)\mathbf{S} & \mathbf{C} - \mathbf{\Omega}^{-1}\mathbf{D}\mathbf{S} \end{bmatrix} \tag{54}$$

It can be shown that the second matrix on the right-hand side of equation (54) should be symplectic. Therefore $\mathbf{T}$ is a symplectic matrix with parameters. Write its eigen-polynomial as follows:

$$\det\left(\mathbf{T} - \lambda \mathbf{I}_{2n}\right) = 0 \tag{55}$$

where $\mathbf{I}_{2n}$ refers to the identity matrix of order $2n$. It is noted that the eigenvalues of the symplectic matrix will appear in pairs, and equation (55) can be treated by matrix partition to arrive at the form

$$\det\left(\mathbf{\Lambda}^2 - 2\mathbf{EC\Lambda} + \mathbf{E}^2\right) = 0 \tag{56}$$

where $\mathbf{\Lambda} = diag(\lambda_1, \cdots, \lambda_n)$. In the deduction of equation (56), the relation $\mathbf{S}^2 + \mathbf{C}^2 = \mathbf{I}_n$ has been used, where $\mathbf{I}_n$ refers to the identity matrix of order n. The eigenvalues of $\mathbf{T}$ can then be obtained from equation (56) as follows:

$$\lambda_k = \exp\left(-\xi_k \omega_k^0 \Delta t\right)\left[\cos(\omega_k \Delta t) \pm i \sin(\omega_k \Delta t)\right],\ (k = 1, \cdots, n) \tag{57}$$

where $i = \sqrt{-1}$. It can be seen from equation (57) that the eigenvalues of $\mathbf{T}$ are all complex in pairs and the spectral radius of $\mathbf{T}$ is smaller than unity for structures with physical damping, i.e.

$$\rho(\mathbf{T}) = \max|\lambda_k| < 1 \tag{58}$$

It is noted that the spectral radius is independent of the size of time step, Δt, so that the algorithm of the second-order PTI method is unconditionally stable. This is as expected, because the algorithm can theoretically yield an analytical solution. But for a structure with physical damping, all the vibrations will decay to zero with the progression of time. Therefore, in order to consider the numerical behavior of the algorithm, it would be more appropriate to consider cases without physical damping. Suppose $\mathbf{B} = \mathbf{0}$ in equation (1), then the amplification matrix and its eigenvalues become

$$\mathbf{T} = \begin{bmatrix} \mathbf{C} & \mathbf{\Omega}_0^{-1}\mathbf{S} \\ -\mathbf{\Omega}_0\mathbf{S} & \mathbf{C} \end{bmatrix} \tag{59}$$

$$\lambda_k = \cos(\omega_k \Delta t) \pm i \sin(\omega_k \Delta t),\ (k = 1, \cdots, n) \tag{60}$$

respectively. The spectral radius of $\mathbf{T}$ becomes unity ($\rho(\mathbf{T}) = 1$) independent of Δt, therefore the algorithm of the second-order PTI method is unconditionally stable. It can be seen that all the eigenvalues are located on the circle with a radius of unity. If equation (53) is rewritten in the form

$$\begin{Bmatrix} \mathbf{u}_{h,j+1} \\ \dot{\mathbf{u}}_{h,j+1} \end{Bmatrix} = \mathbf{T}^{j+1} \begin{Bmatrix} \mathbf{u}_{h,0} \\ \dot{\mathbf{u}}_{h,0} \end{Bmatrix} \tag{61}$$

it is easy to see that the variables (displacement and stress) at any time step, say, step $j+1$, can be evaluated using equation (61), and the change in the magnitude of u is obviously dependent on the properties of matrix T. By checking equation (58), we see that the location of eigenvalues may move around a unit circle but always remains on or within the circle. The system characteristics can thus be kept unchanged unless the system is affected by the accumulation of machine-like truncation errors. It is also evident that for the first-order PTI method, the eigenvalues and spectral radius of the amplification matrix $\exp(\mathbf{H}\Delta t)$ in equation (6) are exactly the same as those of the second-order PTI method, so that the solution features for the two algorithms are the same except for computational efficiency.

It can be found from equation (57) that the algorithm of the PTI method has no numerical damping because the absolute values of all eigenvalues equal unity ($|\lambda_k| = 1, k = 1,\cdots,n$). However, numerical damping is required because some spurious modes of high frequency might be introduced by spatial discretization. Suppose ω_n is the spurious frequency to be damped out, then numerical damping can be introduced into the algorithm by setting $\cos(\omega_n) = 0$ and $\sin(\omega_n) = 0$ in matrices $\mathbf{C}$ and $\mathbf{S}$ simultaneously, that is,

$$\begin{aligned} \mathbf{C} &= \mathbf{Q}\ diag[\cos(\omega_1),\cdots,\cos(\omega_{n-1}),0]\ \mathbf{Q}^{-1}, \\ \mathbf{S} &= \mathbf{Q}\ diag[\sin(\omega_1),\cdots,\sin(\omega_{n-1}),0]\ \mathbf{Q}^{-1} \end{aligned} \tag{62}$$

This is in fact equivalent to simultaneously multiplying $\mathbf{C}$ and $\mathbf{S}$ by a matrix $\mathbf{Q}\ diag(1,\cdots,1,0)\ \mathbf{Q}^{-1}$. In this way, the pair of eigenvalues, $\lambda_n = \cos(\omega_n\Delta t) \pm i\sin(\omega_n\Delta t)$, in the matrix $\mathbf{T}$ are set to zero correspondingly and the contribution of this mode to the solution will be eliminated. Obviously, this technique can be used to study the contribution of each single mode to the numerical solutions. For example, if $\mathbf{C}$ and $\mathbf{S}$ are replaced by $\mathbf{C}_m$ and $\mathbf{S}_m$, defined as

$$\begin{aligned} \mathbf{C}_m &= \mathbf{Q}\ diag[0,\cdots,0,\cos(\omega_m),0,\cdots,0]\ \mathbf{Q}^{-1}, \\ \mathbf{S}_m &= \mathbf{Q}\ diag[0,\cdots,0,\sin(\omega_m),0,\cdots,0]\ \mathbf{Q}^{-1} \\ &(0 \le m \le n) \end{aligned} \tag{63}$$

in the computer program, then the contribution of the m th mode to the numerical solutions can be analyzed. The effects of numerical damping and the contribution of some single modes are studied and presented in the numerical examples in Section 4.

3. Performance and Behavior for Smooth Responses

In this section, several simple numerical examples of one-dimensional problems are presented to illustrate the performance and behavior of the proposed second-order PTI method for the dynamic analysis of smooth responses [15]. The eigenvalues and the corresponding eigenvectors of the system matrix are computed using the FORTRAN packages in reference [21]. $N = 20$ is used in the 2^N algorithm. The first example serves as a benchmark test using the second-order scheme for the solution of a homogeneous equation, which is compared with the first-order scheme. The next two examples are for testing the effect of the inhomogeneous part and the damping term of the differential equation on the solutions with the proposed algorithm. The numerical results of the first three examples are presented in terms of maximum errors in comparison with the theoretical solutions.

3.1. One-dimensional Wave Equation

Consider the vibration of a string clamped at both ends. The governing equation is a one-dimensional wave equation

$$\ddot{u} = \frac{\partial^2 u}{\partial x^2}, \; x \in [0, 2\pi] \tag{64}$$

with the boundary and initial conditions

$$u(0,t) = u(2\pi,t) = 0, \; u(x,0) = \sin(x), \; \dot{u}(x,0) = 0 \tag{65}$$

where $u(x,t)$ is the transverse displacement of the string with unit amplitude of initial displacement. In equation (65), the first two relations stand for the fixed boundary condition and the last two relations represent the initial condition. The solution to equation (64) is given by the D'Alembert integral, being of the form

$$u(x,t) = \sin(x)\cos(t) \tag{66}$$

The spatial coordinate in equation (64) is divided into a series of subregions using the differential quadrature method [16,24]. In this method, the derivatives of $u(x,t)$ with respect to the spatial coordinate x are approximated by a weighted sum of function values at all discrete points within the interval $x \in [0, 2\pi]$ under consideration, i.e.

$$u^{(r)}(x_j,t) = \frac{d^r}{dx^r} u(x,t)\bigg|_{x=x_j} = \sum_{k=0}^{n} A_{jk}^{(r)} u(x_k,t), \; (j = 0,1,\cdots,n+1) \tag{67}$$

For details of the computation of weighting coefficients $A_{jk}^{(r)}$ ($r = 2$), readers can refer to Quan and Chang [25-26] or Shu [27]. In the quadrature formula (67), $j = 0$ and $j = n+1$ correspond to the boundary nodes, respectively, at $x = 0$ and $x = 2\pi$ so that n is the number of the interior nodes. The positions of the interior grid points are located at

Gaussian points, the zeros of the shifted Legendre polynomial, which are computed by the subroutine presented in reference [28]. After discretization incorporated with the boundary condition, equation (64) is transformed to the following form of algebraic and differential equation as

$$\ddot{\mathbf{u}} + \mathbf{A}\mathbf{u} = \mathbf{0} \text{ or } \ddot{\mathbf{u}} + \boldsymbol{\Omega}^2\mathbf{u} = \mathbf{0} \tag{68}$$

where $\boldsymbol{\Omega} = \sqrt{\mathbf{A}}$. The numerical solution of the wave equation with the second-order precision integration algorithm at the $(k+1)$th time step, i.e. $t = t_0 + (k+1)\Delta t$, is written as follows:

$$\mathbf{u}_{k+1} = \mathbf{C}\mathbf{u}_k + \boldsymbol{\Omega}^{-1}\mathbf{S}\dot{\mathbf{u}}_k \tag{69}$$

$$\dot{\mathbf{u}}_{k+1} = -\boldsymbol{\Omega}\mathbf{S}\mathbf{u}_k + \mathbf{C}\dot{\mathbf{u}}_k \tag{70}$$

The maximum errors using both the first- and second-order schemes are compared in Figure 1. Twenty interior grid points and the time step $\Delta t = 1$ were used in the computation. It can be seen that both the algorithms are very stable, but better accuracy is achieved with the second-order scheme. The CPU time used for the second-order algorithm was only about 37.5% of that used for the first-order algorithm for the same amount of computation. This is considered to result mainly from the small size of the system matrix in the second-order algorithm during long-term computation, because the major computational effort comes from the step-by-step integration realized by matrix multiplications.

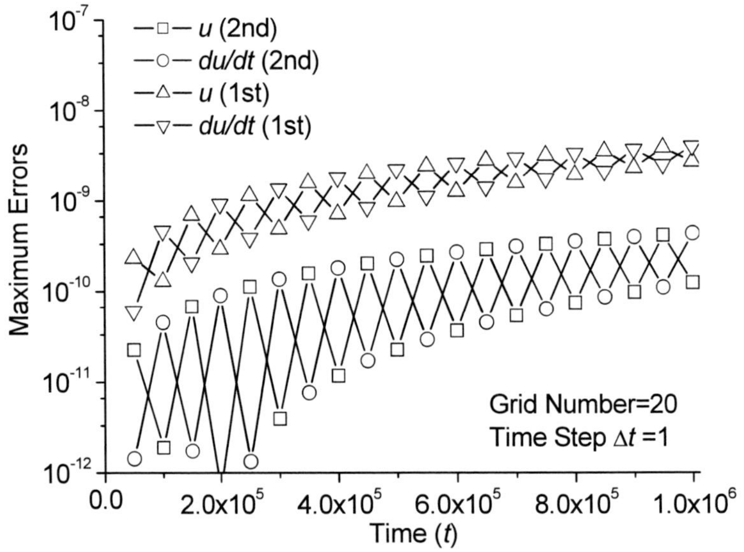

Figure 1. Comparison between the first- and the second-order schemes.

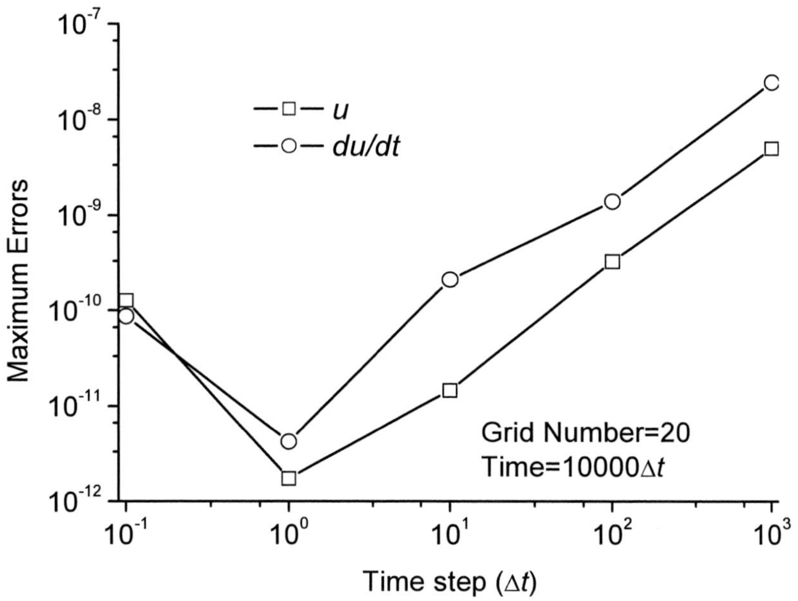

Figure 2. Errors as a function of time step Δt with the second-order scheme.

The effect of the time step Δt on the accuracy with the second-order algorithm is shown in Figure 2. It can be seen that very large time steps can be employed for the solution of homogeneous equations with the PTI method, which verifies the error analysis presented in subsection 2.3.

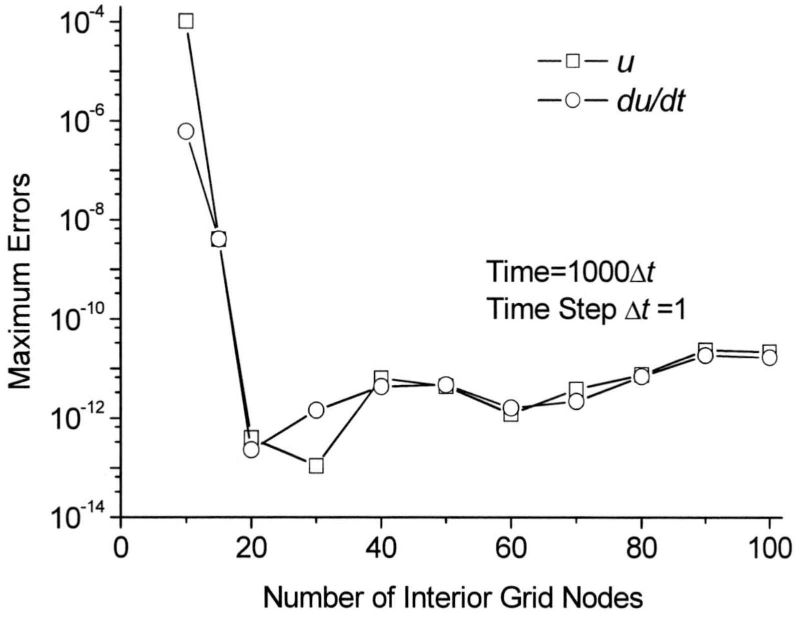

Figure 3. Errors as function of interior grid numbers with the second-order scheme.

The effect of the interior grid number on accuracy is shown in Figure 3. In reference [29], the wave equation was solved, accompanied by differential quadrature discretization, by using the fourth-order Runge-Kutta method with a time step $\Delta t = 10^{-4}$. The solution in [29] becomes unstable at $t = 0.45$ even though the grid number is not too large ($n = 15$) in global discretization. In contrast, the results from the proposed algorithm show that no instability phenomenon occurs with the increase of the grid number after long-term integration ($t = 1000$) as shown in Figure 3. This result demonstrates the superiority of the precision integral compared to the traditional finite-difference-based time-marching method.

3.2. Forced One-dimensional Wave Equation

Consider the forced vibration of a string clamped at both ends. The governing equation is the following inhomogeneous one-dimensional wave equation:

$$\ddot{u} = \frac{\partial^2 u}{\partial x^2} + f(x,t) \tag{71}$$

where

$$f(x,t) = -3\sin(x)\sin(2t) \tag{72}$$

subject to the boundary and the initial conditions

$$u(0,t) = u(2\pi,t) = 0,\ u(x,0) = \sin(x),\ \dot{u}(x,0) = 2\sin(x) \tag{73}$$

The solution to equation (60) is as follows:

$$u(x,t) = \sin(x)[\cos(t) + \sin(2t)] \tag{74}$$

After spatial discretization by the differential quadrature method, equation (71) becomes

$$\ddot{\mathbf{u}} + \mathbf{\Omega}^2\mathbf{u} = \mathbf{f}(t) \tag{75}$$

The numerical solution of the forced wave equation with the second-order precision integration algorithm at the $(k+1)$th time step, i.e. $t = t_0 + (k+1)\Delta t$, is written as follows:

$$\mathbf{u}_{k+1} = \mathbf{C}\mathbf{u}_k + \mathbf{\Omega}^{-1}\mathbf{S}\dot{\mathbf{u}}_k + \mathbf{\Omega}^{-1}\int_0^{\Delta t}\sin(\mathbf{\Omega}\tau)\mathbf{f}(t_k + \Delta t - \tau)d\tau \tag{76}$$

$$\dot{\mathbf{u}}_{k+1} = -\mathbf{\Omega S u}_k + \mathbf{C}\dot{\mathbf{u}}_k + \int_0^{\Delta t} \cos(\mathbf{\Omega}\tau)\mathbf{f}(t_k + \Delta t - \tau)d\tau \tag{77}$$

The integrals for the terms in the Taylor expansion of $\mathbf{f}(t)$ for the forced wave equation can be computed recursively by the following formula:

$$\mathbf{I}_S^0 = \int_0^{\Delta t} \sin(\mathbf{\Omega}\tau)d\tau = \mathbf{\Omega}^{-1}(\mathbf{I} - \mathbf{C}) \tag{78}$$

$$\mathbf{I}_C^0 = \int_0^{\Delta t} \cos(\mathbf{\Omega}\tau)d\tau = \mathbf{\Omega}^{-1}\mathbf{S} \tag{79}$$

$$\mathbf{I}_S^m = \int_0^{\Delta t} \sin(\mathbf{\Omega}\tau)(\Delta t - \tau)^m d\tau = m\mathbf{\Omega}^{-1}\left(\frac{1}{m}\mathbf{I}\Delta t^m - \mathbf{I}_C^{m-1}\right) \tag{80}$$

$$\mathbf{I}_C^m = \int_0^{\Delta t} \cos(\mathbf{\Omega}\tau)(\Delta t - \tau)^m d\tau = m\mathbf{\Omega}^{-1}\mathbf{I}_S^{m-1}, \ (m = 1,2,\cdots) \tag{81}$$

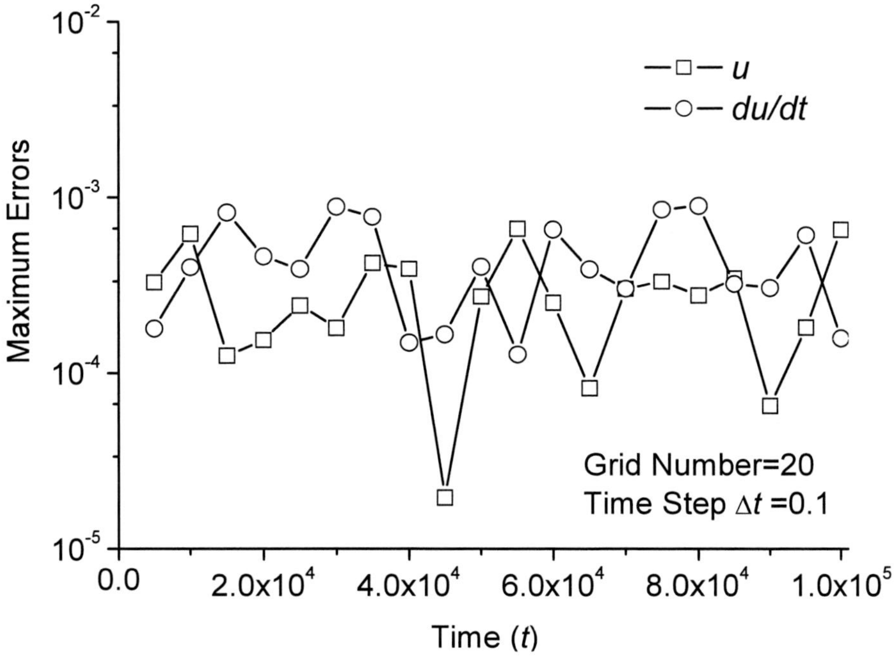

Figure 4. Errors as function of time for the forced wave equation.

The maximum errors as a function of time for the forced wave equation are shown in Figure 4. In the computation, only three terms in the Taylor expansion are necessary to maintain reasonable accuracy, and are retained. Figure 4 shows that the algorithm is long-term stable, although the time step chosen should not be too large ($\Delta t = 0.1$) owing to the existence of the inhomogeneous term. By comparing Figure 4 with Figures 1-3 it is evident that the accuracy of the general solution is predominantly controlled by the accuracy of the particular solution.

3.3. Forced One-dimensional Equation with Damping

The third example to be solved is a forced second-order one-dimensional equation with damping as follows:

$$\ddot{u} + 2\dot{u} + 3u = \frac{\partial^2 u}{\partial x^2} + f(x,t) \tag{82}$$

where

$$f(x,t) = 2\sin(x)\{2\cos(t) - \sin(2t)\} \tag{83}$$

subject to the boundary and initial conditions

$$u(0,t) = u(2\pi,t) = 0,\ u(x,0) = \sin(x),\ \dot{u}(x,0) = 3\sin(x) \tag{84}$$

The theoretical solution to equation (69) is as follows:

$$u(x,t) = \sin(x)\,[\cos(t)\exp(-t) + \sin(2t)] \tag{85}$$

Equation (84) is transformed, after spatial discretization by the differential quadrature method, to the following form of algebraic and differential equation:

$$\ddot{\mathbf{u}} + 2\dot{\mathbf{u}} + \mathbf{G}\mathbf{u} = \mathbf{f}(t) \tag{86}$$

where $\mathbf{G} = 3\mathbf{I} + \mathbf{A}$. The numerical solution of the forced equation with damping using the second-order precision integration algorithm at the $(k+1)$th time step, i.e. $t = t_0 + (k+1)\Delta t$, can be given as follows:

$$\mathbf{u}_{k+1} = e\left[\left(\mathbf{C} + \mathbf{\Omega}^{-1}\mathbf{S}\right)\mathbf{u}_k + \mathbf{\Omega}^{-1}\mathbf{S}\dot{\mathbf{u}}_k\right] + \mathbf{\Omega}^{-1}\int_0^{\Delta t} \exp(-\tau)\sin\mathbf{\Omega}(\tau)\mathbf{f}(t_k + \Delta t - \tau)d\tau \tag{87}$$

$$\dot{\mathbf{u}}_{k+1} = -e\left[\left(\mathbf{\Omega} + \mathbf{\Omega}^{-1}\right)\mathbf{S}\mathbf{u}_k + \left(\mathbf{\Omega}^{-1}\mathbf{S} - \mathbf{C}\right)\dot{\mathbf{u}}_k\right] + \int_0^{\Delta t} \exp(-\tau)\left[\cos\mathbf{\Omega}(\tau) - \mathbf{\Omega}^{-1}\sin\mathbf{\Omega}(\tau)\right]$$
$$\mathbf{f}(t_k + \Delta t - \tau)d\tau \tag{88}$$

where $\mathbf{\Omega} = \sqrt{\mathbf{I} - \mathbf{G}}$ and $e = \exp(-\Delta t)$. The integrals of the terms in the Taylor expansion of $\mathbf{f}(t)$ for the forced equation with damping can also be computed recursively by the following formula:

$$\mathbf{I}_S^0 = \int_0^{\Delta t} \exp(-\tau)\sin(\mathbf{\Omega}\tau)d\tau = \mathbf{G}^{-1}(\mathbf{\Omega} - e\mathbf{S} - e\mathbf{\Omega}\mathbf{C}) \tag{89}$$

$$\mathbf{I}_C^0 = \int_0^{\Delta t} \exp(-\tau)\cos(\mathbf{\Omega}\tau)d\tau = \mathbf{G}^{-1}(\mathbf{I} + e\mathbf{\Omega}\mathbf{S} - e\mathbf{C}) \tag{90}$$

$$\mathbf{I}_S^m = \int_0^{\Delta t} \exp(-\tau)\sin(\mathbf{\Omega}\tau)(\Delta t - \tau)^m d\tau = m\mathbf{G}^{-1}\left(\frac{1}{m}\mathbf{I}\Delta t^m - \mathbf{I}_S^{m-1} - \mathbf{\Omega}\mathbf{I}_C^{m-1}\right) \tag{91}$$

$$\mathbf{I}_C^m = \int_0^{\Delta t} \exp(-\tau)\cos(\mathbf{\Omega}\tau)(\Delta t - \tau)^m d\tau = m\mathbf{G}^{-1}\left(\frac{1}{m}\mathbf{I}\Delta t^m + \mathbf{\Omega}\mathbf{I}_S^{m-1} - \mathbf{I}_C^{m-1}\right)$$
$$(m = 1, 2, \cdots) \tag{92}$$

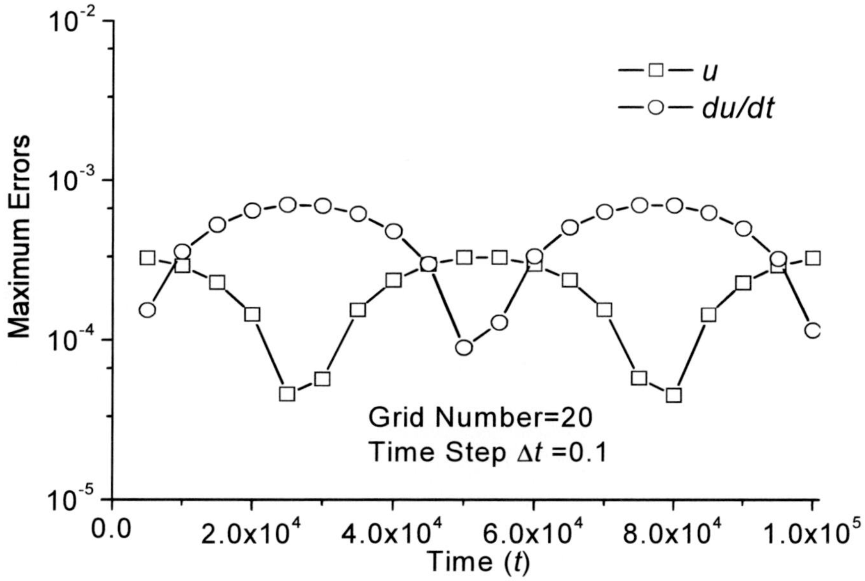

Figure 5. Errors as function of time for the forced equation with damping.

The maximum errors as a function of time for the forced equation are shown in Figure 5. Again, only three terms in the Taylor expansion are retained in the computation. Figure 5 shows that this algorithm is also long-term stable. As the homogeneous part of the solution will decay to be negligibly small after a short period of time, the results in Figure 5 can be considered to be a contribution by the inhomogeneous part only.

3.4. Wave Propagation of a String

The fourth example is the wave propagation in a string having the same governing equation of (51) but with different initial conditions. The initial wave profile is a localized packet, which will sweep over the whole string at subsequent time steps. In this case, the wave profile has a feature of some spatial non-smoothness, a complicated problem as well as a good test for the algorithm. The boundary and initial conditions are as follows:

$$u(0,t)=u(2\pi,t)=0,\ \dot{u}(x,0)=0, \tag{93a}$$

$$u(x,0)=\begin{cases}\sin(2x-1.5\pi) & (0.75\pi\le x\le 1.25\pi)\\ 0 & other\end{cases}, \tag{93b}$$

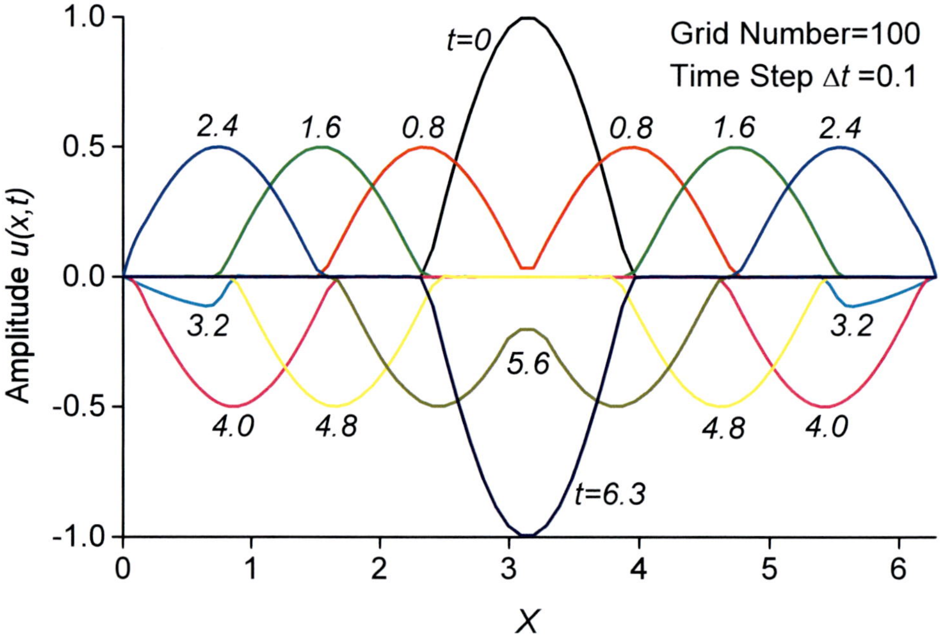

Figure 6. Wave propagation of a string.

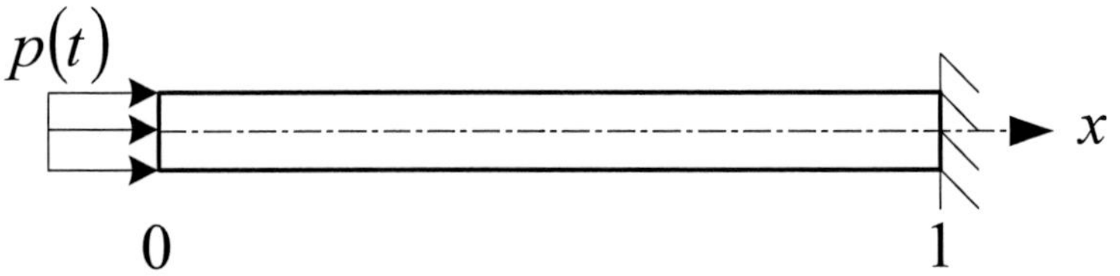

Figure 7. A rod excited by a shock at one end ($x = 0$).

These conditions produce a localized wave concentrated on a small central area. In the numerical computations, the time step $\Delta t = 0.1$ and the grid number 100 were used. The wave profiles at selected time intervals are shown in Figure 6. It is evident from Figure 6 that over time, the wave propagates consecutively outwards towards the right and left ends, respectively. After reaching both ends, the waves are reflected as negative waves and travel consecutively inwards towards the center. The two reflected waves meet again at the center and will repeat the next cycle, which is not presented here in the figure. The results of this example also show that the proposed algorithm can provide a very stable and accurate numerical solution.

4. Performance and Behavior for Transient Responses

For the analysis of shock-excited dynamic responses, it is generally considered that high-frequency behavior in the spatial-discrete model is already in error when compared with the actual physical continuum model. This means that spatial discretization will introduce spurious high-frequency numerical noise. Numerical dissipation must be taken into account in the traditional finite difference-based time-step integration algorithms in order to damp out the effect of this high-frequency noise [17-19], as the responses of interest are in principle provided mainly by the low-frequency modes. On the other hand, the high-frequency modes play an important role in relation to the responses of the structure in the spatial-discrete model. Based on this understanding, Chen and Tanaka indicated that the PTI method is inapplicable for the analysis of shock-excited dynamic problems when using the dual-reciprocity boundary element method for spatial discretization [30]. This is due to the lack of numerical damping. For the analysis of shock-excited dynamic responses in the framework of the PTI method, the questions are: What is the effect of the spatial-discrete format on the high-frequency modes? How do the high-frequency modes influence the numerical results produced via the PTI method? If the high-frequency mode causes spurious responses, how can numerical dissipation be introduced into the PTI method and what is the effect of this introduction? These problems are clarified in this section [20].

4.1. The Computational Model

To address the above-mentioned questions, a simple example of a rod shock-excited at one end (Figure 7) is considered, using the second-order PTI method to study the effect of spatial discretization, numerical damping, and the magnitude of time step on the accuracy of the numerical results. The Newmark- β method is also employed in the numerical tests for the purpose of comparison with regard to efficiency and accuracy of computations. The dimensionless equation governing the longitudinal vibration of the rod without physical damping can be expressed as

$$\frac{\partial^2 u(x,t)}{\partial t^2} = \frac{\partial^2 u(x,t)}{\partial x^2}, \quad x \in [0,1] \tag{94}$$

with the initial and boundary conditions as

$$u(x,0) = \dot{u}(x,0) = 0\,, \; u(1,t) = 0\,, \; \frac{\partial u(0,t)}{\partial x} = p(t) \tag{95}$$

where $u(x,t)$ represents the longitudinal displacement of the rod. The spatial coordinate in equation (94) is also discretized with the differential quadrature method [16,24]. For details of the computation of weighting coefficients $A_{jk}^{(r)}$ ($r = 1,2$) in equation (67) refer to Quan [25-26] or Shu [27]. In the quadrature formula (67), $j = 0$ and $j = n$ correspond to the boundary nodes at $x = 0$ and $x = 1$, respectively, so that $n - 1$ is the number of interior nodes. Equation (94) can be transformed into second-order ordinary differential equations using equation (67) incorporated with the boundary conditions. It can be written in matrix form as

$$\ddot{\mathbf{u}} + \mathbf{\Omega}^2 \mathbf{u} = \mathbf{f}(t) \tag{96}$$

The initial conditions are

$$\mathbf{u}(t_0) = \mathbf{0}\,, \; \dot{\mathbf{u}}(t_0) = \mathbf{0} \tag{97}$$

where

$$\mathbf{\Omega}^2 = -\begin{bmatrix} \sum_{j=1}^{n} A_{0j}^{(1)} A_{j0}^{(1)} & \sum_{j=1}^{n} A_{0j}^{(1)} A_{j1}^{(1)} & \cdots & \sum_{j=1}^{n} A_{0j}^{(1)} A_{j,n-1}^{(1)} \\ A_{10}^{(2)} & A_{11}^{(2)} & \cdots & A_{1,n-1}^{(2)} \\ \vdots & \vdots & & \vdots \\ A_{n-1,0}^{(2)} & A_{n-1,1}^{(2)} & \cdots & A_{n-1,n-1}^{(2)} \end{bmatrix}, \tag{98a}$$

$$\mathbf{f}(t)=\begin{Bmatrix} A_{00}^{(1)} \\ 0 \\ \vdots \\ 0 \end{Bmatrix} p(t)+\begin{Bmatrix} \sum_{j=1}^{n} A_{0j}^{(1)} A_{jn}^{(1)} \\ A_{1n}^{(2)} \\ \vdots \\ A_{n-1,n}^{(2)} \end{Bmatrix} u(1,t) \tag{98b}$$

Obviously the order of the system matrix is n in equation (97). As it is well known that the accuracy of the differential quadrature method is generally sensitive to the sampling grid distribution over the interval under consideration [16], three kinds of grid distribution, i.e., equally-spaced, Chebyshev-Gauss-Lobatto spaced and Legendre-Gauss spaced, are considered in the numerical example. For the Chebyshev-Gauss-Lobatto distribution, the sampling grid points are placed at

$$x_k = \frac{1}{2}\left[1-\cos\left(\frac{k\pi}{n}\right)\right], \ (k = 0,1,\cdots,n) \tag{99}$$

The interior nodes x_k ($k = 1,\cdots,n-1$) are the zeros of the derivative of shifted Chebyshev polynomial $T_n'(2x-1)$ within the interval $x \in [0,1]$ [31]. For the Legendre-Gauss distribution, the interior nodes are placed at the zeros of the shifted Legendre polynomial $P_{n-1}(2x-1)$, computed by the subroutine presented in reference [28]. Three types of impact loads are considered, rectangular, half-triangular and Heaviside, as shown in Figure 8. Equation (96) is solved by the second-order PTI method. The solution behaviors are studied numerically with the aid of the eigen-frequencies and modes of the system matrices associated with the grid distributions and numbers employed in computation as well as the damping effects.

Equation (96) is also solved by the Newmark-β method. The time integration formulae at the $(j+1)$th time step, i.e. $t = t_0 + (j+1)\Delta t$, are written in the following forms incorporated with the quadrature rule and the boundary conditions:

$$\mathbf{u}_j = \left(\mathbf{\Omega}^2 + \alpha_1 \mathbf{I}_n\right)^{-1}\left(\mathbf{f}_j + \alpha_1 \mathbf{u}_{j-1} + \alpha_2 \dot{\mathbf{u}}_{j-1} + \alpha_3 \ddot{\mathbf{u}}_{j-1}\right) \tag{100}$$

$$\ddot{\mathbf{u}}_j = \alpha_1\left(\mathbf{u}_j - \mathbf{u}_{j-1}\right) - \alpha_2 \dot{\mathbf{u}}_{j-1} - \alpha_3 \ddot{\mathbf{u}}_{j-1} \tag{101}$$

$$\dot{\mathbf{u}}_j = \dot{\mathbf{u}}_{j-1} + \alpha_4 \ddot{\mathbf{u}}_{j-1} + \alpha_5 \ddot{\mathbf{u}}_j \tag{102}$$

where $\alpha_1 = \left(\beta\Delta t^2\right)^{-1}$, $\alpha_2 = (\beta\Delta t)^{-1}, \alpha_3 = 0.5\beta^{-1} - 1, \alpha_4 = (1-\gamma)\Delta t, \alpha_5 = \gamma\Delta t$ and the parameters $\gamma = 0.5$ and $\beta = 0.25$ are used in the computation.

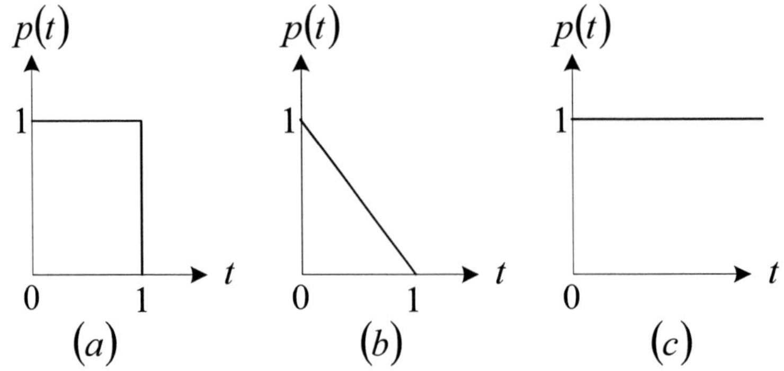

Figure 8. Impact forces applied to the rod: (*a*) Rectangular impact (*b*) Half-triangular impact (*c*) Heaviside impact.

4.2. Effects of Grid Distribution and Grid Number

The solutions of time-displacement curves at the free end ($x = 0$) of the rod subjected to rectangular impact are presented in Figure 9. The results were obtained using small grid numbers ($n = 20$) and different grid distributions. It can be seen from Figure 9 that, for the PTI method, both the unequally-spaced grid distributions, namely the Chebyshev-Gauss-Lobatto and the Legendre-Gauss, give better results than the equally-spaced grid distribution. The displacement solution with equally-spaced grid distribution has small oscillations. For the displacement curve with the Newmark- β method, however, there is a small deviation from the exact solution using the Legendre-Gauss grid points, and a reasonable result cannot be achieved using the equally-spaced grid points. For this reason, only the Legendre-Gauss grid distributions are used for the Newmark- β method in the following analysis.

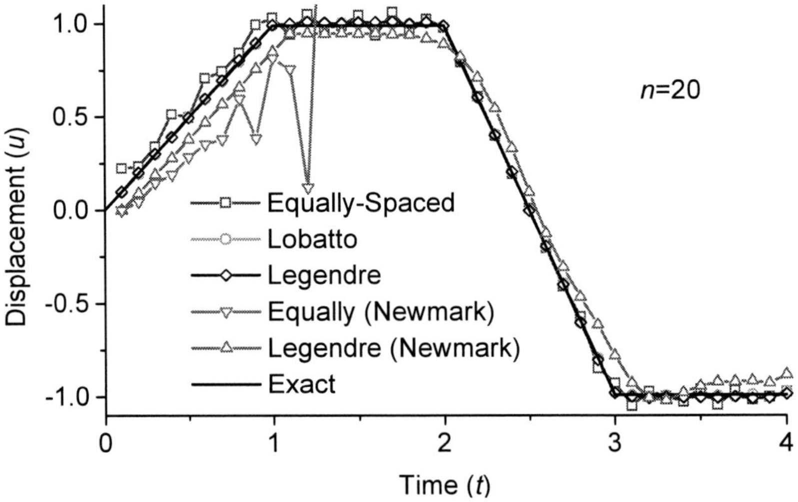

Figure 9. Displacement curves at free end ($x = 0$) of rod subjected to rectangular impact ($\Delta t = 0.1$), with different grid distributions.

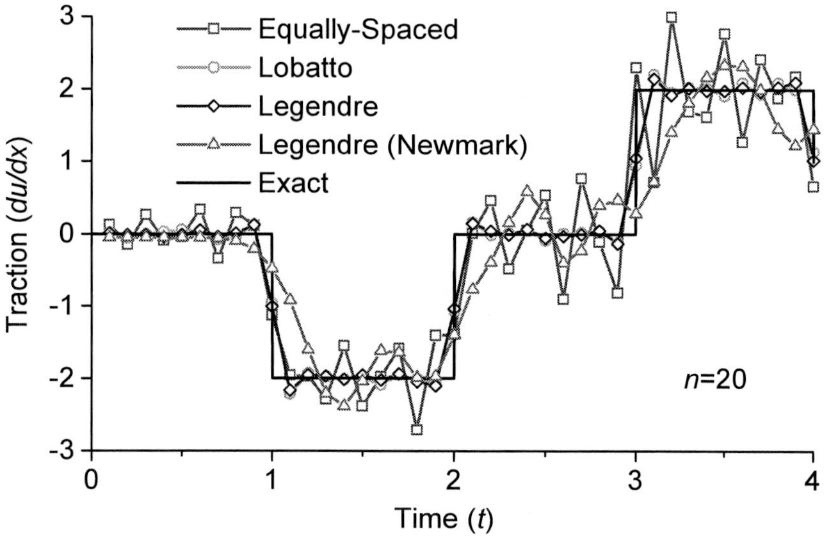

Figure 10. Traction curves at fixed end ($x = 1$) of rod subjected to rectangular impact ($\Delta t = 0.1$), with different grid distributions.

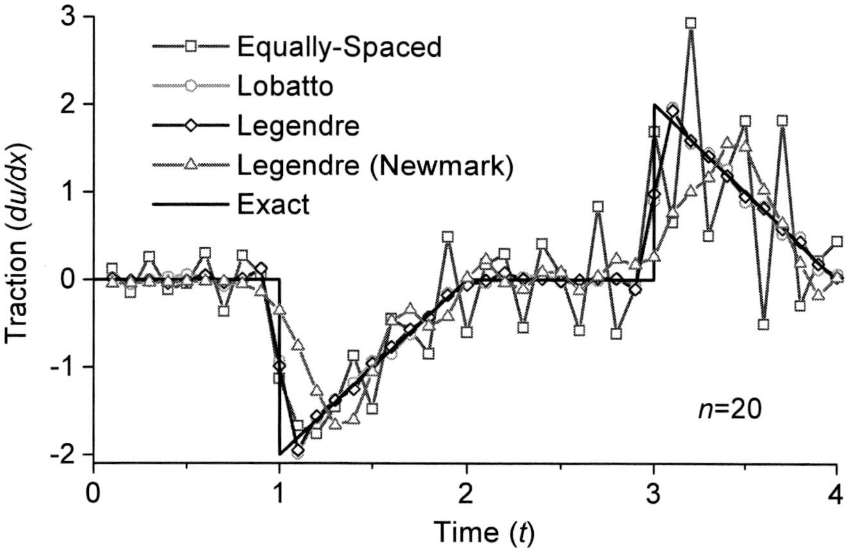

Figure 11. Traction curves at fixed end ($x = 1$) of rod subjected to half-triangular impact ($\Delta t = 0.1$), with different grid distributions.

Figure 10 shows the solution of time-traction curves at the fixed end ($x = 1$) of the rod subjected to rectangular impact. It is obvious from Figure 10 that the unequally-spaced grid distributions give almost the same results, which are far better than that from the equally-spaced grid distribution. It is also evident that the traction solution with the equally-spaced grid distribution has larger oscillations than those with unequally-spaced grid distributions. It is seen also that using the Legendre-Gauss grid points, the traction response with the Newmark- β method is not as good as that with the second-order scheme. The same conclusion can be drawn from Figure 11, where the time-traction curves are presented at the fixed end ($x=1$) of the rod subjected to half-triangular impact. This conforms to the observation by Bert and Malik [16] and is consistent with reports in the literature that unequally-spaced grid distributions perform consistently better than equally-spaced grid distributions. However, the effect of grid distribution is considered from a new point of view in this chapter with respect to the modal types and their corresponding dynamic responses. This is presented in the next subsection under the framework of the PTI method.

In the following analysis, only the Legendre-Gauss grid distribution is considered, since the two unequally-spaced grid distributions yield almost the same results for the second-order PTI method and the equally-spaced grid points give an unacceptable result with the Newmark-β method. Use of the Legendre-Gauss grid distribution here means that the roots of the shifted Legendre polynomials over the interval [0,1] of concern are taken as the grid coordinates. The time-traction curves at the fixed end ($x=1$) of the rod subjected to rectangular and half-triangular impacts are presented in Figures 12 and 13 respectively, using various grid numbers. In order to show clearly the details of the curves with sharp changes, fine time steps ($\Delta t = 0.001$) are used within a small time interval in the two figures. It can be seen from both Figures 12 and 13 that the larger the grid number used, the better can the exact solution be simulated using the second-order PTI method. However, the traction curves with the Newmark-β method not only fail to trace the sharp change as well as those using the PTI method but also show greater overshooting following the sharp change. In fact, it needs to be pointed out from both Figures 12 and 13 that the results with the Newmark-β method using larger numbers of grid points are inferior to those with the PTI method using fewer grid points. For example, the result with the Newmark-β using $n=200$ is not as good as that with the PTI method using $n=100$, and the result with the Newmark-β using $n=100$ is not as good as that with the PTI method using $n=50$, etc.

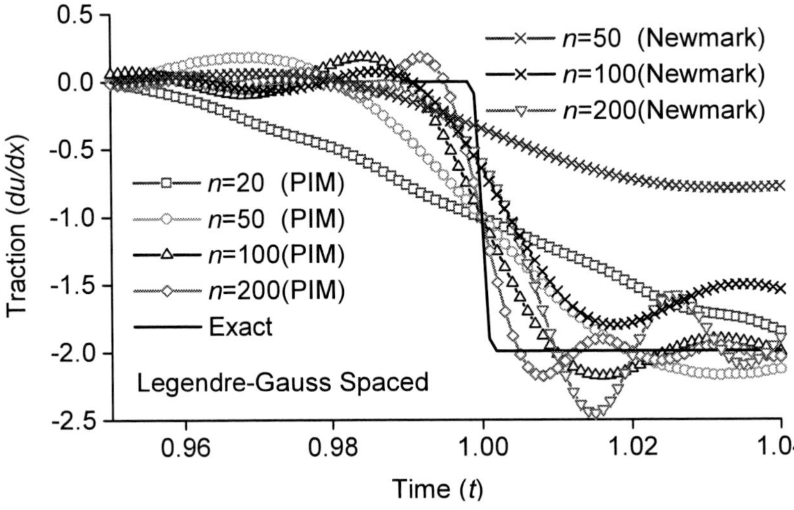

Figure 12. Details of traction curves around $t=1$ at fixed end ($x=1$) of rod subjected to rectangular impact ($\Delta t = 0.001$), with Legendre-Gauss distribution using different grid numbers.

4.3. Eigen-frequency and Mode

The rod under consideration is, in general, a continuum with a continuous distribution of eigen-frequencies. However, the distribution of eigen-frequencies becomes discrete due to spatial discretization with finite mode numbers. It is, moreover, affected by the manner of discretization. Figure 14 shows variations of circular eigen-frequencies, ω_m ($m = 1,\cdots,n$), with the mode number for different grid distributions with the PTI method. It is evident that the frequency distribution of equally-spaced grid distributions is inferior to those from unequally-spaced grid distributions because of the lack of high-frequency modes, although the low-frequency modes are almost identical for the three types of grid distribution. From this result it can be inferred that the large oscillations in time-traction curves shown in Figures 10 and 11 derive from the lack of high-frequency responses when using an equally-spaced grid distribution. However, it should be pointed out that this situation caused by the lack of high-frequency modes cannot be improved by increasing the grid number with the equally-spaced distribution. It is also evident from Figure 14 that the high-frequency modes associated with the Legendre-Gauss-spaced distribution are richer than those associated with the Chebyshev-Gauss-Lobatto-spaced distribution. It seems that the solution using the Legendre-Gauss distribution is slightly better than that using the Chebyshev-Gauss-Lobatto distribution. The distinction between the solutions for the two distributions is, however, relatively small because the contribution of high-frequency modes to the numerical solution does not dominate, as discussed in the next subsection.

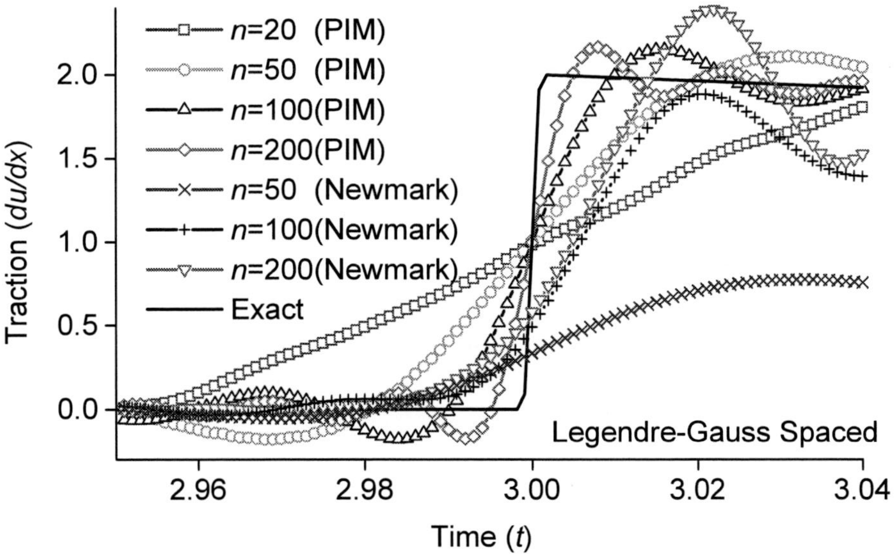

Figure 13. Details of traction curves around $t = 3$ at fixed end ($x = 1$) of rod subjected to half-triangular impact ($\Delta t = 0.001$), with Legendre-Gauss distribution using different grid numbers.

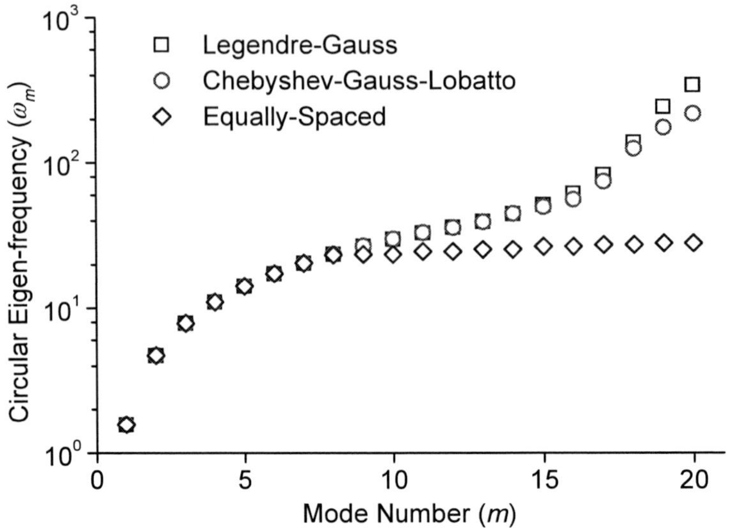

Figure 14. Comparison of circular eigen-frequencies corresponding to the mode number with different grid distributions with the PTI method.

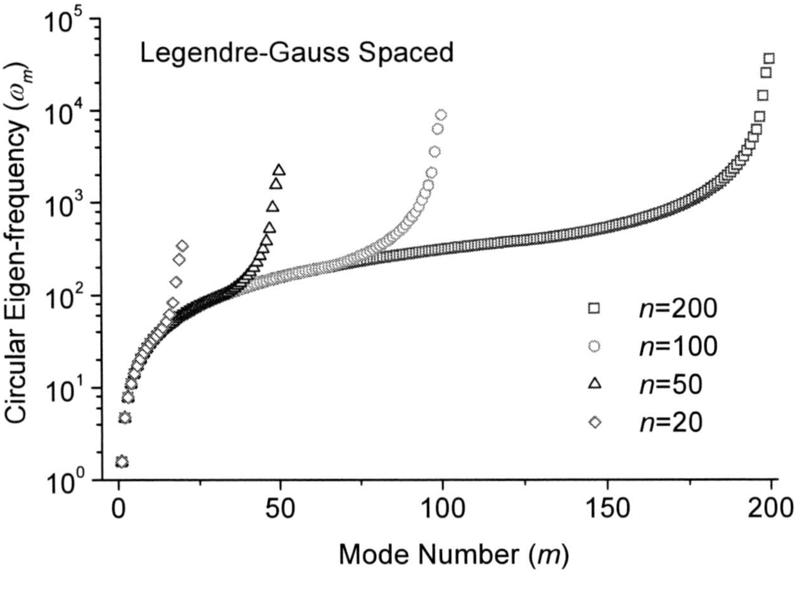

Figure 15. Comparison of circular eigen-frequencies corresponding to the mode number with Legendre-Gauss distributions using various grid numbers with the PTI method.

The effect of grid number on the circular eigen-frequencies with the PTI method is shown in Figure 15. As expected, the higher the grid number used, the greater the values of high-frequency modes. This finding confirms the reason for the phenomenon evident in both Figures 12 and 13. In other words, the contribution of high-frequency responses to the solutions becomes more important along with increases in the grid number, which is important for approximation of the traction curves. Therefore, under the framework of the PTI method, spatial discretization can yield numerically the possible modal types for the structure to be analyzed. An appropriate spatial discretization scheme for a given structure is usually required to obtain an accurate and meaningful numerical solution.

4.4. Effects of Numerical Damping

The process of spatial discretization may introduce spurious high-frequency numerical noise, which can significantly affect the accuracy of the numerical solution. Therefore, numerical dissipation is essential in the time-step integration algorithms to damp out spurious high-frequency noise [17-19]. In particular, the PTI method has been shown to be inapplicable for the analysis of shock-excited dynamic problems, due to the lack of numerical damping [30]. To address this problem, the effect of numerical damping on the accuracy of the PTI method is investigated. Figure 16 shows the time-traction curves around $t = 1$ at the fixed end ($x = 1$) of a rod subjected to a half-triangular impact, with Legendre-Gauss distribution, using $n = 200$. The numerical damping is realized by using equation (62), and the numbers in parentheses in Figure 16 indicate that the modes up to this number are retained in the system matrix whereas all modes with higher numbers are removed from the system matrix. For example, '(20)' means that the modes with numbers 1 through 20 are kept in the system matrix and thus contribute to the solution, while the contributions of modes with numbers 21 through 200 are eliminated. In this case, only the low frequency modes with numbers 1 through 20 have an influence on the resulting time-traction curves. By comparing Figure 16 with Figure 13, it is evident that the behavior of traction curves with different numerical damping is very similar to that using different grid numbers. The results show that high-frequency responses are required in the algorithm to describe the details of traction curves properly, especially for curves with sharp changes. In contrast, numerical damping is unnecessary if the second-order PTI method is used together with an appropriate spatial discretization (see Figures 12, 13 and 16).

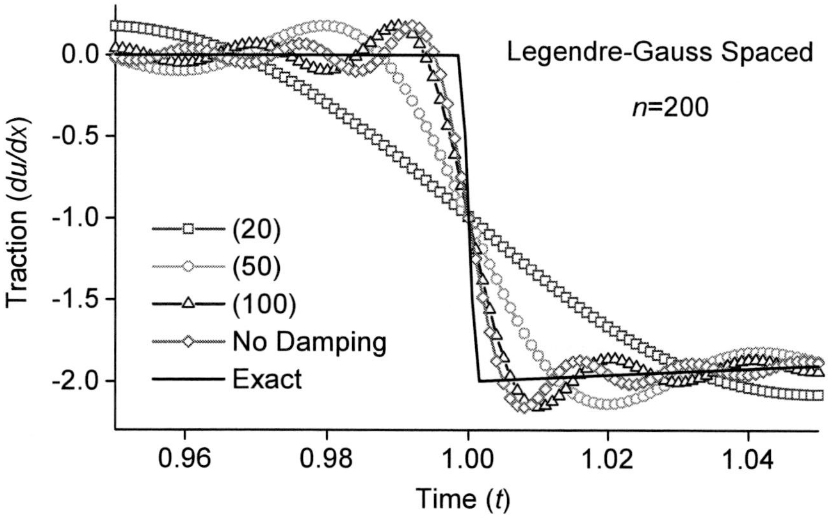

Figure 16. Damping effects on traction curves around $t = 1$ at fixed end ($x = 1$) of rod subjected to half-triangular impact, with Legendre-Gauss distribution, with the PTI method.

The traction responses of some selected single modes (say mode number m) at the fixed end ($x=1$) of the rod subjected to half-triangular impact are presented in Figure 17 for low frequency modes and in Figure 18 for high frequency modes. It can be seen from Figures 17 and 18 that the amplitude of the solution curves decreases along with the increase of mode numbers, as expected, because the second-order PTI method can provide essentially an 'exact' solution of the discrete system with machine-like precision resulting from a homogeneous differential equation.

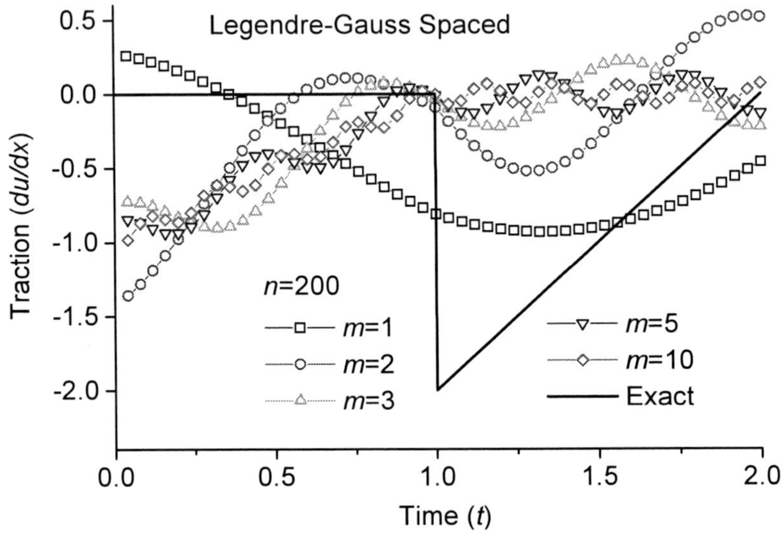

Figure 17. Traction responses of single modes (m) with low frequencies at fixed end ($x=1$) of rod subjected to half-triangular impact, with the PTI method.

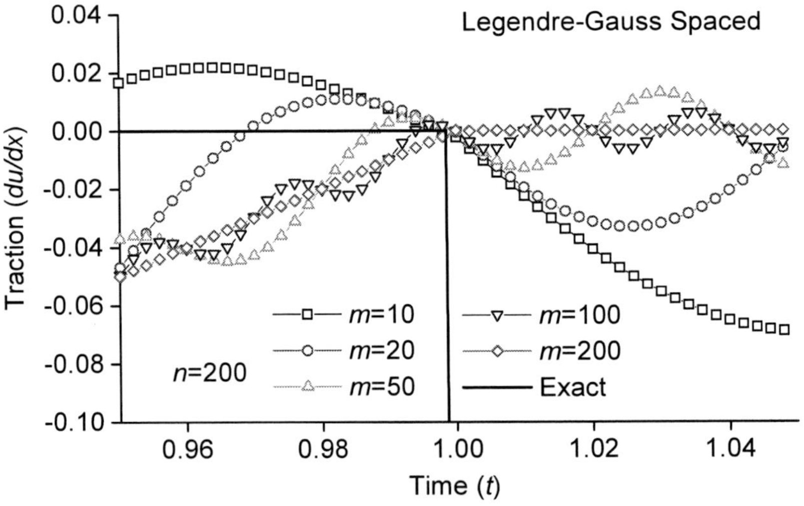

Figure 18. Traction responses of single modes (m) with high frequencies at fixed end ($x=1$) of rod subjected to half-triangular impact, with the PTI method.

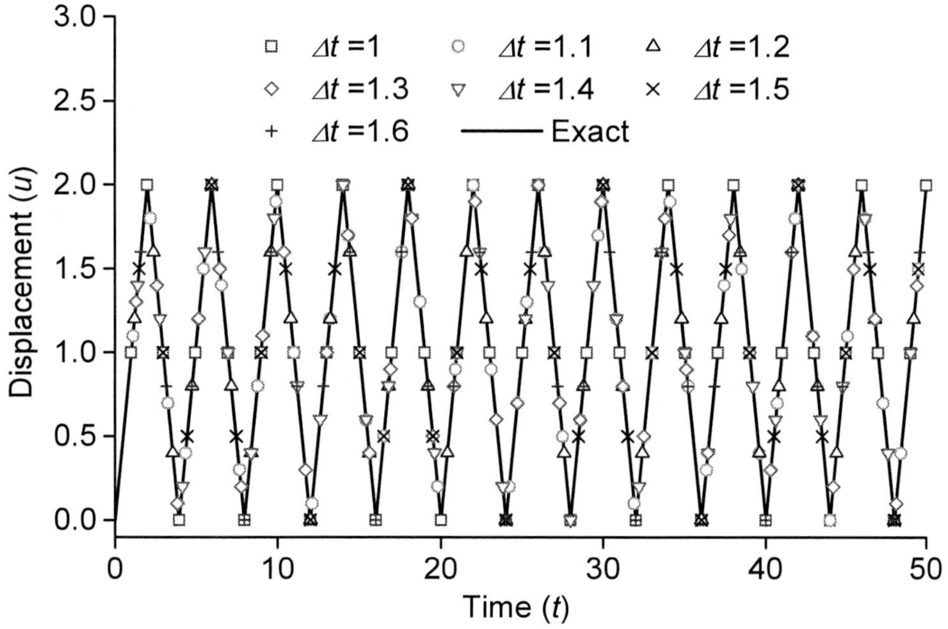

Figure 19. Displacement curves at free end ($x = 0$) of rod using different time steps, Δt , subjected to Heaviside impact, with the PTI method.

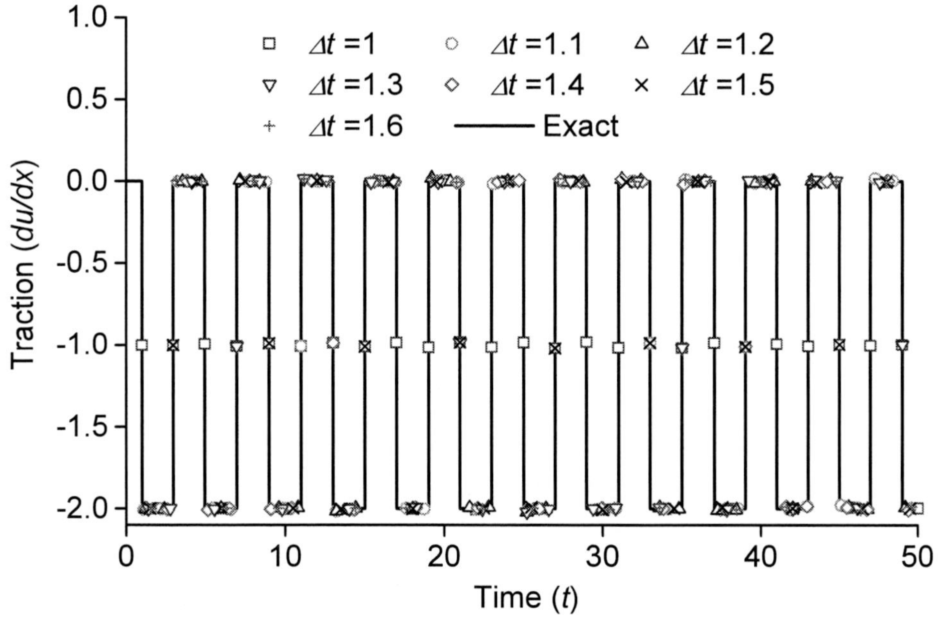

Figure 20. Traction curves at fixed end ($x = 1$) of rod using different time steps, Δt , subjected to Heaviside impact, with the PTI method.

4.5. Effects of Time Step with regard to Accuracy and Efficiency

In this subsection the effect of the time step used on the accuracy of the PTI method is investigated using a grid number $n = 100$. As a comparison, the results with the Newmark-β method are also presented using a grid number $n = 200$. The results with the PTI method are presented in Figures 19 and 20. Figure 19 shows the displacement at the free end of the rod versus time for several time steps, while Figure 20 displays the corresponding traction at the fixed end. It can be seen that the computed results with the PTI method are in good agreement with the exact results in both Figures 19 and 20. In fact, it is possible to use very large time steps with the second-order PTI method without loss of accuracy. In comparison, the results with the Newmark-β method are presented in Figures 21 and 22. It can be seen from Figure 21 that even for a small time step ($\Delta t = 0.1$) the details of the displacement curve cannot be traced satisfactorily with the Newmark-β method. The result becomes unacceptable when a large time step ($\Delta t = 1$) is used. A similar conclusion can be drawn from Figure 22 for the traction responses, where over-shooting is observed, which is considered to stem from the spurious mode imbedded in the Newmark-β method.

Chen and Tanaka have pointed out that the computational efficiency of the first-order PTI method is very low compared with traditional time integration schemes [30]. The efficiency of the PTI method is improved with the second-order scheme since the size of system matrices is reduced in the algorithm [15]. It is well known that there are generally two stages to be performed in a time integration scheme. The first stage is the repair of the system matrix, or the transmission matrix. The second stage is time-marching computations step by step to obtain the desired results. As the 2^N algorithm must be used in the first stage of the PTI method, the efficiency is much lower than that of the Newmark-β method if accuracy is not an essential consideration. In the second stage, by roughly comparing equations (49) and (50) with equations (100)-(102), it is evident that the number of operations of the PTI method would be three to four times those of the Newmark-β method in a single time step if the same number of grid points is used. However, comparing Figures 19 and 20 with 21 and 22 could draw the opposite conclusion if accuracy is an essential consideration. In addition to less restriction in selecting a time step, the results of the PTI method using the lower grid number $n = 100$ are far better than those of the Newmark-β method using the much higher grid number $n = 200$. The CPU times in the two stages are compared in Table 1 for the two algorithms, taking accuracy into consideration, where 10000 steps are computed in the second stage.

Figure 23 shows the displacement and traction responses of the rod using single time steps of very large size from $\Delta t = 100$ to $\Delta t = 104$ with the PTI method. It can be seen from Figure 23 that the computed results are in good agreement with the exact ones. These findings verify the exactness of the second-order PTI method, where very small differences in time steps are used to accurately describe the sharp changes of the traction curve at times around $t = 101$ and $t = 103$, respectively. In contrast, there is a phase shift in the displacement response and over-shooting in the traction response, represented by dotted lines in Figure 23, computed using the Newmark-β method with $\Delta t = 0.1$ and $n = 200$.

Obviously, the second-order PTI method is capable of using a large time step without loss of accuracy, which can significantly save computing time. This is particularly true for long-term integration.

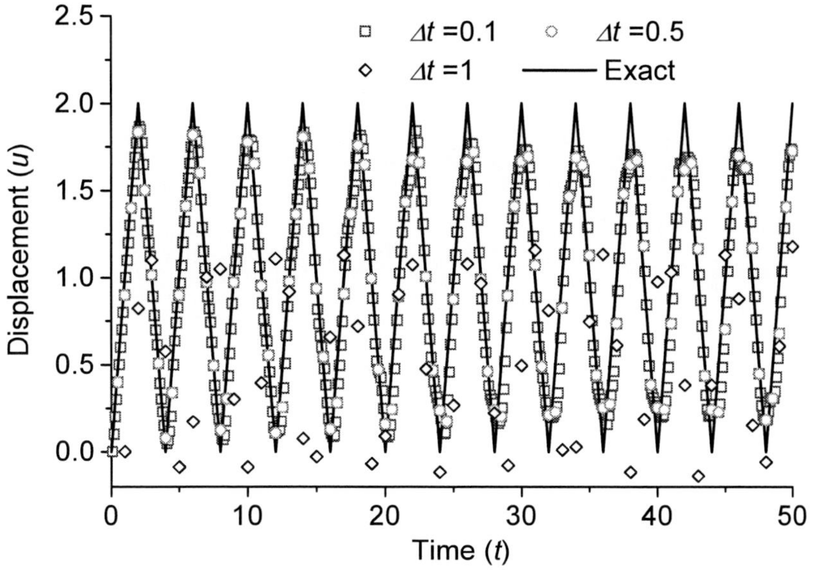

Figure 21. Displacement curves at free end ($x = 0$) of rod using different time steps, Δt , subjected to Heaviside impact, with the Newmark- β method using $n = 200$.

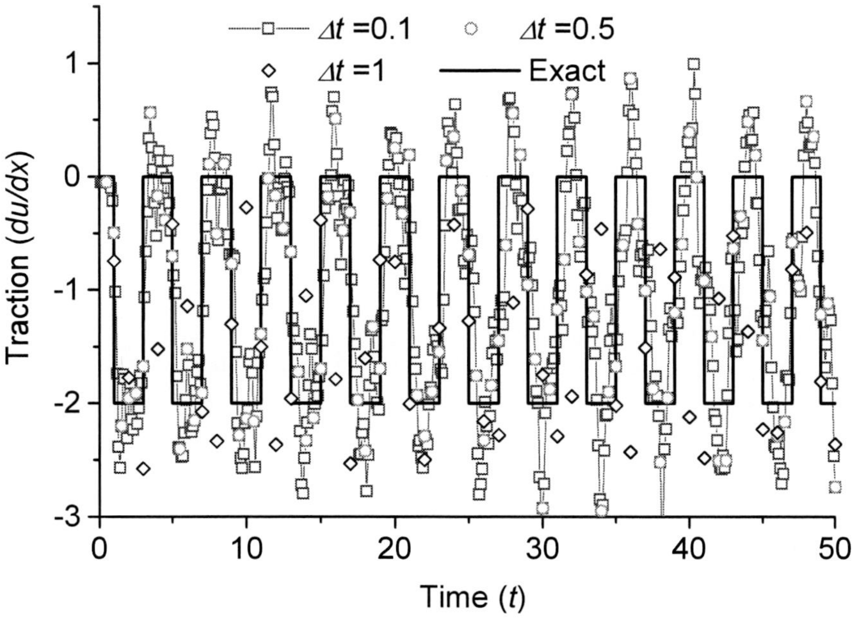

Figure 22. Traction curves at fixed end ($x = 1$) of rod using different time steps, Δt , subjected to Heaviside impact, with the Newmark- β method using $n = 200$.

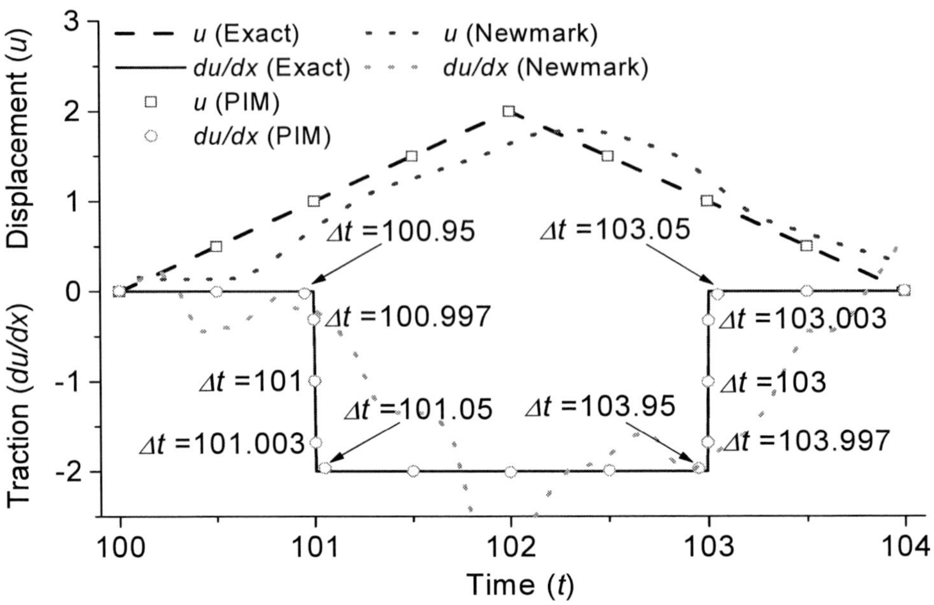

Figure 23. Displacement and traction responses of rod using single time steps with very large sizes subjected to Heaviside impact (dotted lines are computed using Newmark- β method with $\Delta t = 0.1$ and $n = 200$).

Table 1. Comparison of CPU times (μs)

Algorithm	Newmark- β	Newmark- β	PTI
Grid number	100	200	100
1st stage	32	294	297
2nd stage	516	2237	1532
Total	548	2531	1829

5. CONCLUSIONS

A second-order scheme of precise time-step integration (PTI) method is described for dynamic analysis with respect to long-term integration and transient responses. The recursive scheme is presented in detail for the direct solution of the homogeneous part of the second-order algebraic and differential equations. The corresponding particular solution is also presented, incorporated with the second-order scheme where the excitation vector is approximated by the truncated Taylor series. The sine and cosine matrices involved in the scheme are calculated using the so-called 2^N algorithm. It is revealed that, in the analysis of elasto-dynamic impact problems under the framework of the PTI method, the choice of spatial discretization is important to match the corresponding time integration scheme, which, as a result, can provide numerically the possible modal types for the structure to be analyzed. With a proper grid distribution, the higher the grid number, the richer the frequency modes in the system matrix, and the more accurate the solutions. This is particularly true for accurately describing the details of traction responses with sharp changes.

It is noted that the spatial discretization changes a continuum physical modal into a discrete one characterized by the system matrix. With regard to the contribution of each modal type in the system matrix to the solution, the lower the frequency mode, the more it contributes to the solution. However, with appropriate spatial discretization, high-frequency modes play an important role in accurately describing the details of traction responses with sharp changes. This facility is indispensable, as shown in the numerical examples. This is a feature just like that of an analytical solution procedure imbedded in the PTI method, and thus numerical damping, which is often considered essential in traditional time-step integration schemes, is unnecessary in the present scheme, as the high-frequency responses are no longer spurious. For the same reason, there is no time-step restriction in the use of the second-order PTI method, as shown in the numerical examples. It is clear that the second-order PTI method is capable of using a large time step without loss of accuracy, which can significantly save computing time. This is particularly true for long-term integration.

REFERENCES

[1] W.X. Zhong, On the precise time-integration method for structural dynamics. *Journal of Dalian University of Technology* 34(2):131-136, (1994) (in Chinese).

[2] W.X. Zhong and F.W. Williams, A precise time step integration method. *Proceedings of Institution of Mechanical Engineers, Part C: Journal of Mechanical Engineering Science* 208, 427-430, (1994).

[3] J.H. Lin, W.P. Shen and F.W. Williams, A high precision direct integration scheme for structures subjected to transient dynamic loading. *Computers & Structures* 56, 113-120, (1995).

[4] W.P. Shen, J.H. Lin and F.W. Williams, Parallel computing for the high precision direct integration method. *Computer Methods in Applied Mechanics and Engineering* 126, 315-331, (1995).

[5] Y.X. Wang, X.D. Tian and G. Zhou, Homogenized high precise direct integration scheme and its applications in engineering. *Communications in Numerical Methods in Engineering* 18, 429-439, (2002).

[6] X.Q. Zhu and S.S. Law, Precise time-step integration for the dynamic response of a continuous beam under moving loads. *Journal of Sound and Vibration* 240, 962-970, (2001).

[7] X.J. Liu, D.W. Begg, M.A. Devane and W.X. Zhong, High precision integration for dynamic structural systems with holonomic constraints. *Structural Engineering and Mechanics* 5, 283-295, (1997).

[8] J.H. Lin, W.P. Shen and F.W. Williams, A high precision direct integration scheme for non-stationary random seismic responses of non-classically damped structures. *Structural Engineering and Mechanics* 3, 215-228, (1995).

[9] J.H. Lin, W.P. Shen and F.W. Williams, Accurate high-speed computation of non-stationary random structural response. *Engineering Structures* 19, 586-593, (1997).

[10] A.Y.T. Leung, Fast matrix exponent for deterministic or random excitations. *International Journal for Numerical Methods in Engineering* 50, 377-394, (2001).

[11] W.X. Zhong, J. Zhu and X.X. Zhong. On a new time integration method for solving time dependent partial differential equations. *Computer Methods in Applied Mechanics and Engineering* 130, 163-178, (1996).

[12] W.X. Zhong, Combined method for the solution of asymmetric Riccati differential equations. *Computer Methods in Applied Mechanics and Engineering* 191, 93-102, (2001).

[13] W.X. Zhong, On precise integration method. *Journal of Computational and Applied Mathematics* 163, 59-78, (2004).

[14] H.W. Zhang, X.W. Zhang and J.S. Chen, A new algorithm for numerical solution of dynamic elastic-plastic hardening and softening problems. *Computers & Structures* 81, 1739-1749, (2003).

[15] H. Ma and Q.H. Qin, A second order scheme for integration of one-dimensional dynamic analysis. *Computers & Mathematics with Applications* 49, 239-252, (2005).

[16] C.W. Bert and M. Malik, Differential quadrature method in computational mechanics: A review. *Applied Mechanics Review* 49, 1-27, (1996).

[17] T.C. Fung, Complex-time-step Newmark methods with controllable numerical dissipation. *International Journal for numerical methods in Engineering* 41, 65-93, (1998).

[18] T.C. Fung, Complex-time-step methods for transient analysis. *International Journal for numerical methods in Engineering* 46, 1253-1271, (1999).

[19] T.C. Fung, Numerical dissipation in time-step integration algorithms for structural dynamic analysis. *Prog. Struct. Engng. Mater*. 5, 167-180, (2003).

[20] H. Ma, F. Yin and Q.H. Qin, Performance and numerical behaviors of the second-order scheme of precise time-step integration for transient dynamic analysis. *Numerical Methods for Partial Differential Equations* (published online on 2 March 2007).

[21] B.T. Smith, J.M. Boyle, B.S. Garbow, Y. Ikebe, V.C. Klema, and C.B. Moler, Matrix Eigensystem Routines-EISPACK Guide. Springer-Verlag, Berlin: 1976.

[22] A.S. Deif, Advanced Matrix Theory for Scientists and Engineers (2nd Edition). Abacus Press, New York: 1991.

[23] P. Bugl, Differential Equations-Matrices and Models. Prentice Hall, New Jersey: 1995.

[24] R.E. Bellman, B.G. Kashef and J. Casti, Differential quadrature: a technique for rapid solution of nonlinear partial differential equations. *Journal of Computational Physics* 10, 40-52, (1972).

[25] J.R. Quan and C.T. Chang, New insight in solving distributed system equations by the quadrature method-I. Analysis. *Computers in Chemical Engineering* 13, 779-788, (1989).

[26] J.R. Quan and C.T. Chang, New insight in solving distributed system equations by the quadrature method-II. Numerical experiments. *Computers in Chemical Engineering* 13, 1017-1024, (1989).

[27] C. Shu and B.E. Richards, Application of generalized differential quadrature to solve two-dimensional incompressible Navier-Stokes equations. *International Journal of numerical methods in Fluids* 15(7), 791-798, (1992).

[28] H.P. William, A.T. Saul, T.V. William and P.F. Brian, Numerical Recipes in Fortran77 (Vol.1): The Arts of Scientific Computing, Cambridge University Press, Cambridge, 1997.

[29] Z. Zong and K.Y. Lam, A localized differential quadrature (LDQ) method and its application to the 2D wave equation. *Computational Mechanics* 29, 382-391, (2002).

[30] W. Chen and M. Tanaka, A study on time schemes for DRBEM analysis of elastic impact wave. *Computational Mechanics* 28, 331-338, (2002).

[31] A. Quarteroni, R. Sacco and F. Saleri, *Numerical Mathematics*. Springer-Verlag, New York, 2000.

In: Applied Mathematical Modeling
Editor: Eetu N. Virtanen, pp. 185-220
ISBN: 978-1-60021-975-7

Chapter 4

THE EXACT THEORIES OF RECTANGULAR BEAMS BASED ON LINEAR ELASTICITY

Yang Gao [1]***, Bai-xiang Xu*** [2] ***and Bao-sheng Zhao*** [3]
[1]College of Science, China Agricultural University, Beijing 100083, People's Republic of China
[2]State Key Laboratory for Turbulence and Complex Systems and Department of Mechanics and Engineering Science, Peking University, Beijing 100871, People's Republic of China
[3]School of Mechanical Engineering, University of Science and Technology Liaoning, Anshan 114044, People's Republic of China

ABSTRACT

This chapter is designed to present a relatively simple, but realistic exposition of some of the ideas in our previous works. All our calculations are explicit and analytic. By considering an elastic beam of narrow rectangular cross section, the two beam theories, the refined theory and the decomposition theorem, are exact in the sense that one-dimensional (1D) beamlike equations are obtained from the three-dimensional (3D) equations of linear elasticity without any hypotheses or approximations whatsoever. Thus, the exact theories are not asymptotic and no appeal is made to small slenderness ratios. To set up the general framework of the beam theories, the appropriate stress and mixed boundary conditions accurate to all order for the beam of general edge geometry and loadings are formulated explicitly.

1. INTRODUCTION

The beam is the most general and important engineering component, which has been extensively studied for many years. Theories of rectangular beams involve basically the reduction of a 3D problem of elasticity theory to a 1D problem. Since the thickness dimension is much smaller than the longitudinal dimension, it is possible to approximate the distribution of the displacement, strain and stress components in the thickness dimension. Many efforts have been made on seeking for various beam theories with the help of some ad hoc assumptions. Additionally some observations are made on the formulation of the theories of conventional beam due to Timoshenko [1], Levinson [2], Bickford [3], Reddy [4], Tutek and Aganović [5], Lewiński [6], Fan and Widera [7], Tullini and Savoia [8] and the original work of Bernoulli and Euler.

Since the beginning of the history of the theory of elasticity, there are many methods available for reducing the 3D equations of the theory of elasticity to 1D equations in the theory of beams. These methods are as follow: methods based on hypotheses, the method of expansion, the asymptotic method, the method of successive approximation and the mix method. In all the above methods, reduction is carried out with the use of thickness coordinate except for the asymptotic expansion in which, generally, small parameter like thickness to length ratio is used in the expansion. In the method of expansion, generally, Taylor series, MacLaurin series, Legendre polynomials and Bernstein polynomials are used in the expansion. The merits of these methods relative to each other are discussed in detail [9–13].

2. Linear Elasticity

In the absence of body forces, the basic equations of linear elasticity for isotropic materials are of the form

$$\varepsilon_{ij}=\frac{1}{2}\left(u_{i,j}+u_{j,i}\right),\quad \sigma_{ij,i}=0,\quad \sigma_{ij}=\lambda u_{k,k}\delta_{ij}+\mu\left(u_{i,j}+u_{j,i}\right), \tag{2.1}$$

where u_i, ε_{ij} and σ_{ij} are the components of displacement, strain and stress, respectively, λ and μ are Lame constants, and δ_{ij} represents the Kronecker delta symbol. The comma in the subscript denotes the partial derivative with respect to the spatial variable, and the repeated indices imply summation.

For a narrow rectangular straight beam, the width in the y-direction is stress free. Therefore, it is plausible to set the components of stress $\sigma_y=\tau_{xy}=\tau_{yz}=0$. This is a plane stress assumption. We assume that the beam length in x-direction is l, the beam width in y-direction is 1, the beam height in z-direction is h, and $l>>h>>1$. In the plane $x-z$, the equations of deformation geometry

$$\varepsilon_x=\frac{\partial u_x}{\partial x},\quad \varepsilon_z=\frac{\partial u_z}{\partial z},\quad \gamma_{xz}=\frac{1}{2}\left(\frac{\partial u_z}{\partial x}+\frac{\partial u_x}{\partial z}\right). \tag{2.2}$$

The static equilibrium equations in the absence of body forces are

$$\frac{\partial\sigma_x}{\partial x}+\frac{\partial\tau_{xz}}{\partial z}=0,\quad \frac{\partial\tau_{xz}}{\partial x}+\frac{\partial\sigma_z}{\partial z}=0. \tag{2.3}$$

The constitute equations are expressed as

$$\sigma_x=\frac{E}{1-\nu^2}\left(\frac{\partial u_x}{\partial x}+\nu\frac{\partial u_z}{\partial z}\right),\quad \tau_{xz}=\frac{E}{2(1+\nu)}\left(\frac{\partial u_z}{\partial x}+\frac{\partial u_x}{\partial z}\right),\quad \sigma_z=\frac{E}{1-\nu^2}\left(\nu\frac{\partial u_x}{\partial x}+\frac{\partial u_z}{\partial z}\right), \tag{2.4}$$

The compatibility equation in terms of stresses has the form as

$$\nabla^2\left(\sigma_x+\sigma_z\right)=0, \tag{2.5}$$

where E and ν is are the Young's modulus and Poisson's ratio, respectively, and $\nabla^2=\partial^2/\partial x^2+\partial^2/\partial z^2$ is the two-dimensional (2D) Laplacian operator.

Two boundary conditions are considered in this chapter. One is the homogeneous boundary condition on the upper and lower surfaces of the beam,

$$\tau_{xz}=0,\quad \sigma_z=0\quad \left(z=\pm\frac{h}{2}\right), \tag{2.6}$$

and the other is the non-homogenous boundary condition on the upper and lower surfaces of the beam,

$$\tau_{xz}=0,\quad \sigma_z=\pm\frac{q}{2}\quad \left(z=\pm\frac{h}{2}\right). \tag{2.7}$$

In the absence of body forces, the equilibrium equations of elasticity plane stress problem denoted by u_x and u_z are expressed as

$$\nabla^2u_x+\frac{1+\nu}{1-\nu}\frac{\partial}{\partial x}\left(\frac{\partial u_x}{\partial x}+\frac{\partial u_z}{\partial z}\right)=0,\quad \nabla^2u_z+\frac{1+\nu}{1-\nu}\frac{\partial}{\partial z}\left(\frac{\partial u_x}{\partial x}+\frac{\partial u_z}{\partial z}\right)=0. \tag{2.8}$$

Papkovich-Neuber (P-N) solution [14, 15] of the governing equations (2.8) can be obtained as

$$u_x=P_1-\frac{1+\nu}{4}\frac{\partial}{\partial x}\left(P_0+xP_1+zP_3\right),\quad u_z=P_3-\frac{1+\nu}{4}\frac{\partial}{\partial z}\left(P_0+xP_1+zP_3\right). \tag{2.9}$$

where the displacement functions $P_i\left(i=0,1,3\right)$ is a 2D harmonic function, and they satisfy

$$\nabla^2 P_i = \frac{\partial^2 P_i}{\partial z^2} + \partial_x^2 P_i = 0 \quad \left(i = 0,1,3, \ \partial_x = \frac{\partial}{\partial x} \right). \tag{2.10}$$

3. THE REFINED THEORY OF BEAMS

Without employing ad hoc assumptions, Cheng [16] presented a method for the solution of 3D elasticity equations, and with the method deduced directly a refined theory of the plate from Boussinesq-Galerkin solution and Lur'e method [17]. The refined plate theory consists of three parts: the biharmonic equation, the shear equation and the transcendental equation. People may doubt the legitimacy of manipulation performed on differential operators in the derivation; however, the final results obtained by his method can be justified by the satisfaction of all equations in the 3D theory of elasticity. The only approximation in Cheng's plate theory is due to the approximate specification of boundary conditions at the edges of the plate. Therefore, regarding Saint-Venant's principle, Cheng's theory is a very accurate one.

Since the publication of the excellent work of Cheng [16] on deducing the plate theory directly from the 3D theory of elasticity, several extensions have been found in the plate theory of among isotropic plates [18-20], transversely isotropic plates [21, 22], piezoelectric plates [23], thermoelastic plates [24, 25]. Moreover, Gao and coauthors have indicated that applications of Cheng's method are quite successful in various beams, such as elastic beams [26, 27], transversely isotropic beams [28], magnetoelastic beams [29], piezoelectric beams [30], thermoelastic beams [31] and curved beams [32]. The significance of Cheng's method is that it opens a systematic way of developing the exact and approximate lower-dimensional theory from a 3D theory with the aid of the general solution of elasticity and symbolic computation.

3.1. The Refined Theory

As we know, the problem of beams may be decomposed into two fundamental problems: the extension or compression of a beam for symmetric deformation and the bending of a beam for anti-symmetric deformation. In the case of bending of a beam, the beam is subjected only to anti-symmetrical loadings and edge conditions, thus only odd functions of z are required for u_x and even functions of z for u_z. Based on Lur'e method [17] and with these requirements satisfied, treating Eq. (2.10) as an ordinary differential equation in z with constant coefficients, one obtains the following symbolic solution of Eq. (2.10):

$$P_1(x,z) = \frac{\sin(z\partial_x)}{\partial_x} g_1(x), \quad P_3(x,z) = \cos(z\partial_x) g_3(x), \tag{3.1}$$

g_1 and g_3 are unknown functions of x yet to be determined. The operators $\sin(z\partial_x)/\partial_x$ and $\cos(z\partial_x)$, which must be interpreted as representing series in powers of operators $(z\partial_x)^2$, have the following symbolic expressions

$$\frac{\sin(z\partial_x)}{\partial_x}=z\left(1-\frac{1}{3!}z^2\partial_x^2+\frac{1}{5!}z^4\partial_x^4-\cdots\right),\quad \cos(z\partial_x)=\left(1-\frac{1}{2!}z^2\partial_x^2+\frac{1}{4!}z^4\partial_x^4-\cdots\right). \tag{3.2}$$

From Appendix A, we can know that harmonic function P_0 always can satisfy the following expression without loss in generality,

$$P_0+xP_1+zP_3=-z\cos(z\partial_x)f(x), \tag{3.3}$$

where

$$f(x)=\int_0^x g_1(t)\,\mathrm{d}t-g_3(x). \tag{3.4}$$

Substituting Eqs. (3.1) and (3.3) into Eq. (2.9), one obtains

$$\begin{aligned} u_x&=\frac{\sin(z\partial_x)}{\partial_x}g_1+\frac{1+\nu}{4}z\cos(z\partial_x)f',\\ u_z&=\cos(z\partial_x)g_3+\frac{1+\nu}{4}\left[\cos(z\partial_x)-z\partial_x\sin(z\partial_x)\right]f. \end{aligned} \tag{3.5}$$

For the sake of simplicity, the differential symbol "′" denotes differentiation with respect to x. From Eq. (3.5), we can get the angle of rotation and the deflection of the neutral surface,

$$\psi=-\left.\frac{\partial u_x}{\partial z}\right|_{z=0}=-\left(g_1+\frac{1+\nu}{4}f'\right),\quad w=u_z\big|_{z=0}=g_3+\frac{1+\nu}{4}f. \tag{3.6}$$

From Eqs. (3.5) and (3.6), the final expressions for the displacements are

$$\begin{aligned} u_x&=-\frac{\sin(z\partial_x)}{\partial_x}\psi+\frac{1+\nu}{4}\left[z\cos(z\partial_x)-\frac{\sin(z\partial_x)}{\partial_x}\right]f',\\ u_z&=\cos(z\partial_x)w-\frac{1+\nu}{4}z\partial_x\sin(z\partial_x)f \end{aligned} \tag{3.7}$$

with the expression

$$f(x) = -\left[\int_0^x \psi(t)\,\mathrm{d}t + w(x)\right]. \tag{3.8}$$

Using the constitute equations (2.4), from Eq. (3.7) the components of stress can be indicated as

$$\begin{aligned}
\sigma_x &= -\frac{E}{4}\left\{\left[\frac{1-\nu}{1+\nu}\frac{\sin(z\partial_x)}{\partial_x} - z\cos(z\partial_x)\right]f'' + \frac{4}{1+\nu}\frac{\sin(z\partial_x)}{\partial_x}\psi'\right\}, \\
\tau_{xz} &= -\mu\left[\cos(z\partial_x)(\psi - w') + \frac{1+\nu}{2}z\partial_x\sin(z\partial_x)f'\right], \\
\sigma_z &= -\frac{E}{4}\left\{\left[\frac{1-\nu}{1+\nu}\frac{\sin(z\partial_x)}{\partial_x} + z\cos(z\partial_x)\right]f'' + \frac{4}{1+\nu}\frac{\sin(z\partial_x)}{\partial_x}w''\right\},
\end{aligned} \tag{3.9}$$

3.2. Exact Equations: No Transverse Surface Loadings

Making use of the refined theory of beams obtained in the previous section, we will investigate the rectangular straight beam with homogeneous boundary conditions on the upper and lower surfaces in this section. Substituting the stress expressions in Eq. (3.9) into the boundary condition (2.6) of beams, we get the following equations

$$\begin{bmatrix} D_1 - D_2\partial_x^2 & -\left(D_1 + D_2\partial_x^2\right)\partial_x \\ D_3\partial_x & -\dfrac{4}{(1+\nu)h}D_2\partial_x^2 + D_3\partial_x^2 \end{bmatrix}\begin{bmatrix}\psi \\ w\end{bmatrix} = \begin{bmatrix}0 \\ 0\end{bmatrix}. \tag{3.10}$$

The three differential operators D_1, D_2 and D_3 are defined by

$$D_1 = \frac{4}{1+\nu}\cos\left(\frac{h\partial_x}{2}\right),\ D_2 = \frac{h}{\partial_x}\sin\left(\frac{h\partial_x}{2}\right),\ D_3 = \frac{h}{2}\cos\left(\frac{h\partial_x}{2}\right) + \frac{1-\nu}{1+\nu}\frac{1}{\partial_x}\sin\left(\frac{h\partial_x}{2}\right). \tag{3.11}$$

Equations (3.10) are differential equations with respect to unknown deflection and angle of rotation of the neutral surface. Let L_0 be the determinant of the 2×2 operator coefficient matrix of Eq. (3.10), there is

$$L_0 = \frac{4h}{1+\nu}\left\{\frac{1}{\partial_x^2}\left[1 - \frac{\sin(h\partial_x)}{h\partial_x}\right]\right\}\partial_x^4, \tag{3.12}$$

and $L_{ij}\left(i,j=1,2\right)$ be the co-factors of the matrix. The solutions of Eq. (3.10) are

$$\begin{bmatrix}\psi\\ w\end{bmatrix}=\begin{bmatrix}L_{22} & -L_{12}\\ -L_{21} & L_{11}\end{bmatrix}\begin{bmatrix}\xi_1\\ \xi_2\end{bmatrix}, \tag{3.13}$$

and ξ_i satisfies

$$L_0\xi_i=0\quad\left(i=1,2\right). \tag{3.14}$$

In Appendix B, it is proved that the solutions of Eq. (3.14) can be decomposed into two parts which are governed by a fourth-order equation and a transcendental equation, respectively. That is, ξ_i can be rewritten as

$$\xi_i=\xi_i^F+\xi_i^T\quad\left(i=1,2\right), \tag{3.15}$$

where the superscript "F" and "T" indicate the fourth-order part and the transcendental part, respectively, and ξ_i^F and ξ_i^T have to satisfy the following governing differential equations of the beam problem, respectively,

$$\partial_x^4\xi_i^F=0,\quad \frac{1}{\partial_x^2}\left[1-\frac{\sin\left(h\partial_x\right)}{h\partial_x}\right]\xi_i^T=0, \tag{3.16}$$

then the angle of rotation and the deflection of the neutral surface can also be decomposed into two parts, respectively,

$$\psi=\psi^F+\psi^T,\quad w=w^F+w^T. \tag{3.17}$$

The solutions corresponding to these two parts will be given separately in the following discussions.

3.2.1. The Fourth-order Equation and the Fourth-order Solution

ξ_i^F satisfies the following fourth-order equation

$$\partial_x^4\xi_i^F=0. \tag{3.18}$$

From Eq. (3.13), the corresponding solutions of ψ^F and w^F have the form

$$\begin{bmatrix} \psi^F \\ w^F \end{bmatrix} = \begin{bmatrix} L_{22} & -L_{12} \\ -L_{21} & L_{11} \end{bmatrix} \begin{bmatrix} \xi_1^F \\ \xi_2^F \end{bmatrix}. \tag{3.19}$$

By using Eqs. (3.18) and (3.19), we can get

$$\psi^F = \left(1 + \frac{1+\nu}{4} h^2 \partial_x^2\right) \partial_x w^F, \tag{3.20}$$

where

$$\partial_x^4 w^F = 0. \tag{3.21}$$

From Eq. (3.7), the total displacements can be found to be

$$u_x^F = -z\partial_x \left[1 - \frac{1}{6} z^2 \partial_x^2 + \frac{1+\nu}{12} h^2 \partial_x^2 \left(3 - 2\frac{z^2}{h^2}\right)\right] w^F, \quad u_z^F = \left(1 + \frac{\nu}{2} z^2 \partial_x^2\right) w^F, \tag{3.22}$$

and the normal stress and shear stress can be found to be

$$\sigma_x^F = -Ez\left(w^F\right)'', \quad \tau_{xz}^F = -\frac{E}{8}\left(h^2 - 4z^2\right)\left(w^F\right)''', \quad \sigma_z^F = 0. \tag{3.23}$$

Calculating moment and shear force for the present case yields

$$M_x^F = -EI\left(w^F\right)'', \quad Q_x^F = -EI\left(w^F\right)''', \tag{3.24}$$

where $I = h^3/12$.

By the same arguments made in Cheng [16], the Eqs. (3.22)-(3.24) constitute a first-order refined theory for the bending beams with the differential governing equation (3.21), which can satisfy two edge conditions along the boundary of beams and coincide with the corresponding expressions of the classical elasticity. Unlike the customary beam theory, it is deduced directly all the fundamental equations of the refined beam theory.

3.2.2. The Transcendental Equation and the Transcendental Solution

ξ_i^T satisfies the following transcendental equation

$$\frac{1}{\partial_x^2}\left[1 - \frac{\sin(h\partial_x)}{h\partial_x}\right]\xi_i^T = 0, \tag{3.25}$$

and the related solutions of ψ^T and w^T become

$$\begin{bmatrix} \psi^T \\ w^T \end{bmatrix} = \begin{bmatrix} L_{22} & -L_{12} \\ -L_{21} & L_{11} \end{bmatrix} \begin{bmatrix} \xi_1^T \\ \xi_2^T \end{bmatrix}. \tag{3.26}$$

Substituting Eq. (3.26) into the displacement and stress expressions (2.7) and (2.9), respectively, one obtains the following expressions

$$u_x^T = \frac{1}{E}\left[\frac{\partial^2 m}{\partial x^2} - (1+\nu)\frac{\partial^3 \Phi}{\partial x^3}\right], \quad u_z^T = \frac{1}{E}\left[\frac{\partial^2 n}{\partial x^2} - (1+\nu)\frac{\partial^3 \Phi}{\partial x^2 \partial z}\right], \tag{3.27}$$

$$\sigma_x^T = \frac{\partial^4 \Phi}{\partial x^2 \partial z^2}, \quad \tau_{xz}^T = -\frac{\partial^4 \Phi}{\partial x^3 \partial z}, \quad \sigma_z^T = \frac{\partial^4 \Phi}{\partial x^4}, \tag{3.28}$$

where the function $m(x,z)$ and $n(x,z)$ are conjugate harmonic function, and satisfy

$$\frac{\partial m}{\partial x} = \frac{\partial n}{\partial z} = \nabla^2 \Phi. \tag{3.29}$$

Furthermore, the function $\Phi(x,z)$ has the form

$$\begin{aligned} -\frac{4\Phi}{E} = & \left[-\frac{2}{1+\nu} h \cos\left(\frac{h\partial_x}{2}\right)\sin(z\partial_x) + \frac{4}{1+\nu} z \sin\left(\frac{h\partial_x}{2}\right)\cos(z\partial_x)\right]\frac{1}{\partial_x^2}\xi_1^T \\ & + \left[2\cos\left(\frac{h\partial_x}{2}\right)\frac{\sin(z\partial_x)}{\partial_x} - h\sin\left(\frac{h\partial_x}{2}\right)\sin(z\partial_x) - 2z\cos\left(\frac{h\partial_x}{2}\right)\cos(z\partial_x)\right]\frac{1}{\partial_x^2}\xi_2^T, \end{aligned} \tag{3.30}$$

and Φ satisfies the following equations:

$$\nabla^2 \nabla^2 \Phi = 0, \tag{3.31}$$

$$\Phi = 0, \quad \frac{\partial \Phi}{\partial z} = 0 \quad \left(z = \pm\frac{h}{2}\right). \tag{3.32}$$

Therefore, the moment and shear force to be

$$M_x^T = 0, \quad Q_x^T = 0. \tag{3.33}$$

Eq. (3.33) shows that the transcendental solution does not yield moment and shear force which are yielded only from the fourth-order solution. Eqs. (3.27), (3.28) and (3.33) satisfy two edge conditions along the boundary of beams, and yet satisfy exactly all the fundamental equations in theory of elasticity.

Combining the fourth-order solution (3.22)-(3.24) and the transcendental solution (3.27), (3.28) and (3.33), we arrive at a second-order refined theory for the bending beams with the differential governing equations (3.21), (3.31) and (3.32). It is important to note that the equilibrium equations (2.3) are satisfied by any solution of the refined beam theory, and the only approximation in the theory is introduced by the approximate specification of the boundary conditions at the edges of the beam (i.e. the boundary conditions are specified in terms of the stress resultants or some combination of the angle of rotation and the deflection of the neutral surface, instead of the stress or displacement distribution over the thickness $-h/2 \le z \le h/2$). Therefore, in the cases where Saint-Venant's principle holds, the refined beam theory should be a very accurate one.

3.3. Approximate Equations: Transverse Surface Loadings

In this section a rectangular straight beam with transverse surface loadings is considered. Substituting the stress expressions in Eq. (3.9) into the boundary condition (2.7) of the beam, we get the following equations expressed by ψ and w:

$$\left(D_1 - D_2\partial_x^2\right)\psi - \left(D_1 + D_2\partial_x^2\right)\partial_x w = 0,\ \ D_3\partial_x\psi - \left[\frac{4}{(1+\nu)h}D_2 - D_3\right]\partial_x^2 w = \frac{2}{E}q. \quad (3.34)$$

Simplifying Eq. (3.34) leads to

$$\begin{aligned}
&\frac{Eh}{2}\left[1 - \frac{\sin(h\partial_x)}{h\partial_x}\right]w'' = \left[\cos\left(\frac{h\partial_x}{2}\right) - \frac{1+\nu}{4}h\partial_x\sin\left(\frac{h\partial_x}{2}\right)\right]q,\\
&\left[\cos\left(\frac{h\partial_x}{2}\right) - \frac{1+\nu}{4}h\partial_x\sin\left(\frac{h\partial_x}{2}\right)\right]\psi = \left[\cos\left(\frac{h\partial_x}{2}\right) + \frac{1+\nu}{4}h\partial_x\sin\left(\frac{h\partial_x}{2}\right)\right]\partial_x w.
\end{aligned} \quad (3.35)$$

Eq. (3.35) is the exact governing equation for w and ψ in the beam problem subjected to the transverse surface loadings. Since this equation is of infinite order, however, it is not applicable in most cases. Using Taylor series of the trigonometric functions in Eq. (3.2) and then dropping all the terms associated with h^4 or the higher orders, the results finally turn out to be

$$EIw'''' = \left[1 - \left(\frac{1}{5} + \frac{1}{8}\nu\right)h^2\partial_x^2\right]q,\ \ EI\psi''' = \left[1 + \left(\frac{1}{20} + \frac{1}{8}\nu\right)h^2\partial_x^2\right]q. \quad (3.36)$$

From Eq. (3.36), the expressions about the displacements, stresses and stress resultants become

$$u_x = -\frac{z}{EI\partial_x^3}\left(1+\frac{2+5\nu}{40}h^2\partial_x^2 - \frac{2+\nu}{6}z^2\partial_x^2\right)q,$$
$$u_z = \frac{1}{EI\partial_x^4}\left(1-\frac{8+5\nu}{40}h^2\partial_x^2 + \frac{\nu}{2}z^2\partial_x^2\right)q, \tag{3.37}$$

$$\sigma_x = -\frac{1}{I\partial_x^2}z\left(1+\frac{1}{20}h^2\partial_x^2 - \frac{1}{3}z^2\partial_x^2\right)q,$$
$$\tau_{xz} = -\frac{1}{2I\partial_x}\left(\frac{h^2}{4}-z^2\right)q, \tag{3.38}$$
$$\sigma_z = \frac{1}{2I}z\left(\frac{h^2}{4}-\frac{z^2}{3}\right)q,$$

$$M_x'' = -q, \; Q_x' = -q. \tag{3.39}$$

Eqs. (3.36)-(3.39) form the basic equations of an approximate first-order theory for the bending problem of the beam under the transverse loadings, in which boundary conditions can be prescribed at both upper and lower surfaces. It should be pointed out that Eq. (3.9) in the previous section is still valid. Clearly, even if people doubt the legitimacy of the manipulation performed on differential operators, the beam equations obtained above can be justified by comparing their forms with the forms of the corresponding equations in other well-known beam theories.

3.4. Several Examples

By dropping all the terms associated with h^4 or the higher orders in Eq. (3.34), we arrive at the following equations

$$\psi - w' - \frac{h^2}{8}\left[(2+\nu)\psi'' + \nu w'''\right] = 0, \; \frac{2}{3}(\psi' - w'') = \frac{2(1+\nu)}{Eh}q. \tag{3.40}$$

According to the stress expressions in Eq. (3.9), assuming q is a linear function and omitting all the terms associated with h^4 or the higher orders, one can obtain the expressions of the moment and shear force

$$M_x = -\frac{EI(2+5\nu)}{10(1+\nu)}\left(\frac{8+5\nu}{2+5\nu}\psi' + w''\right), \quad Q_x = -\frac{2}{3}\mu h(\psi - w'). \tag{3.41}$$

Through generalizing the method proposed by Gregory and Wan [33] and by invoking P–N solution, Gao et al. [34] derived a set of necessary conditions on the edge-data for the existence of a rapidly decaying solution of the beam problems. It is interesting to note that the new boundary conditions in the stress data case are consistent with the conventional boundary conditions [35]. To illustrate the applications of the refined beam theory developed in the previous sections, we present the following five typical examples by using the boundary conditions given by Gao et al. [34]. Results for the examples are given for both the new theory and Levinson's beam theory, and are compared with the well-known exact solutions and the solutions by Levinson [2].

3.4.1. The end Loaded Cantilever Beam

Considering a cantilever beam of uniform cross-section loaded by a transverse shear force of magnitude Q_0 at $x = 0$ and clamped at $x = l$. For the present theory, the boundary conditions are

$$\frac{8+5\nu}{2+5\nu}\psi'(0) + w''(0) = 0, \quad -\frac{2}{3}\mu h\left[\psi(0) - w'(0)\right] = Q_0, \quad w(l) = \psi(l) = 0. \tag{3.42}$$

From Eqs. (3.40) and (3.42), the solution for the deflection of the neutral surface is obtained

$$w = \frac{Q_0 l^3}{6EI}\left(-\frac{x^3}{l^3} + 3\frac{x}{l} - 2\right) - (1+\nu)\frac{Q_0 l^3}{4EI}\frac{h^2}{l^2}\left(1 - \frac{x}{l}\right). \tag{3.43}$$

The solution based on the new theory is the exact solution of elasticity theory [35] and is same as the solution by Levinson [2].

3.4.2. The Uniformly Loaded Cantilever Beam

In this case, a cantilever beam loaded by a uniformly distributed load of intensity $q = q_0$ is considered. For the present theory, the boundary conditions are

$$\frac{8+5\nu}{2+5\nu}\psi'(0) + w''(0) = 0, \quad -\frac{2}{3}\mu h\left[\psi(0) - w'(0)\right] = 0, \quad w(l) = \psi(l) = 0. \tag{3.44}$$

From Eqs. (3.40) and (3.44), one obtains, as the solution for the deflection curve of the neutral surface

$$w=\frac{q_0 l^4}{24EI}\left(\frac{x^4}{l^4}-4\frac{x}{l}+3\right)+\frac{q_0 l^4}{20EI}\frac{h^2}{l^2}\left[\left(2+\frac{5}{4}\nu\right)\left(1-\frac{x^2}{l^2}\right)+\left(1+\frac{5}{2}\nu\right)\left(1-\frac{x}{l}\right)\right]. \quad (3.45)$$

Whereas the theory of Levinson solution [2] is

$$w=\frac{q_0 l^4}{24EI}\left(\frac{x^4}{l^4}-4\frac{x}{l}+3\right)+\frac{q_0 l^4}{20EI}\frac{h^2}{l^2}\left[(2+2\nu)\left(1-\frac{x^2}{l^2}\right)+(1+\nu)\left(1-\frac{x}{l}\right)\right]. \quad (3.46)$$

The solution (3.45) based on the refined theory is the exact solution of elasticity theory [35] and is same as the solution by Levinson [2] if $\nu=0$.

3.4.3. The Linearly Loaded Cantilever Beam

As a third example, considering a cantilever beam carried a linear distributed load $q(x)=q_0 x$. For the present theory, the boundary conditions are

$$\frac{8+5\nu}{2+5\nu}\psi'(0)+w''(0)=0,\quad -\frac{2}{3}\mu h\left[\psi(0)-w'(0)\right]=0,\quad w(l)=\psi(l)=0. \quad (3.47)$$

Using Eqs. (3.40) and (3.47), we derive

$$w(x)=\frac{q_0 l^5}{120EI}\left(\frac{x^5}{l^5}-5\frac{x}{l}+4\right)+\frac{q_0 l^5}{48EI}\frac{h^2}{l^2}\left[(2+\nu)\left(1-\frac{x^3}{l^3}\right)+3\nu\left(1-\frac{x}{l}\right)\right]. \quad (3.48)$$

Whereas the theory of Levinson solution [2] is

$$w(x)=\frac{q_0 l^5}{120EI}\left(\frac{x^5}{l^5}-5\frac{x}{l}+4\right)+(1+\nu)\frac{q_0 l^5}{120EI}\frac{h^2}{l^2}\left(-4\frac{x^3}{l^3}-3\frac{x}{l}+7\right). \quad (3.49)$$

The solution based on the new theory is the exact solution of elasticity theory [35] and is same as the solution by Levinson [2] if $\nu=2/3$.

3.4.4. The Uniformly Loaded and Simply Supported Beam

The other example is a beam which is simply supported at $x=\pm l$ and which carries a uniformly distributed load of intensity $q=q_0$. The boundary conditions for the present theory are

$$\frac{8+5\nu}{2+5\nu}\psi'(\pm l)+w''(\pm l)=0,\quad w(\pm l)=0. \quad (3.50)$$

From Eqs. (3.40) and (3.50), we obtain the solution for the deflection of the neutral surface

$$w(x)=\frac{q_0 l^4}{24EI}\left(\frac{x^4}{l^4}-6\frac{x^2}{l^2}+5\right)+\left(1+\frac{5}{8}\nu\right)\frac{q_0 l^4}{10EI}\frac{h^2}{l^2}\left(1-\frac{x^2}{l^2}\right). \tag{3.51}$$

The central deflection of Eq. (3.51) is

$$w(0)=\frac{5q_0 l^4}{24EI}\left[1+\frac{12}{25}\left(1+\frac{5}{8}\nu\right)\frac{h^2}{l^2}\right]. \tag{3.52}$$

Whereas the theory of Levinson solution [2] is

$$w(0)=\frac{5q_0 l^4}{24EI}\left[1+\frac{12}{25}(1+\nu)\frac{h^2}{l^2}\right]. \tag{3.53}$$

3.4.5. The Linearly Loaded and Simply Supported Beam

The fourth example is that of a simply supported beam carried a linear distributed load $q(x)=q_0 x$. For the new theory, the boundary conditions are

$$\frac{8+5\nu}{2+5\nu}\psi'(\pm l)+w''(\pm l)=0,\quad w(\pm l)=0. \tag{3.54}$$

From Eqs. (3.40) and (3.54), the solution for the deflection curve of the neutral surface is

$$w(x)=\frac{q_0 l^5}{360EI}\left(3\frac{x^5}{l^5}-10\frac{x^3}{l^3}+7\frac{x}{l}\right)+\left(1+\frac{5}{8}\nu\right)\frac{q_0 l^5}{30EI}\frac{h^2}{l^2}\left(\frac{x}{l}-\frac{x^3}{l^3}\right). \tag{3.55}$$

Whereas the theory of Levinson solution [2] is

$$w(x)=\frac{q_0 l^5}{360EI}\left(3\frac{x^5}{l^5}-10\frac{x^3}{l^3}+7\frac{x}{l}\right)+(1+\nu)\frac{q_0 l^5}{30EI}\frac{h^2}{l^2}\left(\frac{x}{l}-\frac{x^3}{l^3}\right). \tag{3.56}$$

The solutions (3.51) and (3.55) based on the refined theory are the exact solutions of elasticity theory [35]. From Eqs. (3.52), (3.53), Eqs. (3.55) and (3.56), for these problems both of the beam theories equally overestimate the "shear correction" term at the center of the beam less than 14% if $0\le\nu\le 0.5$.

Up to here, Eqs. (3.43), (3.45), (3.48), (3.51) and (3.55) that we give from the new theory are the exact solutions for the neutral surface given for the case of plane stress by the linear theory of elasticity [35]. The new theory provides as good or better results than Levinson's beam theory [2] when compared to those obtained from the linear theory of elasticity.

4. The Decomposition Theorem of Beams

On the basis of his previous experience in the study of the semi-infinite strip [36, 37], without employing ad hoc assumptions, Gregory [38] gave a rigorous proof about the decomposed form of elastic plates when the stresses are anti-symmetrical about the neutral surface. He proved that the exact solution of the plate problem is in fact the sum of the Levy solution (interior solution) and the boundary layer solution (in a decaying form) of the Papkovich-Fadle (P-F) type.

Due to the use of the P-F eigenfunction expansion of biharmonic functions during his derivation, Gregory's work can not be easily extended to the anisotropic elastic body. Wang and Zhao [39], Zhao and Wang [18, 40], Gao and Wang [41], and Gao et al. [42] successfully developed Gregory's decomposition theorem for plates and beams, and gave a new proof about the decomposed form. Results demonstrate that the general stress state of elastic beams can be decomposed into two parts: the interior state and the P-F state. Instead of the P-F eigenfunction expansion of biharmonic functions, only basic mathematical techniques are involved in the derivation.

4.1. The Decomposition Theorem

Definition 1-the interior state: The stresses of the interior state σ_{ij}^I have the form as

$$\sigma_x^I = -Ezw'', \quad \tau_{xz}^I = -\frac{E}{8}\left(h^2 - 4z^2\right)w''', \quad \sigma_z^I = 0, \tag{4.1}$$

The deflection of the neutral surface $w(x)$ is governed by

$$w'''' = 0. \tag{4.2}$$

It is clear that the interior state of the decomposition theorem is consistent with the fourth-order equation of the refined theory.

Definition 2-the P-F state: The stresses of the P-F state σ_{ij}^{PF} are given as

$$\sigma_x^{PF} = \frac{\partial^4 \Phi}{\partial x^2 \partial z^2}, \quad \tau_{xz}^{PF} = -\frac{\partial^4 \Phi}{\partial x^3 \partial z}, \quad \sigma_z^{PF} = \frac{\partial^4 \Phi}{\partial x^4}, \tag{4.3}$$

where $\Phi(x, z)$ is an odd function of z and satisfies

$$\nabla^2\nabla^2\Phi = 0, \tag{4.4}$$

$$\Phi = 0, \quad \frac{\partial \Phi}{\partial z} = 0 \quad \left(z = \pm\frac{h}{2}\right). \tag{4.5}$$

It is again seen that the P-F state of the decomposition theorem corresponds to the transcendental equation of the refined theory.

The decomposition theorem: The stress field of an elastic beam, which is governed by the equilibrium equations (2.3) and the compatibility equations (2.5) and satisfies the boundary conditions (2.6), can be decomposed into two parts: the interior state and the P-F state, i.e.

$$\sigma_{ij} = \sigma_{ij}^I + \sigma_{ij}^{PF}. \tag{4.6}$$

Combining the interior state solution (4.1) and the P-F state solution (4.3), we arrive at a second-order decomposition theorem for the elastic beam with the differential governing equations (4.2), (4.4) and (4.5).

As mentioned in the above context, the interior state and the P-F state are equivalent to the fourth-order equation and the transcendental equation, respectively, which have been derived from the refined theory. Consequently, it is actually proved the equivalence between the decomposition theorem and the refined theory of beams [43].

4.2. Two Lemmas

In the following subsections, we will introduce and prove two lemmas to give a concise and direct proof about the decomposition theorem.

4.2.1. Lemma 1

If σ_{ij} satisfy the equilibrium equations (2.3), the compatibility equation (2.5) and the boundary conditions (2.6), and for any point in the beam

$$\sigma_z = 0, \tag{4.7}$$

then σ_{ij} must have the form $\sigma_{ij} = \sigma_{ij}^I$.

Proof: From Eqs. (2.3), (2.5) and condition (4.7), one obtains

$$\frac{\partial^2 \sigma_x}{\partial x^2} = 0, \quad \frac{\partial^2 \sigma_x}{\partial z^2} = 0, \tag{4.8}$$

and σ_x is an odd function of z. Hence σ_x can be written as

$$\sigma_x = -Ezw'', \tag{4.9}$$

where $w'''' = 0$.

According to the first equation of Eq. (2.3) and the first expression of conditions (2.6), we get

$$\tau_{xz} = -\frac{E}{8}\left(h^2 - 4z^2\right)w'''. \tag{4.10}$$

Namely, $\sigma_{ij} = \sigma_{ij}^I$. This completes the proof of Lemma 1. □

4.2.2. Lemma 2

If σ_{ij} satisfy the equilibrium equations (2.3), the compatibility equation (2.5) and the boundary conditions (2.6), then there exists $\Phi(x,z)$ which is an odd function of z and satisfies Eqs. (4.4) and (4.5), and σ_z can be expressed as

$$\sigma_z = \frac{\partial^4 \Phi}{\partial x^4}. \tag{4.11}$$

Proof: From Eqs. (2.3) and (2.5), one yields

$$\nabla^2\nabla^2\sigma_z = 0. \tag{4.12}$$

Let

$$A_1 = \nabla^2\sigma_z, \tag{4.13}$$

where A_1 is a harmonic odd function of z. Then, there exists A_2, which is an even function of z, such that

$$2\frac{\partial A_2}{\partial z} = A_1, \quad \nabla^2 A_2 = 0, \tag{4.14}$$

Imposition Eq. (4.14) on Eq. (4.13) leads to

$$\nabla^2\sigma_z = 2\frac{\partial A_2}{\partial z} = \nabla^2\left(zA_2\right). \tag{4.15}$$

From (4.15), we have

$$\sigma_z = A_3 + zA_2, \tag{4.16}$$

where A_3 is a harmonic odd function of z.

Employing Eq. (4.14) two times, we know that there exists A_4 and A_5, which satisfy the following equations

$$\frac{\partial^2 A_4}{\partial z^2} = -A_3, \quad \nabla^2 A_4 = 0, \quad \frac{\partial^2 A_5}{\partial z^2} = -A_2, \quad \nabla^2 A_5 = 0. \tag{4.17}$$

Let

$$\Phi_1 = A_4 + zA_5, \tag{4.18}$$

where $\Phi_1(x,z)$ is a biharmonic odd function of z and satisfies

$$\frac{\partial^2 \Phi_1}{\partial x^2} = \frac{\partial^2 A_4}{\partial x^2} + z\frac{\partial^2 A_5}{\partial x^2} = -\frac{\partial^2 A_4}{\partial z^2} - z\frac{\partial^2 A_5}{\partial z^2} = A_3 + zA_2 = \sigma_z. \tag{4.19}$$

Inserting conditions (2.6) into the second equation of Eq. (2.3), we get

$$\frac{\partial \sigma_z}{\partial z} = 0 \quad \left(z = \pm\frac{h}{2}\right). \tag{4.20}$$

From Eqs. (4.19), (4.20) and conditions (2.6), we find

$$\frac{\partial^2 \Phi_1}{\partial x^2} = 0, \quad \frac{\partial^3 \Phi_1}{\partial x^2 \partial z} = 0 \quad \left(z = \pm\frac{h}{2}\right). \tag{4.21}$$

Let

$$\Phi_2 = \Phi_1 - \frac{z}{h}\left(3\Phi_1 - \frac{h}{2}\frac{\partial \Phi_1}{\partial z}\right)_{z=h/2} + \frac{4z^3}{h^3}\left(\Phi_1 - \frac{h}{2}\frac{\partial \Phi_1}{\partial z}\right)_{z=h/2}. \tag{4.22}$$

It is easy to see that Φ_2 is also an odd function of z and satisfies the following equation

$$\frac{\partial^2 \Phi_2}{\partial x^2} = \sigma_z, \tag{4.23}$$

with

$$\nabla^2\nabla^2\Phi_2 = 0, \tag{4.24}$$

$$\Phi_2 = 0, \quad \frac{\partial \Phi_2}{\partial z} = 0 \quad \left(z = \pm\frac{h}{2}\right). \tag{4.25}$$

It is notable from Eqs. (4.12), (4.20), (4.24) and (4.25) that the function Φ_2 is governed by the same equations as σ_z. Therefore, we can repeat the last process from Eq. (4.22) to Eq. (4.25) and get Φ from Φ_2, so that

$$\frac{\partial^2 \Phi}{\partial x^2} = \Phi_2. \tag{4.26}$$

The derivation can show that Φ is an odd function of z and satisfies Eqs. (4.4) and (4.5). By inserting Eq. (4.26) into Eq. (4.23), we can get Eq. (4.11). The lemma is proved. □

4.3. Proof of the Decomposition Theorem

According to Lemma 2, we can get $\Phi(x,z)$ which is a biharmonic odd function of z and satisfies Eqs. (4.4), (4.5) and (4.11), so the P-F state is obtained. Let

$$\tilde{\sigma}_{ij} = \sigma_{ij} - \sigma_{ij}^{PF}. \tag{4.27}$$

From Eqs. (4.11) and (4.27), we get

$$\tilde{\sigma}_z = \sigma_z - \sigma_z^{PF} = \sigma_z - \frac{\partial^4 \Phi}{\partial x^4} = 0. \tag{4.28}$$

Then, it is easy to see that $\tilde{\sigma}_{ij}$ satisfy Lemma 1, such that

$$\tilde{\sigma}_{ij} = \sigma_{ij}^I. \tag{4.29}$$

Namely,

$$\sigma_{ij} = \sigma_{ij}^I + \sigma_{ij}^{PF}. \tag{4.30}$$

Thus, this completes the proof of the theorem. □

5. Boundary Conditions of Elastic Beams

It is generally known that the exact solution of linear elastostatic problems for flat plates consists of an interior component significant throughout the plate and a boundary layer component. Near a cylindrical edge, the interior solution is supplemented by boundary layer solution component which becomes insignificant away from the edge. The prescribed admissible boundary conditions can be satisfied only by a combination of these two components. However, the boundary layer solution, even just a leading term approximation, needed to fit the edge-data is rather intractable except for cases with simple geometries and load symmetries. This and the fact that the solution behavior near the plate edges is often not needed from practical viewpoint have driven people to take efforts over the years to formulate the interior solution of plate problems, by assigning an appropriate portion of the prescribed edge-data to it, without any reference to the boundary layer solution.

By an application of the Betti-Rayleigh reciprocal theorem, Gregory and Wan developed a novel method determining the interior solution successfully and effectively; they provided the results for several plate problems, and derived a set of correct boundary conditions for arbitrarily prescribed admissible edge-data [33, 44-49]. From these results, they have now explicit examples showing that the higher order accuracy offered by the governing differential equations of a higher order plate theory may not be attained unless commensurate boundary conditions are developed and used for these equations. These general results also show that, to be strictly correct, Saint-Venant's principle should generally be applied only to the leading term outer solution, i.e. the classical plate theory.

Through generalizing the method developed by Gregory and Wan, a set of necessary conditions on the edge-data for the existence of a rapidly decaying solution are established by Gao and coauthors, and various extensions have been found among elastic beams [34, 50], piezoelectric beams [51], elastic plates [52], piezoelectric plates [53], 1D quasicrystal plates [54] and 2D quasicrystal plates [55].

5.1. Necessary Conditions for a Decaying States

There are no body forces and the top and bottom faces of the beam are traction-free so that the condition (2.6) is satisfied. The presence of any body or surface loads may be removed by a particular solution. Then the only forcing terms in the problem are prescribed on the end $x = 0$ in terms of stress or displacement edge-data in the form of one of the following four admissible combinations,

Case A:

$$\sigma_x(0,z) = \bar{\sigma}_x(z), \quad \sigma_{xz}(0,z) = \bar{\sigma}_{xz}(z), \tag{5.1}$$

Case B:

$$u_x(0,z) = \bar{u}_x(z), \quad \sigma_{xz}(0,z) = \bar{\sigma}_{xz}(z), \tag{5.2}$$

Case C:

$$\sigma_x(0,z)=\bar{\sigma}_x(z),\ \ u_z(0,z)=\bar{u}_z(z), \tag{5.3}$$

Case D:

$$u_x(0,z)=\bar{u}_x(z),\ \ u_z(0,z)=\bar{u}_z(z). \tag{5.4}$$

Two definitions for two equilibrium states are introduced as follows:

Definition 3: The displacement fields $\boldsymbol{u}$ and stress fields $\boldsymbol{\sigma}$ induced by the prescribed edge-data are said to give rise to a decaying state within the beam if they satisfy the condition

$$\{\boldsymbol{u},\boldsymbol{\sigma}\}=O\left(M_1e^{-\gamma d/h}\right)\ \text{as}\ h\to 0, \tag{5.5}$$

where d is the minimum distance of the observation point from the edge of the beam, M_1 are the maximum modulus of the prescribed edge-data for the decaying state, M_1 and γ are positive constants.

Definition 4: The displacement fields $\boldsymbol{u}$ and stress fields $\boldsymbol{\sigma}$ are said to be a regular state within the beam if they satisfy the condition

$$\{\boldsymbol{u},\boldsymbol{\sigma}\}=O\left(M_2h^{\alpha}\right)\ \text{as}\ h\to 0, \tag{5.6}$$

where M_2 are the maximum modulus for the regular state, $M_2>0$ and $\alpha\geq 0$. That means the fields have at most an algebraic growth as $h\to 0$.

Supposing that the stress data do give rise to the decaying state in the beam, we now apply the reciprocal theorem for a beam, which takes the form

$$\oint\!\!\!\int_S\left(\sigma_{ij}^{(1)}u_j^{(2)}-\sigma_{ij}^{(2)}u_j^{(1)}\right)n_i\mathrm{d}S=0, \tag{5.7}$$

where S is the surface of the beam which consists of two end planes and a lateral surface, n_i is the direction cosine of the outward normal to S. With the foregoing two definitions in mind, now we take the state with a superscript "(1)" to be the exact solution of beam bending problem, and the decaying state induced by the prescribed edge-data $\bar{\sigma}_x$, $\bar{\sigma}_{xz}$, $\bar{u}_x$, $\bar{u}_z$. For the auxiliary state, denoted by superscript "(2)", we take any regular state which fulfills load-free conditions on S. Similar to the derivation of necessary conditions for a decaying state in the plate, generalizing Gregory and Wan's decay analysis technique to a beam, we finally obtain the necessary conditions for a decaying state

Case A:

$$\int_{-h/2}^{h/2}\left[\bar{\sigma}_x(z)u_x^{(2)}+\bar{\sigma}_{xz}(z)u_z^{(2)}\right]_{x=0}\mathrm{d}z=0, \tag{5.8}$$

Case B:

$$\int_{-h/2}^{h/2}\left[\bar{u}_x(z)\sigma_x^{(2)}-\bar{\sigma}_{xz}(z)u_z^{(2)}\right]_{x=0}\mathrm{d}z=0, \tag{5.9}$$

Case C:

$$\int_{-h/2}^{h/2}\left[\bar{u}_z(z)\sigma_{xz}^{(2)}-\bar{\sigma}_x(z)u_x^{(2)}\right]_{x=0}\mathrm{d}z=0, \tag{5.10}$$

Case D:

$$\int_{-h/2}^{h/2}\left[\bar{u}_x(z)\sigma_x^{(2)}+\bar{u}_z(z)\sigma_{xz}^{(2)}\right]_{x=0}\mathrm{d}z=0. \tag{5.11}$$

These necessary conditions for Case A, B and C to induce only a decaying elastostatic state will be translated into the appropriate boundary conditions for the beam later in next section.

5.2. Boundary Conditions of the Beam

The main difficulty in performing the preceding process lies in obtaining suitable regular states which satisfy the appropriate boundary conditions. However, for the case of elastic beam bending, the necessary regular states can be explicitly determined as follows, at least for edge-data in Case A, B and C.

5.2.1. Case A

Our main task lies in obtaining accurate solutions for these regular states. We can take a rigid body translation in the z-direction as the first auxiliary regular state. Then $u_x^{(2)}\equiv 0$ and $u_z^{(2)}=\text{constant}$ so that Eq. (5.8) gives the first condition for the stress data

$$\int_{-h/2}^{h/2}\bar{\sigma}_{xz}\mathrm{d}z=0. \tag{5.12}$$

Now we look for the second auxiliary regular state with the use of P-N solution (2.9) for beam problem. According to the characteristics of bending problem, we assume

$$P_0=C_1\left(z^3-3x^2z\right)+C_2xz,\ \ P_1=0,\ \ P_3=C_3\left(z^2-x^2\right)+C_4x, \tag{5.13}$$

where $C_i\left(i=1,2,3,4\right)$ are unknown constants to be determined later.

As the procedure in the preceding section indicates, any candidate for regular state (2) must meet the requirements stipulated below

$$\sigma_{xz}^{(2)} = \sigma_z^{(2)} = 0 \quad \left(z = \pm\frac{h}{2}\right),$$
$$\sigma_{xz}^{(2)} = \sigma_x^{(2)} = 0 \quad (x = 0). \tag{5.14}$$

Substituting Eq. (2.9) into Eq. (2.4) together with (5.13), then the results into Eq. (5.14), we can determine the relationship among these unknown constants as

$$C_1 = C_3 = 0, \quad C_2 = \frac{1-\nu}{1+\nu} C_4, \tag{5.15}$$

which corresponds to a rigid body rotation. Inserting this auxiliary regular state (5.13) into Eq. (5.8) with the use of Eq. (5.12), after taking account of the relationship (5.15), we obtain the second necessary condition for a decaying state when $\bar{\sigma}_x$ is prescribed

$$\int_{-h/2}^{h/2} \bar{\sigma}_x z \mathrm{d}z = 0. \tag{5.16}$$

The necessary conditions (5.12) and (5.16) are conventional forms of elastic beam theories [35], although they are formulated explicitly by an application of the reciprocal theorem and P-N solution.

5.2.2. Case B

As in Case A, selecting a rigid body translation as the first auxiliary regular state, we certainly must have the corresponding necessary conditions

$$\int_{-h/2}^{h/2} \bar{\sigma}_{xz} \mathrm{d}z = 0. \tag{5.17}$$

To obtain other auxiliary regular state, the regular state (2) must meet the requirements stipulated as follows

$$\sigma_{xz}^{(2)} = \sigma_z^{(2)} = 0 \quad \left(z = \pm\frac{h}{2}\right),$$
$$\sigma_{xz}^{(2)} = 0, \quad u_x^{(2)} = 0 \quad (x = 0). \tag{5.18}$$

Taking the potential functions P_0, P_1 and P_3 in Case B to be of the same form in Case A, after using the condition (5.18) leads to

$$C_1 = \frac{1-\nu}{3(1+\nu)} C_3, \quad C_2 = C_4 = 0. \tag{5.19}$$

With the help of Eqs. (5.17) and (5.19), the second necessary condition for a decaying state is obtained from Eq. (5.9) when $\overline{u}_x$ and $\overline{\sigma}_{xz}$ are prescribed

$$\int_{-h/2}^{h/2}\left(\overline{u}_x z+\frac{\nu}{2E}\overline{\sigma}_{xz}z^2\right)\mathrm{d}z=0. \tag{5.20}$$

5.2.3. Case C

Similar to Case A, we take the regular state (2) as a rigid body rotation, then obtain

$$\int_{-h/2}^{h/2}\overline{\sigma}_x z\mathrm{d}z=0. \tag{5.21}$$

With the features of bending problem, we take the harmonic functions as

$$P_0=0,\ \ P_1=E_1\left(z^3-3x^2z\right)+E_2z,\ \ P_3=E_3\left(x^3-3xz^2\right)+E_4x. \tag{5.22}$$

where $E_i\left(i=1,2,3,4\right)$ are unknown constants yet to be determined. By noting that

$$\begin{aligned}&\sigma_{xz}^{(2)}=\sigma_z^{(2)}=0\ \ \left(z=\pm\frac{h}{2}\right),\\&\sigma_x^{(2)}=0,\ \ u_z^{(2)}=0\ \ \left(x=0\right),\end{aligned} \tag{5.23}$$

we have the relationship among these unknown constants

$$E_1=-\frac{1-\nu}{1+3\nu}E_3,\ \ E_2+E_4=-\frac{6\nu\left(1+\nu\right)}{\left(1-\nu\right)\left(1+3\nu\right)}h^2E_3. \tag{5.24}$$

On substituting the expressions (5.22) into Eq. (5.10), we obtain, after using Eqs. (5.21) and (5.24)

$$\int_{-h/2}^{h/2}\left[\overline{u}_z\left(\frac{h^2}{4}-z^2\right)+\frac{2+\nu}{3E}\overline{\sigma}_x z^3\right]\mathrm{d}z=0. \tag{5.25}$$

More importantly, the appropriate boundary conditions for beam bending with mixed edge-data in Case B and C are obtained for the first time.

5.3. Boundary Conditions of the P-F State

The combination of the two types of solution components is known to be capable of fitting any admissible set of prescribed edge conditions. Hence boundary value problems of the elastostatics of beams are solved in principle. However, the boundary conditions obtained for the interior state do not guarantee that the solution in the beam interior up to exponentially small terms. In other words, we must still prove that the necessary conditions for the P-F state are also sufficient conditions. The sufficiency proof can also offer a general method for determining the correct boundary conditions for the interior state.

For stress or mixed data at an edge $x=0$, the relevant necessary conditions for the absence of the interior state in the induced bending deformations can now be deduced from analytical solution of the decaying state. For the bending problem, it is apparent that Φ has solution of the form

$$\Phi=\left[A\sin\left(\lambda\frac{2z}{h}\right)+Bz\cos\left(\lambda\frac{2z}{h}\right)\right]\frac{h^2}{4\lambda^2}e^{-\lambda\frac{2x}{h}},\tag{5.26}$$

which defines an eigenvalue problem with λ as the positive eigenvalue parameter, and A and B are unknown constants to be determined. Thus, the stresses and displacements of the P-F state can be shown to be

$$\begin{aligned}\sigma_x^{PF}&=-\frac{4\lambda}{h}\left[A\frac{\lambda}{h}\sin\left(\lambda\frac{2z}{h}\right)+B\frac{\lambda}{h}z\cos\left(\lambda\frac{2z}{h}\right)+B\sin\left(\lambda\frac{2z}{h}\right)\right]e^{-\lambda\frac{2x}{h}},\\ \sigma_{xz}^{PF}&=\frac{4\lambda}{h}\left[A\frac{\lambda}{h}\cos\left(\lambda\frac{2z}{h}\right)-B\frac{\lambda}{h}z\sin\left(\lambda\frac{2z}{h}\right)+\frac{B}{2}\cos\left(\lambda\frac{2z}{h}\right)\right]e^{-\lambda\frac{2x}{h}},\\ \sigma_z^{PF}&=\frac{4\lambda^2}{h^2}\left[A\sin\left(\lambda\frac{2z}{h}\right)+Bz\cos\left(\lambda\frac{2z}{h}\right)\right]e^{-\lambda\frac{2x}{h}},\end{aligned}\tag{5.27}$$

$$\begin{aligned}u_x^{PF}&=\frac{2B}{E}\sin\left(\lambda\frac{2z}{h}\right)e^{-\lambda\frac{2x}{h}}-\frac{1+\nu}{E}\frac{2\lambda}{h}\left[A\sin\left(\lambda\frac{2z}{h}\right)+Bz\cos\left(\lambda\frac{2z}{h}\right)\right]e^{-\lambda\frac{2x}{h}},\\ u_z^{PF}&=\frac{2B}{E}\cos\left(\lambda\frac{2z}{h}\right)e^{-\lambda\frac{2x}{h}}-2\frac{1+\nu}{E}\left[A\frac{\lambda}{h}\cos\left(\lambda\frac{2z}{h}\right)-B\frac{\lambda}{h}z\sin\left(\lambda\frac{2z}{h}\right)+\frac{B}{2}\cos\left(\lambda\frac{2z}{h}\right)\right]e^{-\lambda\frac{2x}{h}}.\end{aligned}\tag{5.28}$$

It can find that the fields (5.27) and (5.28) are exponentially small in the interior of the beam as $h\to 0$. As the definition of the decaying state indicates, the fields (5.27) and (5.28) are decaying states in the beam.

We can determine A and B by inserting Eq. (5.26) into the load-free conditions (4.5) on the top and bottom faces of the beam which yields

$$A = \frac{1}{\sin\lambda}, \quad B = -\frac{2}{h\cos\lambda} \tag{5.29}$$

provided that λ is any root of

$$\sin(2\lambda) = 2\lambda. \tag{5.30}$$

For these decaying states, we can easily testify by direct substitution of Eqs. (5.27) and (5.29) that the necessary conditions (5.12) and (5.16) for elastic beams are right because of Eq. (5.30). Now we verify whether the conditions (5.20) and (5.25) are right or not. From the expressions (5.27) and (5.28) for the stresses and displacements and with the help of Eq. (5.29), we integrate the resulting expressions across the beam thickness. Thus it follows after a little algebra that

$$\frac{2E}{h}\int_{-h/2}^{h/2} \bar{u}_x z\mathrm{d}z = -2(1+\nu)\tan\lambda - 2(1+\nu)\cot\lambda + 2(1-\nu)\frac{1}{\lambda} + 4\nu\frac{1}{\lambda^2}\tan\lambda, \tag{5.31}$$

$$\frac{4E}{h^2}\int_{-h/2}^{h/2} \bar{u}_z\left(\frac{h^2}{4} - z^2\right)\mathrm{d}z = 4(1+\nu)\frac{1}{\lambda}\tan\lambda + 4(1+\nu)\frac{1}{\lambda}\cot\lambda + 4(3+\nu)\frac{1}{\lambda^2} - 8(2+\nu)\frac{1}{\lambda^3}\tan\lambda, \tag{5.32}$$

$$\frac{2}{h}\int_{-h/2}^{h/2} \bar{\sigma}_{xz} z^2\mathrm{d}z = 4\tan\lambda + 4\cot\lambda + 4\frac{1}{\lambda} - 8\frac{1}{\lambda^2}\tan\lambda, \tag{5.33}$$

$$\frac{4}{h^2}\int_{-h/2}^{h/2} \bar{\sigma}_x z^3\mathrm{d}z = 2\lambda\tan\lambda + 2\lambda\cot\lambda - 2 - 12\frac{1}{\lambda}\tan\lambda - 12\frac{1}{\lambda}\cot\lambda - 12\frac{1}{\lambda^2} + 24\frac{1}{\lambda^3}\tan\lambda. \tag{5.34}$$

After using Eq. (5.30), the expressions (5.31)-(5.34) together show that the decaying state conditions (5.20) and (5.25) are indeed satisfied by the corresponding decaying stress and displacement fields (5.27) and (5.28). Therefore, the stress and mixed edge-data (5.1)-(5.3) must satisfy the appropriate boundary conditions (5.12), (5.16), (5.20) and (5.25) for elastic beams, if the data give rise only to the P-F state. In other words, they are in fact necessary and sufficient conditions for the elastostatic state induced by the edge-data to be a decaying state, possibly up to a rigid body displacement field.

5.4. Boundary Conditions of the Interior State

These necessary conditions for the P-F state (decaying state) can then be converted into a set of boundary conditions appropriate for the interior state or its various slender and deep beam theory approximations, which do not involve the P-F state components. From the decomposition theorem (4.6), the difference between the exact solution and the interior one is a decaying state. As an immediate consequence, $\bar{u}_i$ and $\bar{\sigma}_{ij}$ must satisfy the conditions

$$\bar{u}_x = \left[u_x - u_x^I\right]_{x=0}, \quad \bar{u}_z = \left[u_z - u_z^I\right]_{x=0}, \tag{5.35}$$

$$\bar{\sigma}_x = \left[\sigma_x - \sigma_x^I\right]_{x=0}, \quad \bar{\sigma}_{xz} = \left[\sigma_{xz} - \sigma_{xz}^I\right]_{x=0}. \tag{5.36}$$

Therefore, the above necessary and sufficient conditions apply to this difference evaluated at an edge of beams. For the stress and mixed edge-data (5.1)-(5.3) to induce only the interior state, the data must satisfy the four conditions

$$\int_{-h/2}^{h/2} \hat{\sigma}_{xz}\mathrm{d}z = \int_{-h/2}^{h/2} \left[\sigma_{xz}^I\right]_{x=0} \mathrm{d}z, \tag{5.37}$$

$$\int_{-h/2}^{h/2} \hat{\sigma}_x z\mathrm{d}z = \int_{-h/2}^{h/2} \left[\sigma_x^I z\right]_{x=0} \mathrm{d}z, \tag{5.38}$$

$$\int_{-h/2}^{h/2} \left[\hat{u}_x z + \frac{\nu}{2E}\hat{\sigma}_{xz} z^2\right]\mathrm{d}z = \int_{-h/2}^{h/2} \left[u_x^I z + \frac{\nu}{2E}\sigma_{xz}^I z^2\right]_{x=0} \mathrm{d}z, \tag{5.39}$$

$$\int_{-h/2}^{h/2} \left[\hat{u}_z\left(\frac{h^2}{4} - z^2\right) + \frac{2+\nu}{3E}\hat{\sigma}_x z^3\right]\mathrm{d}z = \int_{-h/2}^{h/2} \left[u_z^I\left(\frac{h^2}{4} - z^2\right) + \frac{2+\nu}{3E}\sigma_x^I z^3\right]_{x=0} \mathrm{d}z. \tag{5.40}$$

where $\hat{u}_x$, $\hat{u}_z$, $\hat{\sigma}_x$ and $\hat{\sigma}_{xz}$ are the actually prescribed edge-data. In the form, these conditions also conveniently provide the appropriate boundary conditions for slender or deep beam theories.

The interior solutions σ_x^I and σ_{xz}^I have been taken from Eq. (4.1). Since the decomposition theorem and the refined theory of beams are equivalent [43], u_x^I and u_z^I can be obtained from Eq. (3.22), such as

$$\begin{aligned} u_x^I &= -zw' - (1+\nu)\frac{h^2}{4}zw''' + \frac{2+\nu}{6}z^3 w''', \\ u_z^I &= w + \frac{\nu}{2}z^2 w''. \end{aligned} \tag{5.41}$$

After carrying out the integration involved, imposition of Eqs. (4.1) and (5.41) on the conditions (5.37)-(5.40) allows us to determine the following four boundary condtions for the interior state

$$\int_{-h/2}^{h/2} \hat{\sigma}_{xz} \mathrm{d}z = -EI\left[w'''\right]_{x=0}, \tag{5.42}$$

$$\int_{-h/2}^{h/2} \hat{\sigma}_{x} z \mathrm{d}z = -EI\left[w''\right]_{x=0}. \tag{5.43}$$

$$\int_{-h/2}^{h/2} \left(\hat{u}_x z + \frac{\nu}{2E} \hat{\sigma}_{xz} z^2 \right) \mathrm{d}z = -I\left[w' + \frac{4+5\nu}{20} h^2 w''' \right]_{x=0}. \tag{5.44}$$

$$\int_{-h/2}^{h/2} \left[\hat{u}_z \left(\frac{h^2}{4} - z^2 \right) + \frac{2+\nu}{3E} \hat{\sigma}_x z^3 \right] \mathrm{d}z = 2I\left[w - \frac{1}{20} h^2 w'' \right]_{x=0}. \tag{5.45}$$

By superposition, the boundary conditions for the interior state in Cases A-C are formed. To obtain the appropriate edge conditions for a particular (slender or deep) beam theory, we should expand in powers of h all terms in all four necessary conditions deduced in this section for each type of edge-data and retain only a suitable number of terms in each expansion. The above results for Cases A-C of bending beams illustrate the general method for deriving local necessary conditions for a decaying state and therewith the appropriate boundary conditions for beam theories.

Since w must satisfy the fourth-order differential governing equations (4.2), it is evident that conditions of the type (5.42) and (5.43) for Case A or (5.42) and (5.44) for Case B or (5.43) and (5.45) for Case C completely determine the deflection of the neutral surface w, hence the interior state of beams. In effect, they constitute a procedure for assigning an appropriate portion of the edge-data to the interior solution to ensure its correct deformation. Thus, in these cases, the necessary conditions for the P-F state are sufficient for the determination of the interior state.

6. Conclusion

The refined theory of the beam is derived by using P-N solution and Lur'e method without ad hoc assumptions. It is shown that the displacements and stresses of the beam can be represented by the angle of rotation and the deflection of the neutral surface. Based on the refined beam theory, the exact equations for the beam with free faces consist of two governing differential equations: the fourth-order equation and the transcendental equation. The approximate equations for the beam under transverse loadings are accurate up to the second-order terms with respect to beam thickness. In five examples, it is shown that the new theory provides better results than Levinson's beam theory when compared to those obtained from the linear theory of elasticity.

The decomposition theorem of the beam with free faces is obtained. It is shown that the stress state of the beam can be decomposed into two parts: the interior state and the P-F state. In order to provide a concise but strict proof of the decomposition theorem, two basic lemmas are presented as well as their proofs beforehand. Instead of the P-F eigenfunction expansion of biharmonic functions, only basic mathematical techniques are involved in the derivation. In fact, these two states correspond to the two equations derived in the refined theory of the beam: the fourth-order equation and the transcendental equation, respectively. Therefore, this work virtually proves the consistency between the decomposition theorem and the refined theory of the beam.

To obtain the appropriate stress and mixed boundary conditions accurate to all order, the reciprocal theorem and P-N solution are applied in a novel way. Through generalizing the method proposed by Gregory and Wan, a set of necessary conditions on the edge-data for the existence of a rapidly decaying solution are established. When stress and mixed conditions are imposed on the beam edge, these decaying state conditions for the case of bending deformation are derived explicitly. For the stress data, our boundary conditions coincide with those obtained in conventional forms of beam theories. It is important to note that the appropriate boundary conditions with two different sets of mixed edge-data are obtained for the first time. Furthermore, an analytical solution of the decaying state is formulated to verify validity of our boundary conditions. The results in turn show that the necessary conditions for the P–F state are also sufficient conditions. The boundary conditions obtained for the interior state show that the interior solution determined by these conditions is the correct solution in the beam interior up to exponentially small terms.

In the case of extension or compression of a beam, the beam is subjected only to symmetrical loadings and edge conditions. With the first part of this research focusing on the bending of a beam, the similar discussion about the extension or compression of a beam can be made by using the methods developed in this chapter.

APPENDIX A

Next we will prove that when P_0 is defined according to Eq. (3.3), the general solutions (2.9) is complete without loss in generality.

First, from the nonuniqueness of P-N solution [56], $\boldsymbol{P}(x,z)$ and $P_0(x,z)$ in Eq. (2.9) can be changed to $\tilde{\boldsymbol{P}}(x,z)$ and $\tilde{P}_0(x,z)$, respectively, there are

$$\tilde{\boldsymbol{P}} = \boldsymbol{P} + \nabla E,\ \ \tilde{P}_0 = P_0 + \frac{4}{1+\nu}E - \boldsymbol{r}\cdot\nabla E, \tag{A.1}$$

where $\nabla = \boldsymbol{i}\,\partial/\partial x + \boldsymbol{k}\,\partial/\partial z$, $\boldsymbol{r} = (x,z)$, $\boldsymbol{P} = (P_1, P_3)$, in which $\boldsymbol{P}(x,z)$ has the form of expressions (3.1). In the case of bending of a beam, one obtains the following symbolic solution of Eq. (2.10):

$$P_0 = \frac{\sin(z\partial_x)}{\partial_x} g_0(x), \tag{A.2}$$

and $E(x,z)$ is also a harmonic odd function of z. Therefore, we can set

$$E = \frac{\sin(z\partial_x)}{\partial_x} e(x). \tag{A.3}$$

Now we come to prove that it is always possible to choose a function $e(x)$ in Eq. (A.3) so that Eq. (3.3), i.e.,

$$\tilde{P}_0 + \boldsymbol{r} \cdot \tilde{\boldsymbol{P}} = -z\cos(z\partial_x)\tilde{f} \tag{A.4}$$

may hold, in which

$$\tilde{f} = \frac{(g_1 + \partial_x a)}{\partial_x} - (g_3 + e) = f. \tag{A.5}$$

Substituting Eqs. (A.1) and (A.5) into Eq. (A.4), we get the following expression:

$$P_0 + \frac{4}{1+\nu} E + \boldsymbol{r} \cdot \boldsymbol{P} = -z\cos(z\partial_x) f. \tag{A.6}$$

Then inserting Eqs. (A.2), (3.1) and (A.5) into expression (A.6), it will be seen that

$$E = -\frac{1+\nu}{4}\left[\frac{\sin(z\partial_x)}{\partial_x} g_0 + x\frac{\sin(z\partial_x)}{\partial_x} g_1 + z\frac{\cos(z\partial_x)}{\partial_x} g_1\right]. \tag{A.7}$$

Next the expression of $e(x)$ in Eq. (A.3) is to be given. Using the following identity

$$\partial_x^{2n-2}\left(x\partial_x{}^2 g_1\right) = x\partial_x^{2n} g_1 + (2n-2)\left(\partial_x^{2n-2} g_1\right)_{,x}. \tag{A.8}$$

After tedious manipulation by using Eq. (A.8) and Taylor series of the trigonometric functions, the result turns out to be

$$\frac{\sin(z\partial_x)}{\partial_x}\left(x\partial_x^2 g_1 + 3\partial_x g_1\right) = \partial_x^2\left[x\frac{\sin(z\partial_x)}{\partial_x} g_1 + z\frac{\cos(z\partial_x)}{\partial_x} g_1\right]. \tag{A.9}$$

Substituting expression (A.9) into expression (A.7), we get

$$E = -\frac{1+\nu}{4}\frac{\sin(z\partial_x)}{\partial_x}\left[g_0 + \frac{1}{\partial_x^2}\left(x\partial_x^2 g_1 + 3\partial_x g_1\right)\right]. \tag{A.10}$$

From expression (A.10) we know Eq. (A.4) holds when

$$e = -\frac{1+\nu}{4}\left[g_0 + \frac{1}{\partial_x^2}\left(x\partial_x^2 g_1 + 3\partial_x g_1\right)\right] \tag{A.11}$$

in Eq. (A.3).

For convenience $\tilde{\boldsymbol{P}}$ and $\tilde{P}_0$ will still be written as $\boldsymbol{P}$ and P_0, respectively. Thus Eq. (3.3) holds. Consequently, if P_0 is taken according to Eq. (3.3), the general solutions (2.9) losses no generality.

APPENDIX B

Next, we will give and prove a lemma and a theorem.

B.1. Lemma 3

Supposing that $H(x)$ satisfies

$$\frac{1}{\partial_x^2}\left[1 - \frac{\sin(h\partial_x)}{h\partial_x}\right]H = 0, \tag{B.1}$$

then there exists function $D(x)$ which satisfies the following two equations

$$\partial_x^4 D = H, \quad \frac{1}{\partial_x^2}\left[1 - \frac{\sin(h\partial_x)}{h\partial_x}\right]B = 0. \tag{B.2}$$

Proof: Assume function $C(x)$, which satisfies the following equation, can be found

$$\partial_x^2 C = H. \tag{B.3}$$

We can obtain the following equation

$$\partial_x^2 \frac{1}{\partial_x^2}\left[1-\frac{\sin(h\partial_x)}{h\partial_x}\right]C=\frac{1}{\partial_x^2}\left[1-\frac{\sin(h\partial_x)}{h\partial_x}\right]H=0. \tag{B.4}$$

Set

$$B_1=C-\frac{6}{h^2}\frac{1}{\partial_x^2}\left[1-\frac{\sin(h\partial_x)}{h\partial_x}\right]C. \tag{B.5}$$

Using Eqs. (B.3) and (B.4), we can get

$$\partial_x^2 B_1=\partial_x^2 C=H. \tag{B.6}$$

After tedious manipulation by using Eqs. (B.4) and (B.5), the result turns out to be

$$\frac{1}{\partial_x^2}\left[1-\frac{\sin(h\partial_x)}{h\partial_x}\right]B_1=0. \tag{B.7}$$

Because $B_1(x)$ and $H(x)$ satisfy the same equation, so $B_1(x)$ can be used instead of $H(x)$. Repeating Eqs. (B.3)-(B.7), we can obtain $D(x)$ as

$$\partial_x^2 D=B_1,\quad \frac{1}{\partial_x^2}\left[1-\frac{\sin(h\partial_x)}{h\partial_x}\right]D=0. \tag{B.8}$$

From Eqs. (B.6) and (B.8), it is not difficult to verify that $D(x)$ satisfies Eq. (B.2). So the proof of the lemma is finished.

B.2. The Theorem

Supposing that ξ satisfies the following equation

$$\left\{\frac{1}{\partial_x^2}\left[1-\frac{\sin(h\partial_x)}{h\partial_x}\right]\right\}\partial_x^4\xi=0, \tag{B.9}$$

then there exist such ξ^F and ξ^T that

$$\xi=\xi^F+\xi^T, \tag{B.10}$$

where ξ^F and ξ^T satisfying the following two equations

$$\partial_x^4 \xi^F = 0, \quad \frac{1}{\partial_x^2}\left[1 - \frac{\sin(h\partial_x)}{h\partial_x}\right]\xi^T = 0. \tag{B.11}$$

Proof: Let

$$F = \partial_x^4 \xi, \tag{B.12}$$

then

$$\frac{1}{\partial_x^2}\left[1 - \frac{\sin(h\partial_x)}{h\partial_x}\right]F = 0. \tag{B.13}$$

According to Lemma 3, there exists ξ^T satisfying

$$\partial_x^4 \xi^T = F, \quad \frac{1}{\partial_x^2}\left[1 - \frac{\sin(h\partial_x)}{h\partial_x}\right]\xi^T = 0. \tag{B.14}$$

From Eq. (B.12) and the first equation of Eq. (B.14), we get

$$\partial_x^4 \xi^T = F = \partial_x^4 \xi, \tag{B.15}$$

namely,

$$\partial_x^4 \left(\xi - \xi^T\right) = 0. \tag{B.16}$$

Let

$$\xi^F = \xi - \xi^T. \tag{B.17}$$

Eq. (B.16) becomes the first equation of Eq. (B.11), so there are functions ξ^F and ξ^T which satisfy Eq. (B.11). This completes the proof of the theorem.

ACKNOWLEDGEMENTS

The work is supported by the National Natural Science Foundation of China (Nos. 10702077, 10602001 and 10672001) and the National Natural Science Foundation of Beijing (No. 1083012).

REFERENCES

[1] Timoshenko, S. P., On the correction for shear of the differential equation for transverse vibration of prismatic bars, *Phi. Mag.* 41, 744–746 (1921).

[2] Levinson, M., A new rectangular beam theory, J. Sound Vib. 74, 81–87 (1981).

[3] Bickford, W. B., A consistent higher order beam theory, In *Developments in Theoretical and Applied Mechanics,* 11, 137–150 (1982).

[4] Reddy, J. N., *Energy and Variational Methods in Applied Mechanics*, John Wiley, New York (1984).

[5] Tutek, Z., Aganović, I., A justification of the one-dimensional linear model of elastic beam, *Math. Meth. Appl. Sci.* 8, 502–515 (1986).

[6] Lewiński, T., On the twelfth-order theory of elastic plates, *Mech. Res. Commun.* 17, 375–382 (1990).

[7] Fan, H., Widera, G. E. O., Refined engineering beam theory based on the asymptotic expansion approach, *AIAA J.* 29, 444–449 (1991).

[8] Tullini, N., Savoia, M., Elasticity interior solution for orthotropic strips and the accuracy of beam theories, *ASME J. Appl. Mech.* 66, 368–373 (1999).

[9] Gol'denviezer, A. L., Methods for justifying and refining the theory of shells, J. Appl. Math Mech. 32, 704–718 (1968).

[10] Kil'chevskiy, N. A., *Fundamentals of the Analytical Mechanics of Shells*, NASA TTF-292, Washington (1965).

[11] Donnell, L. H., *Beams, Plates and Shells*, McGraw-Hill, New York (1976).

[12] Vlasov, V. Z., Leont'ev, U. N., *Beams, Plates and Shells on Elastic Foundations*, Translated form Russian by Barouch, A., and edited by Pelz, T., Israel Program for Scientific Translations Ltd., Jerusalem (1960).

[13] Sayir, M., Mitropoulos, C., On elementary theories of linear elastic beams, plates and shells, *ZAMP* 31, 1–55 (1980).

[14] Papkovitch, P. F., Solution générale des équations différentielles fondamentales d'élasticité, exprimées par trois fonctions harmoniques, *C. R. Acad. Sci. Paris* 195, 513–515 (1932).

[15] Neuber, H., *Kerbspannungslehre*, J. W. Edwards, Ann Arbor (1944).

[16] Cheng, S., Elasticity theory of plates and a refined theory, *ASME J. Appl. Mech.* 46, 644–650 (1979).

[17] Lur'e, A. I., *Three-Dimensional Problems of the Theory of Elasticity*, Interscience, New York (1964).

[18] Zhao, B. S., Wang, M. Z., The equivalence of the refined theory and the decomposed theorem of an elastic plate, *Appl. Math. Mech.* 26, 486–494 (2005).

[19] Barrett, K. E., Ellis, S., An exact theory of elastic plates, *Int. J. Solids Struct*. 24, 859–880 (1988).

[20] Wang, W., Shi, M. X., Thick plate theory based on general solutions of elasticity, *Acta Mech.* 123, 27–36 (1997).

[21] Wang, F. Y., Two-dimensional theories deduced from three-dimensional theory for a transversely isotropic body-I. Plate problems, *Int. J. Solids Struct.* 26, 445–470 (1990).

[22] Wang, F. Y., Two-dimensional theories deduced from three-dimensional theory for a transversely isotropic body-II. Plane problems, *Int. J. Solids Struct.* 28, 161–177 (1991).

[23] Xu, S. P., Wang, W., A refined theory of transversely isotropic piezoelectric plates, *Acta Mech.* 171, 15–27 (2004).

[24] Gao, Y., Zhao, B. S., The refined theory of thermoelastic rectangular plates, *J. Thermal Stresses* 30, 505–520 (2007).

[25] Gao, Y., Zhao, B.S., The refined theory of thermoelastic plane problems, *J. Thermal Stresses* (accepted).

[26] Gao, Y., Wang, M.Z., A refined beam theory based on the refined plate theory, *Acta Mech.* 177, 191–197 (2005).

[27] Gao, Y., Wang, M. Z., The refined theory of deep rectangular beams based on general solutions of elasticity, *Sci. China Ser. G* 49, 291–303 (2006).

[28] Gao, Y., Xu, B. X., Zhao, B. S., The refined theory of beams for a transversely isotropic body, *Acta Mech.* 191, 109–122 (2007).

[29] Gao, Y., Wang, M. Z., The refined theory of magnetoelastic rectangular beams, *Acta Mech.* 173, 147–161 (2004).

[30] Gao, Y., Wang, M. Z., The refined theory of transversely isotropic piezoelectric rectangular beams, *Sci. China Ser.* G 49, 473–486 (2006).

[31] Gao, Y., Wang, M. Z., A refined theory of thermoelastic beams under steady temperature, *Eng. Mech.* 23, 34–40 (2006) (in Chinese).

[32] Gao, Y., Wang, M. Z., Zhao, B. S., The refined theory of rectangular curved beams, *Acta Mech.* 189, 141–150 (2007).

[33] Gregory, R. D., Wan, F. Y. M., On plate theories and Saint-Venant's principle, *Int. J. Solids Struct.* 21, 1005–1024 (1985).

[34] Gao, Y., Xu, S. P., Zhao, B. S., Boundary conditions for elastic beam bending, *C. R. Mecanique* 335, 1–6 (2007).

[35] Timoshenko, S. P., Goodier, J. C., *Theory of Elasticity*, McGraw-Hill, New York (1970).

[36] Gregory, R. D., The semi-infinite strip $x \geq 0, -1 \leq y \leq 1$; completeness of the Papkovich-Fadle eigenfunctions when $\phi_{xx}(0,y)$, $\phi_{yy}(0,y)$ are prescribed, *J. Elast.* 10, 57–80 (1980).

[37] Gregory, R. D., The traction boundary value problems for the elastostatic semi-infinite strip; existence of solution, and completeness of the Papkovich-Fadle eigenfunctions, *J. Elast.* 10, 295–327 (1980).

[38] Gregory, R. D., The general form of the three-dimensional elastic field inside an isotropic plate with free faces, *J. Elast.* 28, 1–28 (1992).

[39] Wang, M.Z., Zhao, B.S., The decomposed form of the three-dimensional elastic plate, *Acta Mech.* 166, 207–216 (2003).

[40] Zhao, B. S., Wang, M. Z., The decomposed theorem of the transversely isotropic elastic plate, *Acta Mech. Sinica* 36, 57–63 (2004) (in Chinese).
[41] Gao, Y., Wang M. Z., The decomposed form of rectangular beams, *Eng. Mech.* 22 (Supplement), 58–61 (2005) (in Chinese).
[42] Gao, Y., Xu, B. X., Zhao, B. S., The decomposed form of magnetoelastic beams with free faces, *Acta Mech.* (in press).
[43] Gao Y., Wang M. Z., The equivalence of the refined theory and the decomposed theorem of rectangular beams, *Appl. Math. Modell.* 31, 551–563 (2007).
[44] Gregory, R. D., Wan, F. Y. M., Decaying states of plane strain in a semi-infinite strip and boundary conditions for plate theory. *J. Elast.* 14, 27–64 (1984).
[45] Gregory, R. D., Wan, F. Y. M., Edge effect in the stretching of plates, in: *Local Effects in the Analysis of Structures,* Ladeveze, P. (ed.), Elsevier Science Publishers, Amesterdam, 35–54 (1985).
[46] Gregory, R. D., Wan, F. Y. M., On the interior solution for linearly elastic plate, *ASME J. Appl. Mech.* 55, 551–559 (1988).
[47] Lin, Y. H., Wan, F. Y. M., First integrals and the interior solution for orthotropic plates in plane strain or axisymmetric deformations, *Stud. Appl. Math.* 79, 93–125 (1988).
[48] Wan, F. Y. M., Stress boundary conditions for plate bending, *Int. J. Solids Struct.* 40, 4107–4123 (2003).
[49] Wan, F. Y. M, Author's response to Comment by Yang Gao and Min-Zhong Wang on "Stress boundary conditions for plate bending" by F.Y.M. Wan [Int. J. Solids Struct. 40 (2003) 4107–4123], *Int. J. Solids Struct.* 46, 1856 (2006).
[50] Gao Y., Zhao B. S., Xu B. X., The decomposed form and boundary conditions of elastic beams with free faces, *Acta Mech.* (in press).
[51] Gao Y., Xu S. P., Zhao B. S., Boundary conditions for the bending of a piezoelectric beam, *Sci. China Ser. G* (accepted).
[52] Gao, Y., Wang, M. Z., Comment on "Stress boundary conditions for plate bending" by F.Y.M. Wan [Int. J. Solids Struct. 40 (2003) 4107–4123], *Int. J. Solids Struct.* 46, 1854–1855 (2006).
[53] Xu, S. P., Gao, Y., Wang, W., On boundary conditions for transversely isotropic piezoelectric plates, *Mech. Res. Commun.* (in press).
[54] Gao, Y., Xu, S.P., Zhao, B.S., Boundary conditions for plate bending in one-dimensional hexagonal quasicrystals, *J. Elast.* 86, 221–233 (2007).
[55] Gao, Y., Xu, S. P., Zhao, B. S., Stress and mixed boundary conditions for two-dimensional dodecagonal quasi-crystal plates, *Pramana-J. Phys.* 68, 803–817 (2007).
[56] Eubanks, R. A., Sternberg, E., On the completeness of the Boussinesq-Papkovich stress functions, *J. Rat. Mech. Anal.* 5, 735–746 (1956).

In: Applied Mathematical Modeling
Editor: Eetu N. Virtanen, pp. 221-245
ISBN: 978-1-60021-975-7

Chapter 5

A RETRIAL QUEUEING MODEL WITH A SECOND OPTIONAL SERVICE AND WITH BERNOULLI VACATION SCHEDULE

Gautam Choudhury*

Mathematical Sciences Division
Institute of Advanced Study in Science and Technology
Paschim Boragaon, Guwahati-781035, Assam, India

ABSTRACT

The single server queues with repeated attempts have been extended to include several types of extensions and generalizations, to which attention has been paid by several researchers. The pioneering studies of retrial queues present the concept of retrial time as an alternative to classical models of telephone systems. The interest in such types of models have been further enhanced in recent years because of their theoretical structures as well as their applications in many practical situations such as telephone switching systems, telecommunication networks and computer compacting to gain service from a central processing unit. However, our intention in this paper is to look into a mathematical model, that includes many interesting features such as second optional service, Bernoulli schedule vacation and repeated attempts. An extensive stationary analysis of the model has been carried out including discrete time process, continuous time process and calculation of some important performance measures.

1. INRODUCTION

The vast literature of queueing theory abounds in results of considerable theoretical elegance and significance. Nevertheless, it is felt by users of various results in different

* E.mail: choudhuryg@ yahoo.com

situations and contexts, that the theory has, to a large extent, still remained behind the control of service process or service discipline or arrival process. Probabilistic service discipline is a simple method to make a service discipline controllable. A common property of probabilistic disciplines is that each of them assigns a set of user controllable parameters to queue in the system. The parameter assigned to a queue determines the extent to which the queue is served when it is polled by the server. The main advantage of the probabilistic disciplines is that the assigned parameters can be adjusted based on network feedback. By adjusting these parameters carefully, various performance criteria can be satisfied and thus the system performance may be optimized, see for example Ghatir and Silio (1993) and Jiang et.al (2002).

Because of their wide applications, especially in computer networks, several probabilistic disciplines have been proposed and studied to date of literature: Cyclic polling systems with Bernoulli schedule by Keilson and Servi (1986), the Binomial- gated service discipline for cyclic systems by Levi (1991), the Pi – persistent protocol for multi-access communications over unidirectional broadcast bus networks by Mukherjee and Meditch (1988), the priority queue with Bernoulli schedule by Katayama and Takahashi (1992) and second optional service for two phases of service by Madan (2000).

Again many queueing situations have the feature that customer finding the service area busy upon arrival must leave it temporarily and join a group of unsatisfied customers, but they repeat their request after some random amount of time, called retrial time. Between successive trials a customer is said to be in orbit. This type of situation arises in many telecommunication systems, where an arriving call is not allowed to await the termination of a busy signal. The renewed interest on this topic is mainly explained by the advances in telecommunication technology leading to the use of new facilities such as "repeat last number", "ring back when free" etc. For a systematic account of fundamental methods and main results on this topic, we refer the readers to the book written by Falin and Templeton (1987) and Surveys written by Falin (1986), Yang and Templeton (1987) and Kulkarni and Liang (1997). Also, see Artalejo [1999 (a,b)] for bibliographical information.

However, our interest in this paper is to investigate some short of a more generalized model that includes many interesting features such as Bernoulli vacation schedule, second optional service and repeated attempts. The researchers working in this areas have looked at the problem from different angles, depending on applications on hand. This has given rise to a vast and rich literature, from which we list bellow some of the features will be considered in this paper.

1.1. Bernoulli Schedule

The classical vacation scheme with Bernoulli schedule discipline was introduced by Keilson and Servi in 1986 and extended in a few followup papers, Keilson and Servi (1987, 1989), Ramaswami and Servi (1981) and Servi (1986) in which after each service comletion the server takes a vacation with probability p, and with probability $q = 1 - p$ it remains with primary queue to serve the next customer. Kella (1990) suggested a generalized Bernoulli scheme according to which a single server goes on k-consecutive vacations with probability p_i, if the queue upon his return is empty. Infact, various as[pects of Bernoulli vacation models have been investigated recently by a number of authors. We list them bellow.

(a) *Model with Specific distribution:* Madan and Saleh [2001 (a-c)] have been studied as single server queueing model with exponential service and deterministic vacation in [a], deterministic service with exponential vacation in [b] and deterministic service with deterministic vacations in [c], assuming Bernoulli schedule. That is, if an service completion the queue is not empty, the server goes on vacation with probability p $(p > 0)$ and resume service with complementary probability $q = 1 - p$.

(b) *Model with Exhaustive service:* Madan and Abu-Al- Rub [2004 (a)] have studied an $M / D / 1$ queue with optional server vacations based on exhaustive service. That is, the server must complete service of all customers present in the system before exercising his option whether to take a vacation or not. They also assume that whenever the server takes a vacation, it is always a single vacation. Shin and Pearce (1988) treat a single –server vacation model with queue length dependent vacation schedule, which also falls under this category.

(c) *Model with two types of Service:* Madan et.al (2005) analyzed a batch arrival queueing system, where the server providing two kinds of general heterogeneous service. Just before his service starts, a customer may choose one of the services and as soon as a service (of any kind) gets completed, he may take a vacation or continue to stay in the system. They also assume that whenever the server takes a vacation, it is always a single vacation. Anabosi and Madan (2003) studied an $M / M / 1$ queue with optional type service under Bernoulli schedule, which also falls under this category.

(d) *Model with two Servers:* Madan et.al [2003(a)] have studied a two server queue with Bernoulli schedule and a single vacation policy model. They assumed Poisson arrivals waiting in a single queue and two parallel servers who provide constant service to customers on first come first served basis. They consider two models, in one they assume after completion of a service both servers can take a vacation while in the other they assume that only one may take a vacation. The vacation periods of the servers in both models are assume to be exponential. In a subsequent paper Madan et.al [2003(b)] extended this system to the case of exponential service times.

(e) *Model with two queues:* Katayama and Takajashi (1992) analyzed a two queue model where one of the queue follows the 1 – limited regime while the others follows a Bernoulli schedule according to which the server serves the next available job in the queue with probability p $(p > 0)$ or switches to the other queue with complementary probability $q = 1 - p$. The case where both queues are served according to the Bernoulli schedule was investigated by Feny et.al (1998). Another paper in this category is the work of Lee (1997).

(f) *Model with two Phases of Service:* Choudhury and Madan (2004) and Choudhury and Paul (2004 (a)) investigated a batch arrival queueing model, where the server provides two successive phases of heterogenous service to the arriving batches under Bernoulli vacation schedule i.e. after completion of two successive phases of service, the server may take a vacation or decided to stay in the system to provide service to the next unit, if any. This generalized the model studied by Madan [2000 (b), 2001] and Madan and Choudhury (2005).

(g) *Multiple Vacation Model:* Choudhury and Madan (2006) considered a batch arrival queueing model with Bernoulli schedule vacation, where after completion of first vacation or service completion, the server takes sequence of vacations till it finds some units after returning from each vacation. This generalized model considered by Servi (1986) and Ramaswami and Sevi (1988). In subsequent papers Choudhury and Paul [2006 (a)] and Choudhury et.al. (2007) generalized the results for the case of two phase service systems. Another paper in this category is the work of Li et.al.(1997).

(h) *N-Policy Model:* Choudhury and Madan (2005) investigated such a batch arrival queueing model with two phases of service, where the server is turned on when total number of units in the system reaches the threshold value 'N' and turned off when the system becomes empty under Bernoulli vacation schedule. Tadj et.al. [2006 (a,b)] generalized this type of model to the case of batch service system with a random setup time, which corresponds to the preparatory work of the server before server starting actual service.

(i) *Restricted Admissibility Policy Model:* Madan and Choudhury [2004(a)] introduced a new class of Bernoulli vacation model, where all arriving batches are not allowed to join the system at all time. Unlike usual batch arrival queueing system, the restricted admissibility policy differs during a busy period and a vacation period and hence all arriving batches are not allowed to join the system at all times. This extends the model studied in Madan and Abu-dayyeb (2002). In a subsequent paper Madan and Choudhury [2004(b)] extended the result for two phase service system. More recently, Choudhury [2008(a)], Choudhury and Madan (2007) and Madan and Choudhury (2006) generalized models considered in [2004(a,b)] by introducing the concept of a random setup time.

(j) *Retrial Model:* Krishna Kumar and Arivudainambi [2002(b)] were concerned with the analysis of an $M/G/1$ retrial queue with Bernoulli schedules and general retrial times. In a subsequent paper Krishna Kumar et.al. [2002(a)] generalized this model to the case of batch arrival queueing system. More recently, Choudhury [2007] investigates a batch arrival queueing system with two phases of heterogeneous service and Bernoulli schedule, where he introduce the concept of a control to admission mechanism in the form of Bernoulli admission mechanism. Atencia and Moreno (2005) investigated an $M/G/1$ retrial queue with generalized retrial time where the blocked customers either with probability p $(p>0)$ joins the infinite waiting space or with complementary probability $q(=1-p)$ leaves the service area and enter the retrial group. Another paper is in this category is the work of Atencia et.al. (2006).

(k) *Discrete Queueing Model:* Li and Tang (1998) studied the discrete time $Geo/G/1$ retrial queueing model with Bernoulli schedule in which the blocked customers either join the infinite waiting space with probality p $(p>0)$ or leaves the system and join the orbit with complementary probability $q = 1 - p$ .Choi and Park (1990) studied a continuous time version of this model.

(i) *Modified Bernoulli vacation Model:* Madan and Abu Al-Rub [2004(b)] analyzed a single server queue with modified Bernoulli server vacations based on exhaustive service. Unlike other Bernoulli vacation models, they assume that it is only at completion of the service of the system that the server has the option to take a vacation or to remain in the system waiting for next customer to arrive. The service times of the customers have been assumed to be exponential and vacations are phase type exponential.

Next, we will talk at length and survey as comprehensively as possible about another probabilistic service discipline, the second optional service models.

1.2. Second Optional Service

Recently, there have been several contributions considering Poisson input queueing system, in which the server may provide a second phase of service. Madan [2000(a)] studied such an $M/G/1$ queueing model where the server provides a first phase of regular service (FRS) to all the units, whereas only some them received a second phase of optional service (SPS). The FRS time follow general distribution but SPS is assumed to be exponentially distributed random variable. Medhi (2002) generalized the model by considering that SPS is also governed by a general distribution. Choudhury [2003 (a)] investigated the model further this model in more depth and generalized for batch arrival case in [2003(b)]. Recently, Choudhury and Paul [2004(c)] developed a recursive solution for computing system state probabilities of this model. Krishna Kumar and Arivudaimambi (2001) obtained explicit expression for transient probabilities for this type of finite capacity model $M/G/1/1$ Bernoulli feed back queue and $M/G/1/1$ queue with unreliable server in [2002(a)]. Their analysis is based on a transformed approach. Due to various applications in many real life situation there have been several types of extensions and generalizations for this type of queueing model, to which attention has been paid by several researchers. Here we present a brief literature review for such types of models.

(a) *Model with Specific distribution:* Another most practical class of model related to such type of model is "*Coxian type of models*", where both the phases of service are exponentially distributed random variables. The k-phase *Coxian* distribution C_k consist of k-independent exponential phases with rate $'\theta_j'$(say) at phase j $(1 \le j \le k)$ in which after completing service in the phase '*j*', the unit either enters phase $(j+1)$ with probability p_j or completes the service with probability $(1-p_j)$. It covers multistage queueing process where feedback may occur as in manufacturing process and quality control inspection are performed after certain stages and parts that do not meet quality standards are sent back for reprocessing. Betsimas and Papa constantiou [1988 (a,b)] utilized C_2 distribution as service time

to study the bulk service queueing models with applications in the transportation systems. Recently, Sen (2002) and Agarwal (2002) utilized C_2 distribution as approximation of $M/G/1/N$ type and $GI/M/1/N$ type of queueing models respectively, to which references may be made.

(b) *Vacation Model:* There are some situations where as soon as the busy period ends, the server shuts down the service facility. It may get engaged in other works or may just go away and may not available when the next unit arrives to an empty system. The importance of such models have been established in surveys of Doshi [1986,1990], Teghem (1986) and in the monograph of Takagi (1991). Also, see Tadj and Choudhury (2005) for a survey of recent development of vacation models. However, vacation model with optional second service received little attention. Choudhury and Paul [2006(a),2004(b)] investigated batch arrival queueing model under N-policy, D-policy model with second optional service by Choudhury (2006), second optional service with server breakdown by Wang (2004). More recently, Ke (2007) investigates such a model with a random setup time for j-additional options to the server and Choudhury [2008(b)] investigates such a model for exhaustive service policies that includes many interesting features.

(c) *Retrial Model:* Krishna Kumar et.al. [2002(b)] considered an $M/G/1$ retrial queueing model with a second optional service where at the end of FRS, the server may push out the customer who is receiving such a service, to start the service of another priority arriving customer. The interrupted customer join a retrial queue and the customer at the head of this queue is allowed to conduct a repeated attempt in order to start again FRS after some random time. Artalejo and Choudhury (2004) investigated such a model with a classical retrial policy. More recently, Atencia and Moreno [2006(a)] investigate a discrete time version of this model. While discrete time version of a retrial queue with second optional service starting with failures is investigated by Wang and Zhao (2007). Another paper in this category is the work of Li and Wang (2006).

(d) *Related Models:* The papers mentioned in the above three categories are characterized by a common feature: the second phase of service is provided only to a proportion of the original incoming customers. As related literature, next we mention some papers by Doshi (1991), Krishna and Lee (1990) and Selvam and Sivasankaran (1991) arising from distributed system control where all customers received batch mode service in the first phase followed by individual services in the second phase.

Besides the Bernoulli schedule, optional second services, the queueing system of interest to us exhibits another feature that we briefly mention here without surveying any relevant literature since it is well known.

Classical Retrial Policy

Interval between successive repeated attempts of primary customers follow exponential distribution with parameter $n\mu$ (say), when the number of customers in the orbit or retrial group is n [see Falin (1986) and Yang and Templeton (1987)], which is the case in our model.

The fundamental reason for analyzing such a model is that its structure appears in many representations of computer and tele-communication systems. Another reason is that we analyzed the stationary system behaviour in more depth. Most of the previous studies only gives solutions in terms of generating function or the solution of a Riemann-Hillbert boundary value problem [see for example Lee (1997) and Feng et.al (1998)]. However, in the present study, we develop a more detailed study even for a more generalized model which includes the recursive solution of limiting probabilities, calculations of some important performance measures and stochastic decomposition. To this, the mathematical methodology will be based on a combination of embedded Markov chain analysis and Markov renewal theory.

The rest of the paper is organized as follows. Section-2 describe the mathematical model under consideration. Discrete time process and continuous time process for the model are considered in section-3 and section-4 respectively. Finally some important performance measures are derived in section-5.

2. Mathematical Model

We consider a single server queueing system in which primary customers arrive according to a Poisson process with arrival rate $\lambda(>0)$. There is a single server who provides preliminary first phase of service (FPS) denoted by B_1 to all arriving units. As soon as the FPS of a unit is completed, then the customer may leave the system with probability $p_1\ (0 \le p_1 \le 1)$ (say) or may be provide a second phase of optional service (SPS) denoted by B_2 with probability $p_2 = 1 - p_1$. The service times follow general law of probability with probability distribution function $B_i(x), i = 1,2; (B_i(0) = 0)$, Laplace Stieltjes transform (LST) $\beta_i(\theta) = E\left[e^{-\theta B_i}\right]$ and finite moments $\beta_i^{(k)}; k = 1,2$; where the sub index $i = 1$ (respectively 2) denotes FPS (respectively the SPS). On completion of the total service (i.e. FPS or SPS) the server may decided to go for a vacation denoted by V with probability $q_1\ (0 \le q_1 \le 1)$ (say), otherwise it may continue to serve the next customer, if any, with probability $q_2 = 1 - q_1$. Next, we assume that the vacation time random variable follows a general law of distribution with probability distribution function $V(x), (V(0) = 0)$; LST $\gamma^*(\theta) = E\left[e^{-\theta V}\right]$ and finite moment $\gamma^{(k)}; k = 1,2$ and independent of service time random variable B_i and arrival process. The primary units finding the server free automatically starts their FPS. However, if a primary customer finds the server busy (attending FPS or SPS) or the server is on vacation (attending other secondary job), then he joins the orbit (i.e. a group of unsatisfied customers) in order to seek service again and again until he finds the server free. The time between two successive repeated attempts of each customer in the orbit is assumed to be exponentially distributed random variable with rate $n\mu$ when the orbit size is $n \in Z^+$.

It should be noted here that the same server serve both the phases of service. We may define the modified service time required by the customer to complete the service cycle as:

$$G = \begin{cases} B = \begin{cases} = B_1 \text{ with probability } p_1 \\ = B_1 + B_2 \text{ with probability } p_2 \end{cases} \text{ with probability } q_2 \\ = B + V \text{ with probability } q_1 \end{cases}$$

Now if we define $l_{i,j}$ and m_j as the probabilities that '*j*' individual units arrive during the *i*-th phase of service time for $i \in \{1,2\}$ and vacation time respectively, then for $j = 0,1,2,\ldots$; we have $l_{i,j} = \int_0^\infty p_j(t)dB_i(t)$ and $m_j = \int_0^\infty p_j(t)dV(t)$; respectively. Here $p_j(t) = e^{-\lambda t}(\lambda t)^j / j!$ is the probability that '*j*' arrivals occur during [0,t] and therefore we can write

$$l_{i,j} = \int_0^\infty \frac{e^{-\lambda t}(\lambda t)^j}{j!} dB_i(t) = \frac{(-\lambda)^j \beta_i^{(j)}(\lambda)}{j!} \text{ ; j = 0,1,2,..., for } i \in \{1,2\}$$

and $m_j = \int_0^\infty \frac{e^{-\lambda t}(\lambda t)^j}{j!} dV(t) = \frac{(-\lambda)^j \gamma^{(j)}(\lambda)}{j!}$; j = 0,1,2,..., respectively, where $\beta_i^{(j)}(\theta)$ and $\gamma^{(j)}(\theta)$ are the j-th derivative of the LST of B_i and V with respect to θ.

3. Discrete Time Process

First of all we are interested with the embedded process i.e. discrete time process of the model. Let t_n be the time instant at which n-th service completion occurs i.e. we consider the epochs at which the modified service time requested by a customer expires. Then the sequence $N_n = N(t_n+)$, where $N(t_n)$ represents the number of customers in the orbit at time instant t_n, forms a Markov chain which is the embedded Markov renewal process.

The sequence $\{ N_n; n \geq 1 \}$ is a homogeneous Markov chain and it is owing to the following transition

$$\left(N_n \middle| N_{n-1} = j\right) = \begin{cases} j - 1 + S_n \text{ with probability } \dfrac{j\mu}{\lambda + j\mu} \\ j + S_n \text{ with probability } \dfrac{\lambda}{\lambda + j\mu} \end{cases}$$

where S_n is the number of customers that arrived during the n-th modified service time.

It is not difficult to see that $\{ N_n; n \geq 1 \}$ is irreducible and aperiodic. Now by using classical (Foster's) criteria bassed on mean drift [see Pakes (1969) and Sennott et.al (1983)], it can be shown that $\{ N_n; n \geq 1 \}$ is positive recurrent if and only if $\rho = \lambda\left[\beta_1^{(1)} + p_2\beta_2^{(1)} + q_1\gamma^{(1)}\right] < 1$.Next, we assume that $\{ N_n; n \geq 1 \}$ is positive recurrent to guarantee that the limiting probabilities

$$\pi_j = \lim_{n\to\infty} P_r\{ N_n = j \};\ j \geq 0,$$

exist and are positive.

The one step transition matrix, $P = (p_{i,j})$, associated to $\{ N_n; n \geq 1 \}$ has the elements

$$p_{i,j} = \begin{cases} \left(\dfrac{j\mu}{\lambda + j\mu}\right)\{q_2(p_1 h_{1.0} + p_2 l_{1,0}) + q_1(p_1 g_0 + p_2 l_0)\};if \quad i \geq 1, \quad j = i-1 \\ \left(\dfrac{\lambda}{\lambda + j\mu}\right)\{q_2(p_1 h_{1,j-i} + p_2 l_{1,j-i}) + q_1(p_1 g_{j-i} + p_2 l_{j-i})\} \\ + \left(\dfrac{j\mu}{\lambda + j\mu}\right)\{q_2(p_1 h_{1,j-i+1} + p_2 l_{1,j-i+1}) + q_1(p_1 g_{j-i+1} + p_2 l_{j-i+1})\};if \quad 0 \leq i \leq j \end{cases}$$

where $h_{i,j} = \sum_{k=0}^{j} m_k l_{i,j-k}$, $l_j = \sum_{i=0}^{j} l_{1,i} l_{2,j-i}$ and $g_j = \sum_{i=0}^{j} l_i m_{j-i}$ are the probabilities that 'j' customers arrive during the time intervals $[B_i + V], [B_1 + B_2]$ and $[B_1 + B_2 + V]$ respectively.

Then the Kolmogorov equations associated with the Markov chain $\{ N_n; n \geq 1 \}$ can be written as

$$\pi_j = \sum_{i=0}^{j}\left(\frac{\lambda\pi_i}{\lambda + i\mu}\right)\left[q_2(p_1 h_{1,j-i} + p_2 l_{1,j-i}) + q_1(p_1 g_{j-i} + p_2 l_{j-i})\right]$$
$$+ \sum_{i=1}^{j+1}\left(\frac{\lambda}{\lambda + i\mu}\right)\left[q_2(p_1 h_{1,j-i+1} + p_2 l_{1,j-i+1}) + q_1(p_1 g_{j-i+1} + p_2 l_{j-i+1})\right];\ j \geq 0. \qquad (3.1)$$

We now introduce the following generating functions:

$$\pi(z) = \sum_{j=0}^{\infty} z^j \pi_j$$

and $\psi(z)=\sum_{j=0}^{\infty}\frac{\pi_j z^j}{\lambda+j\mu}$; respectively.

Such that

$$\pi(z)=\mu z\psi'(z)+\lambda\psi(z). \tag{3.2}$$

Now because of convolution, the equation (3.1) can be transformed with the help of the following generating functions $L_i(z)=\sum_{j=0}^{\infty} z^j l_{i,j}$, $L(z)=\sum_{j=0}^{\infty} z^j l_j$, $H_i(z)=\sum_{j=0}^{\infty} z^j h_{i,j}$ and $G(z)=\sum_{j=0}^{\infty} z^j g_j$; respectively.

Note that $L_i(z)=\beta_i^*(\lambda-\lambda z)$, $H_i(z)=L_i(z)\gamma^*(\lambda-\lambda z)$ for $i\in\{1,2\}$, $L(z)=L_1(z)L_2(z)$ and $G(z)=L(z)\,\gamma^*(\lambda-\lambda z)$.

Then the equation (3.1), becomes

$$\pi(z)=\lfloor\lambda\psi(z)+\mu\psi'(z)\rfloor\lfloor p_1+p_2\beta_2^*(\lambda-\lambda z)\rfloor\lfloor q_2+q_1\gamma^*(\lambda-\lambda z)\rfloor\beta_1^*(\lambda-\lambda z). \tag{3.3}$$

We now combine equations (3.2) and (3.3) to get

$$\psi'(z)=\frac{\lambda[1-\{q_2+q_1\gamma^*(\lambda-\lambda z)\}\{p_1+p_2\beta_2^*(\lambda-\lambda z)\}\beta_1^*(\lambda-\lambda z)]}{\mu[\{q_2+q_1\gamma^*(\lambda-\lambda z)\}\{p_1+p_2\beta_2^*(\lambda-\lambda z)\beta_1^*(\lambda-\lambda z)-z]}\psi(z) \tag{3.4}$$

and $$\pi(z)=\frac{\lambda(1-z)[q_2+q_1\gamma^*(\lambda-\lambda z)][p_1+p_2\beta_2^*(\lambda-\lambda z)]\beta_1^*(\lambda-\lambda z)}{\{q_2+q_1\gamma^*(\lambda-\lambda z)\}\{p_1+p_2\beta_2^*(\lambda-\lambda z)\}\beta_1^*(\lambda-\lambda z)-z}\psi(z). \tag{3.5}$$

Now we can solve the differential equation (3.4), which yields the following solution in terms of generating function

$$\psi(z)=\psi(1)\exp\left\{-\frac{\lambda}{\mu}\int_z^1\frac{[1-\{q_2+q_1\gamma^*(\lambda-\lambda u)\}\{p_1+p_2\beta_2^*(\lambda-\lambda u)\}\beta_1^*(\lambda-\lambda u)]}{[\{q_2+q_1\gamma^*(\lambda-\lambda u)\}\{p_1+p_2\beta_2^*(\lambda-\lambda u)\}\beta_1^*(\lambda-\lambda u)-u]}du\right\};$$

where $\psi(1)$ can be determined easily by using the normalizing condition $\pi(1)=1$ and therefore we have

$$\psi(1)=\frac{(1-\rho)}{\lambda}.$$

The preceeding results are summed up in the following theorem:

Theorem 3.1: If $\rho < 1$, the stationary distribution of the discrete time process $\{ N_m; m \geq 1 \}$ has the following probability generating function:

$$\pi(z) = \frac{(1-\rho)(1-z)[p_1 + p_2 L_2(z)][q_2 + q_1 M(z)]L_1(z)}{[\{p_1 + p_2 L_2(z)\}\{q_2 + q_1 M(z)\}L_1(z) - z]}$$
$$\times \exp\left\{ -\frac{\lambda}{\mu} \int_z^1 \frac{[1 - \{p_1 + p_2 L_2(u)\}\{q_2 + q_1 M(u)\}L_1(u)]}{[\{p_1 + p_2 L_2(u)\}\{q_2 + q_1 M(u)\}L_1(u) - u]} du \right\} \tag{3.6}$$

where $L_i(z) = \beta_i^*(\lambda - \lambda z)$ and $M(z) = \gamma^*(\lambda - \lambda z)$; respectively.

Remark 3.1: Now by putting z = 0 in (3.6), we get

$$\pi_0 = (1-\rho)\exp\left\{ -\left(\frac{\lambda}{\mu}\right) \int_0^1 \frac{[1 - \{q_2 + q_1\gamma^*(\lambda - \lambda u)\}\{p_1 + p_2\beta_2^*(\lambda - \lambda u)\}\beta_1^*(\lambda - \lambda u)]}{[\{q_2 + q_1\gamma^*(\lambda - \lambda u)\}\{p_1 + p_2\beta_2^*(\lambda - \lambda u)\}\beta_1^*(\lambda - \lambda u) - u]} du \right\};$$

… (3.7)

which is the stationary probability that no unit is waiting in the system at a departure epoch of the system. It should be pointed out that limiting probabilities $\{\pi_j; j \geq 0\}$ can be computed recursively from equation (3.1) and the expressions (3.7) for π_0. The computation of the integrals $l_{i,j}$ and m_j is reduced to explicit expression in the case of many standard service time and vacation time distributions.

Remark 3.2: The result obtained in this section is quite general and it covers many particular cases. For example, let us consider the situation where $q_1 = 0$ (i.e. there is no Bernoulli schedule in the system), then $\rho = \lambda(\beta_1^{(1)} + p_2\beta_2^{(1)}) < 1$ and therefore equation (3.6) yields

$$\pi(z) = \frac{(1-\rho)(1-z)[p_1 + p_2\beta_2^*(\lambda - \lambda z)]\beta_1^*(\lambda - \lambda z)}{[\{p_1 + p_2\beta_2^*(\lambda - \lambda z)\}\beta_1^*(\lambda - \lambda z) - z]}$$
$$\times \exp\left\{ -\frac{\lambda}{\mu} \int_z^1 \frac{[1 - \{p_1 + p_2\beta_2^*(\lambda - \lambda u)\}\beta_1^*(\lambda - \lambda u)]}{[\{p_1 + p_2\beta_2^*(\lambda - \lambda u)\}\beta_1^*(\lambda - \lambda u) - u]} du \right\};$$

which is consistent with the result obtained by Artalejo and Choudhury (2004).

Similarly, consider the case when $p_2 = 1$ (i.e. there is no optional service in the two phase service system), then $\rho = \lambda(\beta_1^{(1)} + \beta_2^{(1)} + q_1\nu^{(1)}) < 1$ and therefore equation (3.6) yields

$$\pi(z) = \frac{(1-\rho)(1-z)[q_2 + q_1\gamma^*(\lambda - \lambda z)]\beta_1^*(\lambda - \lambda z)\beta_2^*(\lambda - \lambda z)}{[\{q_2 + q_1\gamma^*(\lambda - \lambda z)\}\beta_1^*(\lambda - \lambda z)\beta_2^*(\lambda - \lambda z) - z]}$$

$$\times \exp\left\{-\frac{\lambda}{\mu}\int_z^1 \frac{[1-\{q_2 + q_1\gamma^*(\lambda - \lambda u)\}\beta_1^*(\lambda - \lambda u)\beta_2^*(\lambda - \lambda u)]}{[\{q_2 + q_1\gamma^*(\lambda - \lambda u)\}\beta_1^*(\lambda - \lambda u)\beta_2^*(\lambda - \lambda u) - u]}du\right\};$$

which is consistent with the result obtained by Choudhury [2007(b)] for single unit arrival case.

Further, if we consider the limiting case of the model when $\mu \to \infty$, then equation (3.6) yields

$$\lim_{\mu\to\infty} \pi(z) = \frac{(1-\rho)(1-z)[q_2 + q_1\gamma^*(\lambda - \lambda z)][p_1 + p_2\beta_2^*(\lambda - \lambda z)]\beta_1^*(\lambda - \lambda z)}{[\{q_2 + q_1\gamma^*(\lambda - \lambda z)\}\{q_1^* + p_2\beta^*{}_2(\lambda - \lambda z)\}\beta_1^*(\lambda - \lambda z) - z]}$$

$$= \zeta(z) \text{ (say).} \tag{3.8}$$

Note that for $p_2 = 0$ (i.e. there is no optional service in the two phase service system) and deterministic vacation the above result (3.8) is consistent with the result obtained by Madan (1999) for single unit arrival case. Similarly for $p_1 = 0$ the above formula (3.8) is consistent with the result obtained by Choudhury and Madan (2004) for single unit arrival case. Further, when $q_1 = 0$ the above formula (3.8) is consistent with Medhi (2002) [also see Choudhury (2003(a))]. Also, we note that when $\mu = 0$ this model is reduces to the Erlang loss formula.

4. Continuous Time Process

We now turn to the continuous time process where state of the system at time 't' for this model can be described by means of Markov process $X(t) = \{C(t), N(t), \xi(t)\}$, where $C(t) = 0,1,2 \text{ or } 3$ according to whether the system is idle, server is busy with *FPS*, server is busy with *SPS* or the server is on vacation at time '*t*', $N(t)$ represents number of customer in the orbit at time '*t*'. If $C(t) \in \{1,2,3\}$, then $\xi(t)$ represents the corresponding elapsed *FPS*, elapsed *SPS* and elapsed vacation times are in progress.

We neglect the elapsed times $\xi(t)$ and consider only the pair $Y(t) = \{C(t), N(t)\}$, whose state space is $\Omega = \{0,1,2,3\} \times Z^+$ to study the limiting behaviour of the process $Y(t) = \{C(t), N(t)\}$ as $t \to \infty$. Let $Q_{i,,j} = \lim_{t\to\infty} P_r\{Y(t) = (i,j)\}$, $(i,j) \in \Omega$ be the limiting distribution of the system state and number in the orbit.

Now since the arrival process is Poisson, it follows from Burke's theorem that the stationary probabilities $\{Q_{i,,j}, (i,j) \in \Omega\}$ exist and are positive, under the same conditions given in section-3 of the limiting probabilities $\{\pi_j; j \geq 0\}$ of embedded Markov chain $\{N_m; m \geq 1\}$.

Theorem 4.1: Under the stability conditions, we have

The limiting probabilities are given by

$$Q_{0,j} = \frac{\lambda \pi_j}{\lambda + j\mu}; j \geq 0 \tag{4.1}$$

$$Q_{1,j} = \lambda \left[\sum_{i=0}^{j} \frac{\lambda \pi_i}{\lambda + i\mu} k_{1,j-i} + \sum_{i=1}^{j+1} \frac{i\mu \pi_i}{\lambda + i\mu} k_{1,j-i+1} \right] \; ; j \geq 0 \tag{4.2}$$

$$Q_{2,j} = \lambda p_2 \left[\sum_{i=0}^{j} \frac{\lambda \pi_i}{\lambda + i\mu} \sum_{n=0}^{j-i} h_{1,n} k_{2,j-i-n} + \sum_{i=1}^{j+1} \frac{i\mu \pi_i}{\lambda + i\mu} \sum_{n=1}^{j-i+1} h_{1,n} k_{2,j-i+1-n} \right] \; ; j \geq 0 \tag{4.3}$$

and $$Q_{3,j} = \frac{(j+1)\mu}{\lambda + (j+1)\mu} \pi_{j+1} - Q_{1,j} - Q_{2,j} \quad ; j \geq 0, \tag{4.4}$$

where $k_{i,j} = (1 - \sum_{n=0}^{j} h_{i,n})\lambda^{-1} \quad ; j \geq 0, \text{for } i \in \{1,2\}$.

The partial generating functions $Q_i(z) = \sum_{j=0}^{\infty} z^j Q_{i,j}, 0 \leq i \leq 3$, are given by

$$Q_0(z) = (1-\rho)\exp\left\{ -\left(\frac{\lambda}{\mu}\right) \int_z^1 \frac{[1 - \{p_1 + p_2 H_2(u))\}\{q_2 + q_1 M(u)\}H_1(u)]}{[\{p_1 + p_2 H_2(u)\}\{q_2 + q_1 M(u)\}H_1(u) - u]} du \right\}$$
$$\ldots \tag{4.5}$$

$$Q_1(z) = \frac{[1 - \beta_1^*(\lambda - \lambda z)]Q_0(z)}{\{p_1 + p_2 \beta_2^*(\lambda - \lambda z)\}\{q_2 + q_1 \gamma^*(\lambda - \lambda z)\}\beta_1^*(\lambda - \lambda z) - z} \tag{4.6}$$

$$Q_2(z) = \frac{p_2[1 - \beta_2^*(\lambda - \lambda z)]\beta_1^*(\lambda - \lambda z)Q_0(z)}{\{p_1 + p_2 \beta_2^*(\lambda - \lambda z)\}\{q_2 + q_1 \gamma^*(\lambda - \lambda z)\}\beta_1^*(\lambda - \lambda z) - z} \tag{4.7}$$

and $$Q_3(z) = \frac{q_1[1 - \gamma^*(\lambda - \lambda z)][p_1 + p_2 \beta_2^*(\lambda - \lambda z)]\beta_1^*(\lambda - \lambda z)Q_0(z)}{\{p_1 + p_2 \beta_2^*(\lambda - \lambda z)\}\{q_2 + q_1 \gamma^*(\lambda - \lambda z)\}\beta_1^*(\lambda - \lambda z) - z} \tag{4.8}$$

Proof: Here we note that $\{Y(t); t \geq 0\}$ is a Markov regenerative process with the embedded Markov renewal process $\{N_m; m \geq 1\}$, so we may apply main convergence theorem for Markov regenerative process and thus we may write

$$Q_{i,j} = \frac{\sum_{n=0}^{\infty} \pi_n \tau_n(i,j)}{\sum_{n=0}^{\infty} \pi_n \varpi_n} \quad ;(i,j) \in \Omega, \tag{4.9}$$

where

$\tau_n(i,j) =$ The expected amount of time spent by the process $\{Y(t); t \geq 0\}$ in the state (i,j) during a service cycle (i.e. the interval of time between two consecutive total completion epochs) given that at the beginning of this interval the number of customers in the orbit was 'n'

and $\varpi_n =$ The expected length of the service cycle given that at the beginning of this interval the number of customer in the orbit was 'n'.

Now for this model, we have

$$\varpi_n = \frac{1}{\lambda + n\mu} + \beta_1^{(1)} + p_2 \beta_2^{(1)} + q_1 \nu^{(1)}; \text{ for } n \geq 0$$

and therefore $\sum_{n=0}^{\infty} \varpi_n \pi_n = \lambda^{-1}$; is the mean service cycle.

$$\text{Since } \tau_n(0,j) = \frac{1}{\lambda + j\mu} \delta_{j,n}; \quad n \geq 0, j \geq 0, \tag{4.10}$$

where $\delta_{j,n}$ denotes kronecker's delta function.

Therefore utilizing (4.10) in (4.9), we get finally

$$Q_{0,j} = \frac{\lambda}{\lambda + j\mu} \pi_j; \quad j \geq 0,$$

as we claimed in the formula (4.1).

We next prove the expression (4.2) for $Q_{1,j}$. In this case, the quantity $\tau_n(1,j)$ for $j > \max(0, n-1)$, is given by

$$\tau_n(1,j) = \frac{\lambda}{\lambda + n\mu}(1-\delta_{j,n-1})k_{1,j-n} + \frac{n\mu}{\lambda + n\mu}k_{1,j-n+1} \tag{4.11}$$

To understand the above expression we suppose that a modified service time ends leaving 'n' customer in the pool (orbit). We may distinguish two cases according to the origin of the customer who received the next service. Let us assume that this customer is a primary customer, then its *FPS* starts at time (we say) $t = 0$. Then we observe that the time interval $(t, t+\Delta t)$ contributes to $\tau_n(1,j)$ if

(i) the *FPS* has not been completed before time 't' (with probability $(1 - B_1(t))$, and
(ii) (j-n) primary customers arrive during (0,t].

The case in which the customer in service proceeds from the orbit can be analyzed analogously leading to the second term in the right hand side of the expression (4.11).

Initially, it is convenient to express $k_{1,j}$ in the following form

$$k_{1,j} = \int_0^\infty P_j(t)(1-B_1(t))dt = \int_0^\infty \frac{e^{-\lambda t}(\lambda t)^j}{j!}(1-B_1(t))dt\,.$$

After this, it is easy to get the relation $k_{1,j} = [1 - \sum_{i=0}^{j} h_{1,i}]\lambda^{-1}$ connecting the quantities $k_{1,j}$ and $h_{1,j}$ .Now formula (4.2) follows by inserting (4.11) in to (4.9).

Similarly, the quantity $\tau_n(2,j)$ for $j > \max(0, n-1)$ can also be calculated as follows:

$$\tau_n(2,j) = \frac{p_2\lambda}{\lambda + n\mu}(1-\delta_{j,n-1})\sum_{i=0}^{j-n} h_{i,i}k_{2,j-n-i} + \frac{p_2 n\mu}{\lambda + n\mu}\sum_{i=1}^{j-n+1} h_{1,i}k_{2,j-n-i+1}\,, \tag{4.12}$$

where $k_{2,j} = [1 - \sum_{i=0}^{j} h_{2,i}]\lambda^{-1}$.

Hence, formula (4.3) follows by inserting (4.12) in to (4.9).

The stationary probabilities $Q_{3,j}$ can be obtained along the same lines. However, we next offer an alternative approach. To obtain it, we may apply rate of equality principle according to which the number of transitions in a cycle at which the number of customers in the orbit increases from (j-1) to j equals the number of transition at which the number of customers in the orbit decreases from j to (j-1). After an appeal to *PASTA* property (e.g. see [26]) this readily implies that

$$(j+1)\mu Q_{0,j+1} = \lambda[Q_{1,j} + Q_{2,j} + Q_{3,j}];\, j \geq 0. \tag{4.13}$$

Combining (4.1) and (4.14), we get expression (4.4).

A stable recursive scheme for the computation of limiting probabilities $\{Q_{i,j}; (i, j) \in \Omega\}$ in terms of $\{\pi_j; j \geq 0\}$ follows by combining (3.1) with (4.1)-(4.4).

The proof of the formulas (4.5) – (4.8) for the partial generating functions follow straight forwardly after routine algebraic manipulations of the equations (4.1)-(4.4).

Our next objective is to investigate the distribution of number of customers in the orbit and the total number of customers in the system. These results are summarized in the following corollary.

Corollary 4.1: Let us consider the marginal distribution of the number of units in the orbit $Q_j = \sum_{i=0}^{3} Q_{i,j}; j \geq 0$. Then the corresponding generating function is given by

$$Q(z) = \sum_{j=0}^{\infty} z^j Q_j = \frac{(1-z)Q_0(z)}{\{p_1 + p_2\beta_2(\lambda - \lambda z)\}\{q_2 + q_1\gamma(\lambda - \lambda z)\}\beta_1(\lambda - \lambda z) - z}$$

Corollary 4.2: Let P_j be the stationary distribution of the total number of customers in the system i.e. $P_j = Q_{0,j} + (1 - \delta_{j,0})(\sum_{i=1}^{3} Q_{i,j}); j \geq 0$, then its corresponding generating function is given by

$$P(z) = \sum_{j=0}^{\infty} z^j P_j = Q_0(z) + z\{Q_1(z) + Q_2(z) + Q_3(z)\} = \zeta(z)\frac{Q_0(z)}{Q_0(1)}; \tag{4.14}$$

where $Q_0(1) = (1 - \rho)$ and $\zeta(z)$ denotes the generating function of the number of customers present in the system in an *M/G/1* queue with second optional service and with Bernoulli vacation schedule (for instance see formula (3.8) of section -3)

Remark 4.1: It is important to note here that the stationary distribution of the system size distribution of an *M/G/1* queue with second optional service and with Bernoulli vacation schedule given in equation (4.14) in terms of generating functions decomposes in to the distributions of two independent random variables viz-

(i)The system size distribution of an *M/G/1* queue with a second optional service and with Bernoulli vacation schedule [represented by the first term of the equation (4.14)].

and (ii) The conditional distribution of the number of customers in the orbit given that the system is idle [represented by the second term of the equation (4.14)]

This confirms the stochastic decomposition property of Fuhrmann and Cooper (1985). It should be noted here that our retrial model can also be viewed as a non exhaustive vacation model where the server may take a vacation even though there are some units remain in the system which starts after completion of either FPS or SPS. Further, some of useful applications of stochastic decomposition property such as. moments relationship could also be derived along with the lines described in Artalejo and Falin (1994).

5. Some Performance Measures

Our next objective is to provide explicit expression for the system state probabilities and some important performance measures of the system. The results are summaries in the following corollaries.

Corollary 5.1: The stationary probabilities of the server states are given by

Pr {The system is idle} $= Q_0(1) = (1-\rho)$

Pr { The server is busy with *FPS*} $= Q_1(1) = \lambda\beta_1^{(1)}$

Pr { The server is busy with *SPS*} $= Q_2(1) = p_2\lambda\beta_2^{(1)}$

Pr { The server is on vacation} $= Q_3(1) = q_1\lambda\nu^{(1)}$.

Next we investigate the first partial factorial moments

Corollary 5.2: Under the stability conditions,

(i) The first partial moments $M_i = \sum_{j=1}^{\infty} jQ_{i,j}$, for $i \in \{0,1,2,3\}$ are found to be:

$$M_0 = \frac{\lambda\rho}{\mu}$$

$$M_1 = \frac{\lambda^2\beta_1^{(1)}\rho}{(1-\rho)\mu} + \frac{\lambda^2\beta_1^{(2)}}{2} + \frac{\lambda^3\beta_1^{(1)}[\beta_1^{(2)} + p_2\{\beta_2^{(2)} + 2\beta_1^{(1)}\beta_2^{(1)}\} + q_1\{\nu^{(2)} + 2\nu^{(1)}(\beta_1^{(1)} + p_2\beta_2^{(1)})\}]}{2(1-\rho)}$$

$$M_2 = p_2\left[\frac{\lambda^2\beta_2^{(1)}\rho}{(1-\rho)\mu} + \frac{\lambda^2[\beta_2^{(2)} + 2\beta_1^{(1)}\beta_2^{(1)}]}{2}\right]$$

$$+ p_2\left[\frac{\lambda^3\beta_1^{(1)}[\beta_1^{(2)} + p_2\{\beta_2^{(2)} + 2\beta_1^{(1)}\beta_2^{(1)}\} + q_1\{\nu^{(2)} + 2\nu^{(1)}(\beta_1^{(1)} + p_2\beta_2^{(1)})\}]}{2(1-\rho)}\right]$$

and $M_3 = q_1\left[\frac{\lambda^2\nu^{(1)}\rho}{(1-\rho)\mu} + \frac{\lambda^2[\nu^{(2)} + 2\nu^{(1)}(\beta_1^{(1)} + p_2\beta_2^{(1)})]}{2}\right]$

$$+ q_1\left[\frac{\lambda^3\beta_1^{(1)}[\beta_1^{(2)} + p_2\{\beta_2^{(2)} + 2\beta_1^{(1)}\beta_2^{(1)}\} + q_1\{\nu^{(2)} + 2\nu^{(1)}(\beta_1^{(1)} + p_2\beta_2^{(1)})\}]}{2(1-\rho)}\right].$$

(ii) The expected number of customers is the orbit is found to be

$$E(N) = Q'(1) = M_0 + M_1 + M_2 + M_3$$

$$= \frac{\lambda\rho}{\mu(1-\rho)} + \frac{\lambda^2[\beta_1^{(2)} + p_2\{\beta_2^{(2)} + 2\beta_1^{(1)}\beta_2^{(1)}\} + q_1\{\nu^{(2)} + 2\nu^{(1)}(\beta_1^{(1)} + p_2\beta_2^{(1)})\}]}{2(1-\rho)}$$

(iii) The expected number of customers in the system is found to be

$E(L) = P'(1) = \rho + E(N)$; as expected. (5.1)

The proof of both the corollaries follows easily from the result given in Theorem 4.1 so it is omitted.

Remark 5.1: The result obtained in this section is quite general and it covers many practical situations. For example, let us consider the situation where $q_1 = 0$ (i.e. there is no Bernoulli schedule in the system), then $\rho = \lambda(\beta_1^{(1)} + p_2\beta_2^{(1)}) < 1$ and therefore equation (5.1) yields

$$E(L) = \rho + \frac{\lambda\rho}{\mu(1-\rho)} + \frac{\lambda^2[\beta_1^{(2)} + p_2\{\beta_2^{(2)} + 2\beta_1^{(1)}\beta_2^{(1)}\}]}{2(1-\rho)};$$

which is consistent with the result obtained by Artalejo and Choudhury (2004).

Again when $p_2 = 1$ (i.e. there is no optional service in the two phase service system), then $\rho = \lambda(\beta_1^{(1)} + \beta_2^{(1)} + q_1\nu^{(1)}) < 1$ and therefore equation (5.1) yields

$$E(L) = \rho + \frac{\lambda\rho}{\mu(1-\rho)} + \frac{\lambda^2[\beta_1^{(2)} + \beta_2^{(2)} + 2\beta_1^{(1)}\beta_2^{(1)} + q_1\{\nu^{(2)} + 2\nu^{(1)}(\beta_1^{(1)} + \beta_2^{(1)})\}]}{2(1-\rho)};$$

which is the expression for mean number of customer present in the system for a retrial queue with two phases of service.

Similarly by taking limit $\mu \to \infty$ in equation (5.1), we get

$$E(L) = \rho + \frac{\lambda^2[\beta_1^{(2)} + p_2\{\beta_2^{(2)} + 2\beta_1^{(1)}\beta_2^{(1)}\} + q_1\{\nu^{(2)} + 2\nu^{(1)}(\beta_1^{(1)} + p_2\beta_2^{(1)})\}]}{2(1-\rho)}. \quad (5.2)$$

Note that for $q_1 = 0$ the above expression (5.2) is consistent with the result obtained by Medhi (2002) and for $p_2 = 1$ the above expression (5.2) is consistent with the result obtained by Choudhury and Madan (2004) for single unite arrival case.

Moreover, we can estimate the following measure of proximity between the stationary distribution of this type of queueing models with and without retrials

$$D = \sum_{i=0}^{3}\sum_{j=0}^{\infty}\left|Q_{i,j}(\mu) - Q_{i,j}(\infty)\right|$$

as follows:

$$2(1-\rho)\left(1 - \frac{\pi_0}{(1-\rho)}\right) \le D \le 2\left(1 - \frac{\pi_0}{(1-\rho)}\right).$$

Next, we derive mean busy period and expected length of busy cycle in corollary 5.3

Corollary 5.3: (i) let T_b be the length of an ordinary busy period of this model, then under the stability condition the expected length of the busy period is found to be

$$E(T_b)=\frac{(1-\pi_0)}{\lambda\pi_0}$$

$$=\frac{1}{\lambda(1-\rho)}\exp\left\{\left(\frac{\lambda}{\mu}\right)\int\frac{[1-\{p_1+p_2\beta_2^*(\lambda-\lambda u)\}\{q_2+q_1\gamma^*(\lambda-\lambda u)\}\beta_1^*(\lambda-\lambda u)du}{\{p_1+p_2\beta_2^*(\lambda-\lambda u)\}\{q_2+q_1\gamma^*(\lambda-\lambda u)\}\beta_1^*(\lambda-\lambda u)-u}\right\}$$

$$-\frac{1}{\lambda}$$

(ii) let T_c be the length of the busy cycle of this model, then under the stability condition the expected length of the busy cycle is found to be

$$E(T_c)=\frac{1}{\lambda(1-\rho)}\exp\left\{\left(\frac{\lambda}{\mu}\right)\int\frac{[1-\{p_1+p_2\beta_2^*(\lambda-\lambda u)\}\{q_2+q_1\gamma^*(\lambda-\lambda u)\}\beta_1^*(\lambda-\lambda u)du}{\{p_1+p_2\beta_2^*(\lambda-\lambda u)\}\{q_2+q_1\gamma^*(\lambda-\lambda u)\}\beta_1^*(\lambda-\lambda u)-u}\right\}$$

These two results follow directly by applying the argument of alternating renewing renewal process.

CONCLUDING REMARK

In this paper, we have studied an $M/G/1$ retrial queue with second optional service under Bernoulli vacation schedule. Our investigations include the steady state behaviour of both discrete time parameter and continuous time parameter processes, along with some system characteristics, which generalized many interesting features of existing models e.g. see Remark.3.2.

A practical generalization of this model would also be possible by introducing the concept of Bernoulli admission mechanism along with its prescriptive use of linear retrial policy. Various scenarios have been discussed in Artalejo and Atencia (2004) only for bulk arrival queueing systems but neither with second optional service nor with Bernoulli vacation schedule. A clever solution to these types of problems could lead to significant enhancement over the proposed models.

ACKNOWLEDGEMENT

This work is supported by Department of Science and Technology, Govt. of India through Project No.:SR/S4/MS:229/04.

REFERENCES

[1] Anaposi, R.F and Madan.K.C (2003): "A single server queue with two types of service, Bernoulli schedule server vacations and a single vacation policy", *Pakistan Journal of Statistics*, 19(3),331-342.

[2] Artalejo, J.R (1999(a)): "Accessible bibliography of research on retrial queues"; *Mathematical and Computer Modelling*, 30, 1-6.

[3] Artalejo, J.R (1999(b)): "A classified bibliography of research on retrial queues", *TOP*, 7, 187-211.

[4] Artalejo, J.R and Atencia, I (2004): "On the single server retrial queue with batch arrivals", *Sankhyá*, 66, 140-158.

[5] Artalejo, J.R and Choudhury. G. (2004): "Steady state analysis of an $M/G/1$ queue with repeated attempts and two phase service", *Quality Technology and Quantitative Management*, 1 (2), 189-199.

[6] Artalejo, J.R and Falin. G. (1994): "Stochastic decomposition for retrial queues", *TOP*, 2, 329-342.

[7] Atencia. I. And Moreno. P. (2006): "$Geo/G/1$ retrial queue with 2nd optional service", *International Journal of Operational Research*, 1 (4), 340-362.

[8] Atencia. I. And Moreno. P. (2005): "A single server retrial queue with general retrial times and Bernoulli schedule", *Applied Mathematics and Computation*, 162, 855-880.

[9] Atencia. I., Fortes. I., P. Moreno.and S. Sanchez. (2006(b)): "An $M/G/1$ retrial queue with active breakdowns and Bernoulli schedule in the server", *International Journal of Information and Management Sciences*,17(1), 1-17.

[10] Agarwal. M. (2002): "Distribution of number served during a busy period of $GI/M/1/N$ queues- Lattice path approach", *Journal of Statistical Planning and Inference*, 101, 7-21.

[11] Bertsimas. D. and Papa Constatiou. X. (1988(a)): "On the steady state solution of $M/C_2(a,b)/S$ queueing system", *Transportation Sciences*, 22, 125-138.

[12] Bertsimas. D. and Papa Constatiou. X. (1988(b)): "Analysis of the stationary $E_k/C_2/S$ queueing system", *European Journal of Operational Research*, 37, 272-282.

[13] Choudhury. G. (2008 (a)): "A note on the $M^x/G/1$ queue with a random setup time under a restricted admissibility policy with a Bernoulli vacation schedule", *Statistical Methodology* (to appear).

[14] Choudhury. G. (2007): "A two phase batch arrival retrial queueing system with Bernoulli vacation schedule", *Applied Mathematics and Computation*, 188(2), 1455-1466.

[15] Choudhury. G. (2007 (b)): "A single server queueing system with two phases of service and vacations", *Quality Technology and Quantitative Technology* (5(1), 33-49).

[16] Choudhury. G. (2005): "An $M/G/1$ queueing system with two phase service under D- policy", *International Journal of Information and Management Sciences*, 16(4), 1-17.

[17] Choudhury. G. (2003(a)): "Some aspects of an $M/G/1$ queueing system with optional second service", *TOP*, 11, 141-150.

[18] Choudhury. G. (2003(b)): "A batch arrival queueing system with an additional service channel", *International Journal of Information and Management Sciences*, 14(2), 17-30.

[19] Choudhury. G., Tadj. L and Paul. M. (2007): "Steady state analysis of an $M^X/G/1$ queue with two phase service and Bernoulli vacation schedule under multiple vacation policy", *Applied Mathematical Modelling, 31*, 1079-1091.

[20] Choudhury. G. and Madan. K. C.(2007): "A batch arrival Bernoulli vacation queue with a random setup time under restricted admissibility policy", *International Journal of Operational Research*, 2(1), 81-97.

[21] Choudhury. G. and Madan. K. C.(2006): "A batch arrival queue with Bernoulli vacation schedule under multiple vacation policy", *International Journal of Management Sciences*, 12(2), 1-18.

[22] Choudhury. G. and Paul. M. (2006(a)): "A two phases queueing system with Bernoulli vacation schedule under multiple vacation policy", *Statistical Methodology,* 3, 174-185.

[23] Choudhury. G and Paul. M. (2006(b)): "A batch arrival queue with second optional service channel under N-policy", *Stochastic Analysis and Applications*, 24, 1-21.

[24] Choudhury. G. and Madan. K. C.(2005): "A two stage batch arrival queueing system with Modified Bernoulli schedule vacation under N- policy", *Mathematical and Computer Modelling*, 42, 71-85.

[25] Choudhury. G. and Madan. K. C.(2004): "A two phase batch arrival queueing system with a vacation time under Bernoulli vacation schedule", *Applied Mathematics and Computation,* 149, 337-349.

[26] Choudhury. G and Paul. M. (2004(a)): "Analysis of a two phases batch arrival queueing model with Bernoulli vacation schedule", *Revista Investigation Operational*, 25(3), 217-228.

[27] Choudhury. G and Paul. M. (2004(b)): "A batch arrival queue with an additional service channel under N-policy", *Applied Mathematics and Computation*, 156, 115-130.

[28] Choudhury. G and Paul. M. (2004(c)): "A batch arrival queue with an additional service channel: A simple alternative approach", *Journal of Indian Statistical Association*, 42, 63-73.

[29] Choi. B.D and Park. K.K (1990): "The $M/G/1$ retrial queue with Bernoulli schedule", *Queueing Systems*, 7, 219-228.

[30] Doshi. B.T (1991): "Analysis of two phase queueing system with general service time", *Operations Research Letters*, 10, 265-277.

[31] Doshi. B.T (1990): "Single server queues with vacations, in Stochastic Analysis of Computer and Communication systems", H.Takagi (Ed.), Elsivier, Amsterdam, 217-265.

[32] Doshi. B.T (1986): "Queueing Systems with vacations – A survey", *Queueing Systems*, 1, 29-66.

[33] Fuhrmann. S.W and Cooper. R.B (1985): "Stochastic decomposition in $M/G/1$ queue with generalized vacations", *Operations Research*, 32, 1117-1129.

[34] Falin.G (1986): "A survey of retrial queues", *Queueing Systems*, 7, 127-168.

[35] Falin. G and Templeton. J.G.C.(1997): "Retrial Queues", Chapman and Hall, London.

[36] Feng. W, Kowada. M and Adahi..K (1998): "A two queue model with Bernoulli service schedule and switching times", *Queueing Systems,* 30, 405-434.

[37] Ghatir. H. M. and Silio. C.B. (1993): "Performance analysis of a multiple access ring network", *IEEE Transection on Communications*, 41, 1494-1506.

[38] Jaing. Y, Tham. C. K and Ko. C.(2002): "Delay analysis of a probabilistic priority discipline", *European Transaction of Telecommunications*, 13(6), 563-572.

[39] Katayama. T and Takajashi. Y (1992): "Analysis of a two class priority queue with Bernoulli schedule", *Journal of Operational Research Society of Japan*, 35(3), 231-249.

[40] Ke. J. C.(2007): "An $M^X/G/1$ system with start up server and J additional options for service", *Applied Mathematical Modelling* (to appear).

[41] Keilson. J and Servi. L.D (1986(a)): "Oscillating random walk models for $GI/G/1$ vacation systems with Bernoulli schedule", *Journal of Applied Probability*, 23, 790-802.

[42] Keilson. J and Servi. L.D (1987): "Dynamics of the $M/G/1$ queue", *Operations Research*, 31, 575-582.

[43] Keilson. J and Servi. L.D (1989): "Blocking Probability for $M/G/1$ vacations systems with occupancy level dependent schedule", *Operations Research*, 37, 134-140.

[44] Krishna Kumar B and Arivudainambi D(2001): "An $M/G/1/1$ feed back queue with regular and optional services"; *International Journal of Information and Management Sciences*, 12(1), 67-73.

[45] Krishna Kumar B and Arivudainambi D(2002(a)): "An $M/G/1$ queue with unreliable server and no waiting capacity", *International Journal of Information and Management Sciences*, 13(2), 35-50.

[46] Krishna Kumar B and Arivudainambi D(2002(b)): "The $M/G/1$ retrial queue with Bernoulli schedules and general retrial time", *Computer and Mathematics with Applications*,43, 15-30.

[47] Krishna Kumar B, Arivudainambi D and Vijay Kumar. A (2002(a)): "On the $M^X/G/1$ retrial queue with Bernoulli schedules and general retrial time", *Asia Pacific Journal of Operational Research*, 19(c),177-194.

[48] Krishna Kumar B, Vijay Kumar. A and Arivudainambi D (2002(b)): "An $M/G/1$ retrial queueing system with two phase service and preemptive resume", *Annals of Operations Research,* 113, 61-79.

[49] Krishna. C. M and Lee. Y. H(1987): "A study of two phase service", *Operations Research Letter,*9, 265-272.

[50] Kella. O. (1990): "Optimal control of the vacation in an $M/G/1$ queue", *Operations Research*, 38, 724-728.

[51] Lee. D.S (1997): "Analysis of a two queue model with Bernoulli schedules", *Journal of Applied Probability*, 34, 176-191.

[52] Levy. H.(1991): Binomeal -gated service: "A method for effective operations and optimization of polling systems", *IEEE Transaction on Communication*, 39(4), 1341-1350.

[53] Li. H and Yang. T. (1998): "$Geo/G/1$ discrete time retrial queue with Bernoulli schedule", *European Journal of Operational Research*, 111, 629-649.

[54] Li. J. and Wang J. (2006): "An $M/G/1$ retrial queue with second multi optional service, Feedback and unreliable server", *Appl. Math. J. Chinese University*, Sec-B, 21(3), 252-262.

[55] L. W, Shi. D and Chao. X (1997): "Reliability analysis of $M/G/1$ queueing systems with server breakdowns and vacations", *Journal of Applied Reliability*, 34, 546-555.

[56] Madan. K. C. (2006(a)): "An $M/G/1$ queue with second optional service", *Queueing Systems*, 34, 37-46.

[57] Madan. K. C. (2006(b)): "On a single server queue with two stage general heterogeneous service and Binomial schedule server vacations", *The Egyptian Statistical Journal*, 44, 39-55.

[58] Madan. K. C. (2001): "On a single server queue with two stage general heterogeneous service and deterministic server vacations", *International Journal of System Sciences*,32, 837-844.

[59] Madan K.C, Al-Rawi. Z.R and Al- Nassel.A.D (2005): "On $M^X/\binom{G_1}{G_2}/1$ vacation queue with two types of general heterogeneous service", *Journal of Applied mathematics and Decision Sciences*, 3, 123-135.

[60] Madan.K.C. and Choudhury.G(2006): "Steady state analysis of an $M^X/(G_1, G_2)/1$ queue with restricted admissibility and random setup time", *International Journal of Information and Management Sciences*, 17(2), 33-56.

[61] Madan.K.C. and Choudhury.G(2004(a)): "An $M^X/G/1$ queue with a Bernoulli vacation schedule under restricted admissibility policy", *Sankhya*, 66, 175-193.

[62] Madan.K.C. and Choudhury.G(2004(b)): "Steady state analysis of an $M^X/(G_1, G_2)/1$ queue with restricted admissibility and Modified Bernoulli schedule server vacations", *Journal of Probability and Statistical Sciences*, 2, 165-185.

[63] Madan.K.C. and Al-Rub, A.Z. Abu (2004(a)): "On a single server queue with optional phase type server vacations based on exhaustive deterministic service and a single vacation policy", *Applied Mathematics and Computation*, 149, 723-734.

[64] Madan.K.C. and Al-Rub, A.Z. Abu (2004(b)): "Transient and steady state solution of a single server queue with modified Bernoulli schedule vacation based on exhaustive service and a single vacation policy", *Revista Investigation Operational*, 25(2), 158-165.

[65] Madan. K.C., Abu-Dayyeh. W. and Tayyan. F.(2003(a)): "Steady state analysis of an $M/D/2$ queue with Bernoulli schedule server vacations", *Journal of Modern Applied Statistical Method*, 2(1), 202-209.

[66] Madan. K.C., Abu-Dayyeh. W. and Tayyan. F.(2003(b)): "A two server queue with Bernoulli schedules and a single vacation policy", *Applied Mathematics and Computation*, 145, 59-71.

[67] Madan. K.C.and Abu-Dayyeh. W.(2002): "Restricted admissibility of batches into an $M^X/G/1$ type bulk queue with modified Bernoulli schedule server vacations", *ESSAIM; Probability and Statistics*, 6, 113-125.

[68] Madan. K.C. and Saleh. M.F.(2001(a)): "On $M/D/1$ queue with general server vacations", *International Journal of Information and Management Sciences*, 12(2), 25-37.

[69] Madan. K.C. and Saleh. M.F.(2001(b)): "On single server vacation queues with deterministic service or deterministic vacations", *Calcutta Statistical Association Bulletin*, 51, 225-241.

[70] Madan. K.C. and Saleh. M.F.(2001(c)): "On an $M/D/1$ queue with deterministic server vacations", *Systems Science Journal*, 27, 107-118.

[71] Medhi. J.(2002): "A single server Poisson input queue with a second optional channel", *Queueing Systems*, 42, 239-242.

[72] Mukherjee. B and Meditch. J. S.(1988): "The Pi-persistent protocol for unidirectional broadcast bus networks", *IEEE Transactions on Communications*, 36(12), 1277-1286.

[73] Pakes. A.G.(1969): "Some conditions for ergodicity and recurrence of Markov chains", *Operations Research*, 17, 1058-1061.

[74] Ramaswami. R and Servi. L. D (1988): "The busy period of the $M/G/1$ vacation model with a Bernoulli schedule", *Stochastic Models*, 4, 171-179.

[75] Sen Kanwar (2002): "Latice path approach to transient analysis of $M/G/1/N$ non Markovian queues using Cox distribution", *Journal of Statistical Planning and Inference*, 101, 133-147.

[76] Sennot. L.I, Humblet. P. A and Tweedie. R.L (1983): "Mean drilts and the non ergodicity of Markov chain", *Operations Research*, 31, 783-789.

[77] Selvam. D. D and Sivasankaran. V.(1994): "A two phase queueing system with server vacations", *Operations Research Letters*, 15, 163-168.

[78] Servi. L.D.(1986): "Average delay approximation of $M/G/1$ cyclic service queue with Bernoulli schedule", *IEEE Trans. Selected Area of Communication*, CAS-4, 813-820 Correction(1987), CAS-5, 547.

[79] Shin. Y. W and Pearce. C.E.M. (1988): "The $BMAP/G/1$ Vacation queue length dependent vacation schedule", *Journal of the Australian Mathematical Society*, Ser. B, 40, 207-221.

[80] Tadj. L and Choudhury. G (2005): "Optimal design and control of queues", *TOP*, 13(2), 359-412.

[81] Tadj. L, Choudhury. G and Tadj. C. (2006(a)): "A quorum system with a random setup time under N- policy and with Bernoulli vacation schedule", *Quality Technology and Quantitative Management*, 3(2), 145-160.

[82] Tadj. L, Choudhury. G and Tadj. C. (2006(b)): "A bulk quorum queueing system with a random setup time under N- policy and with Bernoulli vacation schedule", *Stochastics*, 78(1), 1-11.

[83] Takagi. H.(1991): Queueing Analysis: "A Foundation of Performance Evaluation, Vol-1, *Vacation and Priority Systems*", Part-1, North Holland, Amsterdam.

[84] Teghem. J. (1986): "Control of service process in a queueing system", *European Journal of Operational Research*, 23, 141-158.

[85] Wang. J (2004): "An $M/G/1$ queue with second optional service and server breakdown", *Computer and Mathematics with Applications*, 47, 1713-1723.

[86] Wang. J and Zhao. Q (2007): "A discrete-time $Geo/G/1$ retrial queue with starting with failures and second optional service", *Computer and Mathematics with Applications*, 53, 115-127.

[87] Wolff. R. W. (1982): "Poisson arrivals see time averages", *Operations Research*, 30, 223-231.

In: Applied Mathematical Modeling
Editor: Eetu N. Virtanen, pp. 247-272
ISBN: 978-1-60021-975-7

Chapter 6

Discrete Particle Simulation of Particulate Systems

H. P. Zhu and A. B. Yu *
Centre for Simulation and Modelling of Particulate Systems
School of Materials Science and Engineering
The University of New South Wales, Sydney, NSW 2052, Australia

Abstract

Particle science and technology is a rapidly developing interdisciplinary research area with its core being the understanding of the relationships between micro- and macroscopic properties of particulate/granular matter – a state of matter that is widely encountered but poorly understood. The macroscopic behaviour of particulate matter is controlled by the interactions between individual particles as well as interactions with surrounding fluids. Understanding the microscopic mechanisms in terms of these interaction forces is therefore key to leading to truly interdisciplinary research into particulate matter and producing results that can be generally used. This aim can be effectively achieved via particle scale research based on detailed microdynamic information such as the forces acting on and trajectories of individual particles in a considered system. In recent years, such research has been rapidly developed worldwide, mainly as a result of the rapid development of discrete particle simulation technique and computer technology. This article introduces such simulation techniques and their application to the study of some typical particulate systems.

* Corresponding author: Phone: + 61 2 93854429; Fax: + 61 2 9385 5956; Email: a.yu@unsw.edu.au

1. Introduction

Particle science and technology is a rapidly developing interdisciplinary research area with its core being the understanding of the relationships between micro- and macroscopic properties of particulate/granular matter – a state of matter that is widely encountered but poorly understood. Previous studies are largely at a macroscopic or global scale, the resulting information being helpful in developing a broad understanding a particulate matter of particular interest. However, the lack of quantitative fundamental understanding makes it difficult to generate a general method for reliable scale-up, design and control/optimization of processes of different types.

The macroscopic behaviour of particulate matter is controlled by the interactions between individual particles as well as interactions with surrounding gas or liquid and wall. Understanding the microscopic mechanism in terms of these interactions is therefore the key leading to truly interdisciplinary research into particulate matter and producing results that can be generally used. This aim can be effectively achieved via particle scale research based on detailed microdynamic information. In recent years, such research has been rapidly developing worldwide, mainly as a result of the rapid development of discrete particle simulation technique and computer technology. An important discrete model is the so-called discrete element method (DEM) originally developed by Cundall and Strach (1979). The method considers a finite number of discrete particles interacting by means of contact and non-contact forces, and every particle in a considered system is described by Newton's equations of motion related to translational and rotational motions. DEM simulations can provide dynamic information, such as the trajectories of and transient forces acting on individual particles, which is extremely difficult, if not impossible, to obtain by physical experimentation at this stage of development. Consequently, it has been increasingly used to study various phenomena, such as particle packing (Yang et al, 2000, 2003a), hopper flow (Langston et al, 1994, 1995; Zhu and Yu, 2005), mixing (Stewart et al, 2001; Zhou et al, 2004), and grinding/granulation (Yamane et al, 2004) in the past two decades or so.

This article presents a brief introduction of this method. It covers the development of simulation techniques and their application to the study of some typical particulate systems such as packing of cohesive particles, hopper flow and mixer.

2. Theoretical Treatments

2.1. Governing Equations

A particle in a granular flow can have two types of motion: translational and rotational. During its movement, the particle may interact with its neighbouring particles or walls and interact with its surrounding fluid, through which the momentum and energy are exchanged. Strictly speaking, this movement is affected not only by the forces and torques originated from its immediate neighboring particles and vicinal fluid but also the particles and fluids far away through the propagation of disturbance waves. The complexity of such a process has defied any attempt to model this problem analytically. In DEM approach, it is generally assumed that this problem can be solved by choosing a numerical time step less than a critical value so that during a single time step the disturbance cannot propagate from the particle and fluid farther than its immediate neighbouring particles and vicinal fluid (Cundall and Strack, 1979). Thus, at all times the resultant forces on a particle can be determined exclusively from its interaction with the contacting particles and vicinal fluid for coarse particle system. For a fine particle system, non-contact forces such as the van der Waals and electrostatic forces should also be included. Based on these considerations, Newton's second law of motion can be used to describe the motion of individual particles. If the effect of interstitial fluid is ignored, the governing equations for the translational and rotational motion of particle i with mass m_i and moment of inertia I_i can be written as

$$m_i \frac{d\mathbf{v}_i}{dt} = \sum_j \mathbf{F}_{ij}^c + \sum_k \mathbf{F}_{ik}^{nc} + \mathbf{F}_i^g, \tag{1}$$

$$I_i \frac{d\boldsymbol{\omega}_i}{dt} = \sum_j \mathbf{M}_{ij}, \tag{2}$$

where $\mathbf{v}_i$ and $\boldsymbol{\omega}_i$ are the translational and angular velocities of particle i, respectively, $\mathbf{F}_{ij}^c$ and $\mathbf{M}_{ij}$ are the contact force and torque acting on particle i by particle j or walls, $\mathbf{F}_{ik}^{nc}$ is the non-contact force acting on particle i by particle k or other sources, and $\mathbf{F}_i^g$ ($=m_i\boldsymbol{g}$) is the gravitational force. Figure 1 schematically shows the forces and torques. Various models have been proposed to calculate these forces and torques, which will be discussed below. Once the forces and torques are known, Eqs. (1) and (2) can be readily solved numerically. Thus, the trajectories, velocities and the transient forces of all particles in a system considered can be determined.

2.2. Forces Governing Particle Flow

2.2.1. Contact Forces between Particles

In general, the contact between two particles is not at a single point but on a finite area due to the deformation of the particles, which is equivalent to the contact of two rigid bodies allowed to overlap slightly in the DEM. The contact traction distribution over this area can be decomposed into a component in the contact plane (or tangential plane) and one normal to the plane, thus a contact force has two components: normal and tangential. It is very difficult to accurately and generally describe the contact traction distribution over this area and then the total force and torque acting on a particle, as it is related to many geometrical and physical factors such as the shape, material properties and movement state of particles. Alternatively, to be computationally efficient and hence applicable to multi-particle systems, the DEM generally adopts simplified models or equations to determine the forces and torques resulting from the contact between particles. Various approaches have been proposed for this purpose. The most popular force models are developed based on the consideration of contact elastic force and viscous contact damping force, including linear and nonlinear ones.

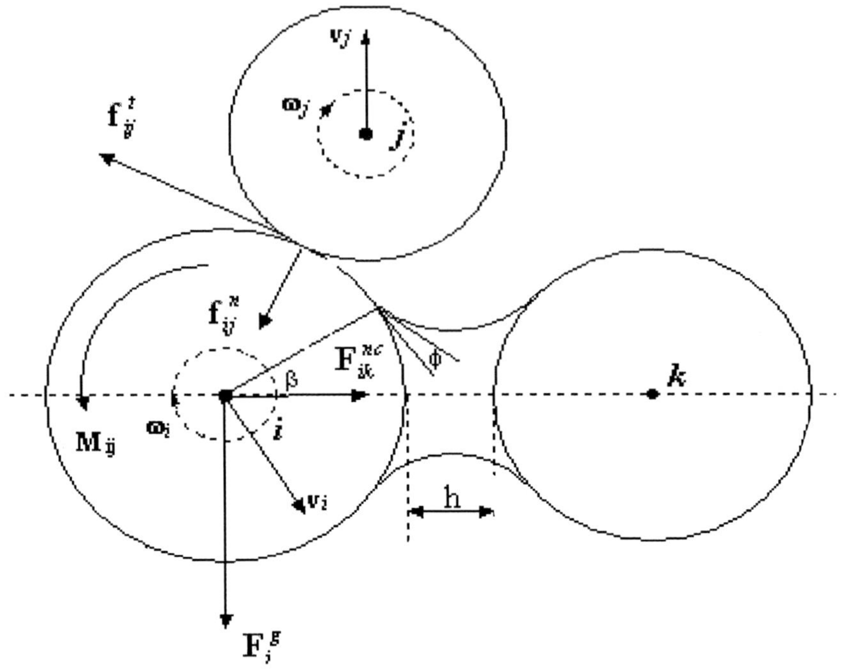

Figure 1. Schematic illustration of the forces acting on particle i from contacting particle j and non-contacting particle k (capillary force here; Yang et al. 2003b).

Linear Force Model

The linear models are the most intuitive and simple models. The most common linear model is the so-called linear spring-dashpot model proposed by Cundall and Strack (1979), where the spring is used for the elastic deformation while the dashpot accounts for the viscous dissipation. In the model, the normal force, $\mathbf{f}_{ij}^n$, includes elastic and damping components, whilst the tangential force, $\mathbf{f}_{ij}^t$, includes frictional and damping components. The normal elastic force between particles i and j, $\mathbf{f}_{ij}^{ne}$, which is expressed by the Hooke's linear spring relationship, is given by

$$\mathbf{f}_{ij}^{ne} = -K_n \delta_{ij}^n \mathbf{n}_{ij}, \tag{3}$$

where K_n is the normal spring coefficient, δ_{ij}^n is the relative normal displacement at contact, and $\mathbf{n}_{ij}$ is the unit normal vector at contact. δ_{ij}^n and $\mathbf{n}_{ij}$ are respectively given by $\delta_{ij}^n = R_i + R_j - |\mathbf{r}_j - \mathbf{r}_i|$, and $\mathbf{n}_{ij} = (\mathbf{r}_j - \mathbf{r}_i)/|\mathbf{r}_j - \mathbf{r}_i|$, where R_i and R_j are the radii of particles i and j, and $\mathbf{r}_i$ and $\mathbf{r}_j$ are the position vectors of particles i and j. The normal damping force between particles i and j, $\mathbf{f}_{ij}^{nd}$, which is modeled as a dashpot that dissipates a proportion of the relative kinetic energy, is given by

$$\mathbf{f}_{ij}^{nd} = -C_n(\mathbf{v}_{ij} \cdot \mathbf{n}_{ij})\mathbf{n}_{ij}, \tag{4}$$

where C_n is the damping coefficient, $\mathbf{v}_{ij}$ is the relative velocity of particles i and j at contact, given by $\mathbf{v}_{ij} = \mathbf{v}_i - \mathbf{v}_j + \boldsymbol{\omega}_i \times \mathbf{C}_{ij} - \boldsymbol{\omega}_j \times \mathbf{C}_{ji}$, $\mathbf{v}_i$ and $\mathbf{v}_j$ are the velocities of particle i and j, and $\mathbf{C}_{ij}$ is a vector from the mass centre of the particle to the contact point. The frictional component of the tangential force is given by

$$\mathbf{f}_{ij}^{te} = -K_t \mathbf{v}_{ij}^t, \tag{5}$$

where K_t is the tangential spring coefficient, and $\mathbf{v}_{ij}^t$ is the vector of the relative tangential displacement of particles i and j at contact. The tangential damping force is given by

$$\mathbf{f}_{ij}^{td} = C_t(\mathbf{v}_{ij} \times \mathbf{n}_{ij}) \times \mathbf{n}_{ij}, \tag{6}$$

where C_t is the tangential damping coefficient. The linear force model has been extensively used to investigate the dynamical behavior of various particulate systems (see, for example, Yang et al, 2000, 2003a; Di Renzo and Di Maio, 2004).

Nonlinear Force Model

More complex and theoretically sound model, Hertz-Mindlin and Deresiewicz model, has also been developed (see the review of Di Renzo and Di Maio (2004)). Hertz (1882) proposed a theory to describe the elastic contact between two spheres in the normal direction. He considered that the relationship between the normal force and normal displacement was nonlinear. Mindlin and Deresiewicz (1953) proposed a general tangential force model. They demonstrated that the force-displacement relationship depends on the whole loading history and instantaneous rate of change of the normal and tangential force or displacement. Due to its complication, however, the complete Hertz-Mindlin and Deresiewicz model is time-consuming for DEM simulations of granular flows often involving a large number of particles, and is therefore not so popular in the application of DEM. Instead, various simplified models based on the Hertz, and Mindlin and Deresiewicz theories have been developed for DEM modelling. An intuitive model was adopted by Langston et al. (1994). They used a direct force-displacement relation for the tangential force and the Hertz theory for the normal force. Due to its simplicity and intuitiveness, the model has been extensively used to study the dynamic behavior of granular matter (Langston et al., 1995; Zhou et al., 1999, Zhu and Yu, 2002). In this model, the normal elastic force, $\mathbf{f}_{ij}^{ne}$, described by the Hertz theory, is given by

$$\mathbf{f}_{ij}^{ne} = -\frac{4}{3}E_i^* \sqrt{R_{ij}^*}\left(\delta_{ij}^n\right)^{3/2} \mathbf{n}_{ij}, \tag{7}$$

where E_i^* and R_{ij}^* are the reduced modulus of elasticity and particle radius, respectively given by $E_i^* = E_i / \left[2(1-\nu_i^2)\right]$, $R_{ij}^* = \left(R_i R_j\right) / \left(R_i + R_j\right)$, where E_i and ν_i are the modulus of elasticity and Poisson's ratio of particle i. The normal damping force, $\mathbf{f}_{ij}^{nd}$ is given by

$$\mathbf{f}_{ij}^{nd} = -c_n \left(8 m_{ij} E_i^* \sqrt{R_{ij}^* \delta_{ij}^n}\right)^{1/2} \left(\mathbf{v}_{ij} \cdot \mathbf{n}_{ij}\right) \mathbf{n}_{ij}, \tag{8}$$

where c_n is the normal damping coefficient, and $m_{ij} = m_i m_j / (m_i + m_j)$. The frictional component of the tangential force is given by

$$\mathbf{f}_{ij}^{te} = -\mu \left|\mathbf{f}_{ij}^{ne}\right| \left(1 - \left(1 - \min\left(\left|\mathbf{v}_{\mathrm{ij}}^{\mathrm{t}}\right|, \delta_{ij}^{\max}\right) / \delta_{ij}^{\max}\right)^{3/2}\right) \hat{\mathbf{v}}_{\mathrm{ij}}^{\mathrm{t}}, \tag{9}$$

where μ is the sliding friction coefficient, $\hat{\mathbf{v}}_{ij}^t = \mathbf{v}_{ij}^t / \left|\mathbf{v}_{ij}^t\right|$, and $\delta_{ij}^{\max}$ is the relative tangential displacement when the sliding starts, given by $\delta_{ij}^{\max} = \mu \delta_{ij}^n (2 - \nu_i) / [2(1-\nu_i)]$. Eq. (9) suggests that the friction is described by the theory proposed by Mindlin and Deresiewicz (1953) prior to the sliding (i.e., when $\left|\mathbf{v}_{ij}^t\right| < \delta_{ij}^{\max}$) and the Coulomb friction model when the particles at contact start to slide relatively. The tangential damping force is given by

$$\mathbf{f}_{ij}^{td} = 2 c_t \left(1.5 \mu m_{ij} \left|\mathbf{f}_{ij}^{ne}\right| \sqrt{1 - \left|\mathbf{v}_{ij}^t\right| / \delta_{ij}^{\max}} \Big/ \delta_{ij}^{\max}\right)^{1/2} (\mathbf{v}_{ij} \times \mathbf{n}_{ij}) \times \mathbf{n}_{ij}, \tag{10}$$

where c_t is the tangential damping coefficient. Note that Eq. (10) implies that there is no contribution of the tangential damping to the tangential force when sliding occurs.

The inter-particle forces act at the contact point between particles rather than the mass centre of the particle and they will generate a torque causing particles to rotate. The torque $\mathbf{m}_{ij}$ causing particle i to rotate is contributed by the tangential and normal components of the traction distribution, $\mathbf{m}_{ij}^t$ and $\mathbf{m}_{ij}^n$, respectively. Implemented in a discrete simulation, $\mathbf{m}_{ij}^t$ is usually expressed as

$$\mathbf{m}_{ij}^t = \mathbf{C}_{ij} \times \mathbf{f}_{ij}^t. \tag{11}$$

Compared with the contribution of the tangential component, the determination of the contribution of the normal component, usually called as rolling friction torque, is very difficult and is still an active research area (Johnson, 1985; Brillianton and Pöschel, 1998). One of the popular explanations is beaded on the consideration of the asymmetry of normal traction distribution. For many inelastic bodies such as viscoelastic particles, the normal traction distribution on the contact area between two contacting particles is asymmetrical due to the relative rotation of particles, as shown in Figure 2. The asymmetry leads to a non-zero torque, $\mathbf{m}_{ij}^{n}$, given by

$$\mathbf{m}_{ij}^{n} = -\min\left\{\mu_{r,ij}\left|\mathbf{f}_{ij}^{n}\right|,\ \mu'_{r,ij}\left|\boldsymbol{\omega}_{ij}^{n}\right|\right\}\hat{\boldsymbol{\omega}}_{ij}^{n}, \tag{12}$$

where $\mu_{r,ij}$ is the rolling friction coefficient, $\mu'_{r,ij}$ is the rotational stiffness and $\hat{\boldsymbol{\omega}}_{ij}^{n} = \boldsymbol{\omega}_{ij}^{n} / \left|\boldsymbol{\omega}_{ij}^{n}\right|$, $\boldsymbol{\omega}_{ij}^{n}$ is the component of the vector of the relative angular velocity of particles i and j in their contact plane. $\mathbf{m}_{ij}^{n}$ is often referred to as the rolling friction torque which provides a resistance to relative rolling motion between particles (Johnson, 1985; Brillianton and Pöschel, 1998; Zhu and Yu, 2003). The rolling friction is considered to be negligible in many DEM models. However, it has been shown that the torque plays a significant role in some cases such as the formations of shear band (Iwashita and Oda, 1998) and heaping (Zhou et al., 1999), and movement of a single particle on a plane (Zhou et al., 1999; Zhu and Yu, 2006).

As discussed above, linear and nonlinear contact force models have been developed for DEM simulations. Theoretically, the more complex nonlinear models simplified based on the Hertz and Mindlin-Deresiewicz theories should be more accurate than the linear model. However, in contrast, the numerical investigations conducted by Di Renzo and and Di Maio (2004) showed that the simple linear model sometimes gives better results. This may be because theoretical models are often based on geometrically ideal particles, whereas there are no such perfect particles in practical applications. Selection of proper parameters also plays an important role in generating accurate results. Complicated models may also consume computational time with insignificant gain in DEM simulations. Moreover, as recently demonstrated by Zhu and Yu (2006), most of the force models were developed focused on one or two aspects or based on simplified conditions, their combination in DEM simulation may lead to theoretical or conceptual problems. These issues should be considered in the further development of force models.

The above models are suitable for spherical particles. The treatment of the contact force between non-spherical particles is far more complicated. Two techniques have been proposed to cope with particles of irregular shapes. One is to model a non-spherical particle as a collection of spherical particles (Gallas and Sokolowski, 1993; Bertrand et al., 2005). The advantage of the strategy is that it can be used to handle particles with very complex shapes, and only a contact model for spherical particle is required. The other technique considers that particles are of a given shape, such as ellipsoid, polygon and cylinder, and determines whether or not there is a contact between two such neighboring particles by solving the underlying mathematical equations (Kohring et al., 1995; Favier et al., 2001; Dziugys and Peters, 2001; Munjiza et al., 2003). The contact force model used in this approach is one for spherical particles (Favier et al., 2001) or its modification (Kohring et al., 1995). This approach is more accurate than the first one, but is computationally more demanding.

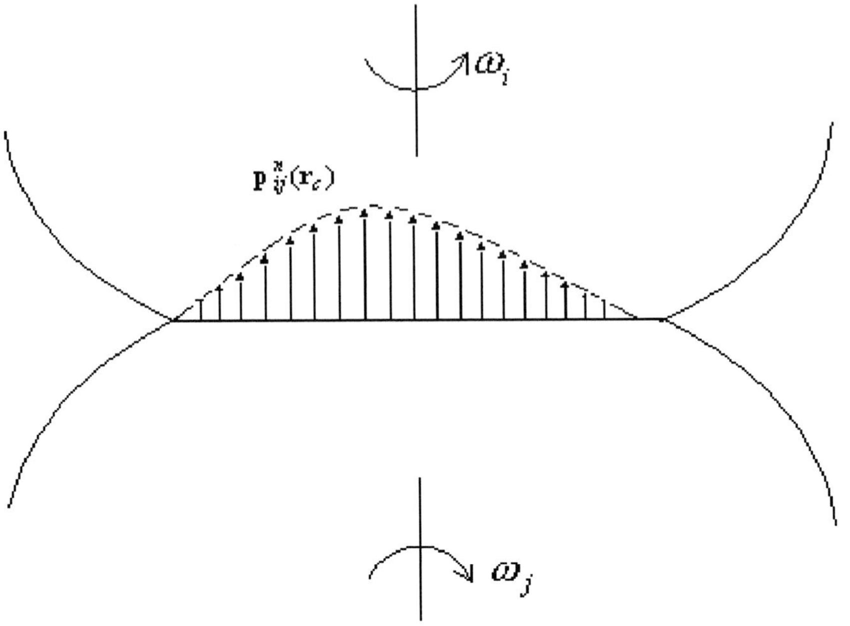

Figure 2. Normal traction distribution exerted on particle *i* due to the collision with particle *j* (Zhu and Yu, 2003).

2.2.2. Non-contact Forces between Particles

When fine particles are involved and/or moisture exists, non-contact interparticle forces may affect the flow behaviour of particles significantly. Often the non-contact forces involve a combination of three fundamental forces, i.e. van der Waals force, capillary force and electrostatic force, which can act concurrently or successively to different extents.

The van der Waals force is the force between molecules having closed shell and consists of several types of interactions (Israelachvili, 1991). The van der Waals force between molecules is proportional to h^{-6}, where h is the separation distance. For two particles, the van der Waals force, obtained by integration of all force for all pair of molecules, decays much more slowly with distance than for two molecules, and is proportional to h^{-2}. The Hamaker theory (Hamaker, 1937) is usually used to calculate the van der Waals force in DEM simulation. This theory is based on the assumption of "pairwise additivity" and starts from the interactions between individual atoms and molecules and postulates their additive so that the van der Waals attraction between macroscopic bodies can be calculated by integration over all pairs of atom and molecules (Hamaker, 1937). For smooth spheres, the van der Waals force can be calculated according to the equation as follows (Hamaker, 1937; Israelachvili, 1991)

$$\mathbf{F}^{v} = -\frac{Ad_p}{24h^2}\hat{\mathbf{n}}_{ij}, \tag{13}$$

where A is the Hamaker constant, d_p is the particle diameter, $\hat{\mathbf{n}}_{ij}$ is the unit vector connecting the mass centers of particle i and j. It can be seen that the so-called Hamaker constant is the only parameter to be determined for given geometry and materials. Theoretically speaking, this constant depends on many variables related to physical and chemical properties such as particle surface roughness or asperity, medium chemistry and so on. It is difficult to evaluate the effects of these variables comprehensively. In numerical modelling, this constant is often treated as a lumped parameter and determined empirically (Yang et al., 2000). The equation from the Hamaker theory shows the van der Waals force becomes infinite as two particles get into contact ($h = 0$), which induces a singularity problem in DEM simulation. To solve this problem, a "cut-off" distance is assumed in the calculation of this force, and this distance ranges from $0.165 \sim 1\text{nm}$ (Krupp, 1967; Israelachvili, 1991; Yang et al., 2000).

The electrostatic force exists between charged particles. It can be catagorised into three types: Coulombic force, image-charge force and space charge force. The electrostatic forces between particles are usually approximated by the classical Coulomb equation, given by (Krupp, 1967)

$$\mathbf{F}^{e} = -\frac{Q^2}{16\pi q_0 h^2}\left(1-\frac{h}{\sqrt{(R^2+h^2)}}\right)\hat{\mathbf{n}}_{ij}, \tag{14}$$

where Q is the particle charge, q_0 is the permittivity of vacuum, and R is the particle radius. Rumpf (1962) compared van der Waals attractive force with electrostatic forces produced by contact potential or by excess charges. The results showed that van der Waals force is greater by one order of magnitude than the electrostatic force for micro-sized particle.

The capillary force is mainly due to the surface tension at solid/liquid/gas interfaces. For the cases shown in Figure 1, the force due to the reduced hydrostatic pressure in the bridge itself can be given by (Fisher, 1926; Princen, 1968)

$$\mathbf{F}^{l} = -[2\pi\gamma R\sin\phi\sin(\theta+\phi)+\pi R^2\Delta p\sin^2\phi]\hat{\mathbf{n}}_{ij}, \tag{15}$$

where γ is the liquid surface tension, ϕ is the half-filling angle, and θ is the contact angle. The reduced hydrostatic pressure within the bridge, Δp, can be given by the Laplace-Young equation, which states that the mean curvature of the meniscus profile is constant and proportional to Δp. To implement the capillary force in DEM simulation, liquid distribution among particles has to be determined. There are three treatments on the liquid distribution based on different assumptions: 1) liquid can transport among particles and is distributed evenly among all gaps smaller than the rupture distance (Muguruma et al., 2000); 2) liquid is distributed evenly among particles and liquid transport between particles can be neglected if liquid viscosity is sufficiently small (Mikami et al., 1998); 3) liquid is distributed evenly and not transferable among particles (Yang et al., 2003b). Once the particle gap is smaller than the rupture distance, a liquid bridge is formed and the liquid assigned to a particle will be evenly distributed to its liquid bridges.

In real systems, the relative importance of these forces varies, depending on a range of variables as illustrated in Figure 3.

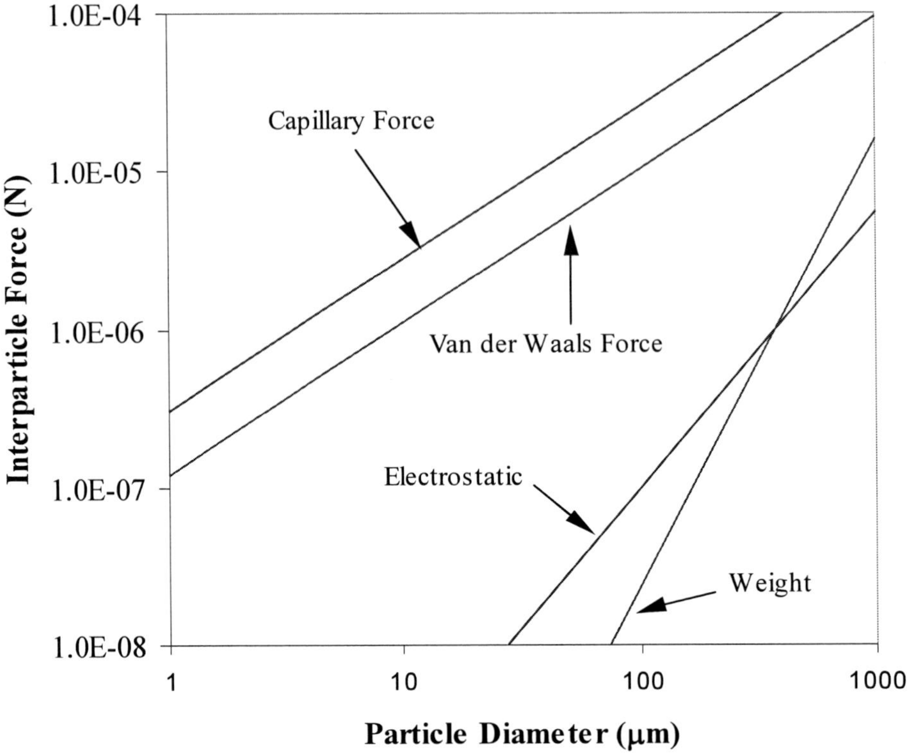

Figure 3. Comparison of the magnitude of inter-particle forces calculated based on: (i) van der Waals forces: $H_a = 6.5\times10^{-20}J$, $h = 1.65$ Å, (ii) capillary force: $\gamma = 72.8\times10^{-2}Nm^{-1}$ (water); (iii) electrostatic forces: $\varepsilon_r = 1$; and (iv) weight: $\rho_g = 3\times10^3 kgm^{-3}$ (Rumpf, 1962).

2.3. Transition from Discrete to Continuum

By use of a proper averaging procedure, the discrete particle system can be transferred into a corresponding continuum system in terms of local average variables. Assume that the mass density ρ, the velocity $\mathbf{u}$, the angular velocity $\boldsymbol{\omega}$ are respectively

$$\rho = \int_{T_t}\sum_i h_i m_i \mathrm{d}s, \tag{16}$$

$$\mathbf{u} = \frac{1}{\rho}\int_{T_t}\sum_i h_i m_i \mathbf{v}_i \mathrm{d}s, \tag{17}$$

$$\boldsymbol{\omega} = \frac{1}{\lambda}\int_{T_t}\sum_i h_i I_i \boldsymbol{\omega}_i \mathrm{d}s, \tag{18}$$

where $h_i = h(\mathbf{r}_i - \mathbf{r}, s - t)$, $h = h(\mathbf{r}, t)$ is the weighting function, which is positive in the limited domain $\Omega = \{(\mathbf{r}, t) | \mathbf{r} \in \Omega_p \subset R^3, t \in T_t \subset R\}$; otherwise, zero. $\mathbf{r}$ and t are, respectively, the position vector and time to which the average value of the considered quantities is assigned, and $\lambda = \int_{T_t} \sum_i h_i I_i \mathrm{d}s$. Differentiating ρ, $\rho\mathbf{u}$ and $\lambda\boldsymbol{\omega}$ by use of Eqs. (1), (2), (16)-(18), we can obtain the balance equations of mass and linear momentum and angular momentum of the continuum system, respectively given by (Zhu and Yu, 2002)

$$\mathrm{D}(\rho) + \rho\nabla \cdot \mathbf{u} = 0, \tag{19}$$

$$\mathrm{D}(\rho\mathbf{u}) + \rho\mathbf{u}\nabla \cdot \mathbf{u} = \nabla \cdot \mathbf{T} + \rho\mathbf{g}, \tag{20}$$

$$\mathrm{D}(\lambda\boldsymbol{\omega}) + \lambda\boldsymbol{\omega}\nabla \cdot \mathbf{u} = \nabla \cdot \mathbf{M} + \mathbf{M}', \tag{21}$$

where $\mathrm{D}(\cdot)$ represents the material derivative of tensor. $\mathbf{T}$ is the stress tensor, $\mathbf{M}$ is the couple stress tensor, and $\mathbf{M}'$ is the rate of supply of internal spin to particles, respectively given by

$$\mathbf{T} = -\int_{T_t} \sum_i h_i m_i \mathbf{v}_i' \otimes \mathbf{v}_i' \mathrm{d}s + \int_{T_t} \sum_i \sum_{j>i} g_{ij} \mathbf{d}_{ij} \otimes \mathbf{f}_{ij} \mathrm{d}s, \tag{22}$$

$$\mathbf{M} = -\int_{T_t} \sum_i h_i I_i \mathbf{v}_i' \otimes \boldsymbol{\omega}_i \mathrm{d}s + \frac{1}{2} \int_{T_t} \sum_i \sum_{j>i} g_{ij} \mathbf{d}_{ij} \otimes (\mathbf{m}_{ij} - \mathbf{m}_{ji}) \mathrm{d}s, \tag{23}$$

$$\mathbf{M}' = \frac{1}{2} \int_{T_t} \sum_i \sum_{j>i} (\mathbf{m}_{ij} + \mathbf{m}_{ji})(h_i + h_j) \mathrm{d}s, \tag{24}$$

where $\mathbf{v}_i' = \mathbf{v}_i - \mathbf{u}$ is the fluctuant velocity of particle i with respect to the average velocity, $\mathbf{d}_{ij}$ is the part of the branch vector connecting the mass centres of particles i and j within domain Ω_p. g_{ij} is the weighting coefficient determined by the weighting function $h(\mathbf{r}, t)$, given by $g_{ij} = \int_0^1 h(\bar{\mathbf{r}}_i + r\mathbf{d}_{ij} - \mathbf{r}, s - t)\mathrm{d}r$, $\bar{\mathbf{r}}_i = \mathbf{r}_i$ if $\mathbf{r}_i - \mathbf{r} \in \Omega_p$; otherwise, $\bar{\mathbf{r}}_i$ is the position vector of the point of intersection of vector $\mathbf{d}_{ij}$ and boundary $\partial\Omega_p$ of domain Ω_p.

A particle may collide with a wall. Therefore, the interaction between particles and walls should be included in the continuum description. The contribution of the interaction between particles and walls to stress and couple stress can be respectively given by

$$\mathbf{T}^b = \int_{T_t} \sum_i g_i^b \mathbf{d}_i^b \otimes \mathbf{f}_i^b \mathrm{d}s, \tag{25}$$

$$\mathbf{M}^b = \int_{T_t} \sum_i g_i^b \mathbf{d}_i^b \otimes \mathbf{m}_i^b \mathrm{d}s, \tag{26}$$

where $\mathbf{d}_i^b$ is the ray from the center of mass of particle i to boundary $\partial\Omega_p$ of domain Ω_p, via a point on the wall and perpendicular to the tangential plane of the point. g_i^b is the weighting coefficient corresponding to walls, given by $g_i^b = \int_0^1 g(\mathbf{r}_i + r\mathbf{d}_i^b - \mathbf{r}, s-t)\mathrm{d}r$.

Eqs. (16)-(18), (22)-(26) give the relationship between microscopic and macroscopic quantities of granular materials. Therefore, the DEM results can be used to calculate the macroscopic variables such as mass density, velocity, stress and couple stress according to these equations. However, these macroscopic quantities depend on the weighting function $h(\mathbf{r}, s)$. So, the selection of a proper weighting function is important in applying these equations. The need to find out a suitable weighting function has been noticed, a weighting function has been recommended, and given by (Zhu and Yu, 2002, 2005)

$$h(\bar{\mathbf{r}}, \bar{t}) = \begin{cases} \dfrac{1}{2\pi\sqrt{2\pi}} \dfrac{cL_t}{(L_t^2 - \bar{t}^2)L_p(L_p^2 - \bar{r}^2)} \exp\left[-\dfrac{1}{2}\left(\ln^2\dfrac{L_t + \bar{t}}{L_t - \bar{t}} + \ln^2\dfrac{L_p + \bar{r}}{L_p - \bar{r}}\right)\right], & (\bar{\mathbf{r}}, \bar{t}) \in \Omega \\ 0, & \text{others} \end{cases} \tag{27}$$

where c is the normalized constant of the weighting function. Parameter L_p or L_t determines the amount of the contributing particles to a probe point or time, and the magnitude of contribution of every particle as well, as illustrated in Figure 4. The specification of appropriate values of L_p and L_t is a significant part of the application of the averaging method. If L_p and L_t are too small, the resulting macroscopic quantities fluctuate, giving unreasonable average results. On the other hand, if they are too large, the local properties of these quantities may be eliminated. Our previous studies on a hopper flow showed that, to balance the friction, we can choose $L_p = 4.0d$ and $L_t = 4.0\sqrt{d/g}$. Such selection can ensure the averaged quantities smoothly vary in a considered domain and retain the local properties of these quantities as largely as possible. However, for different systems, the values may be different.

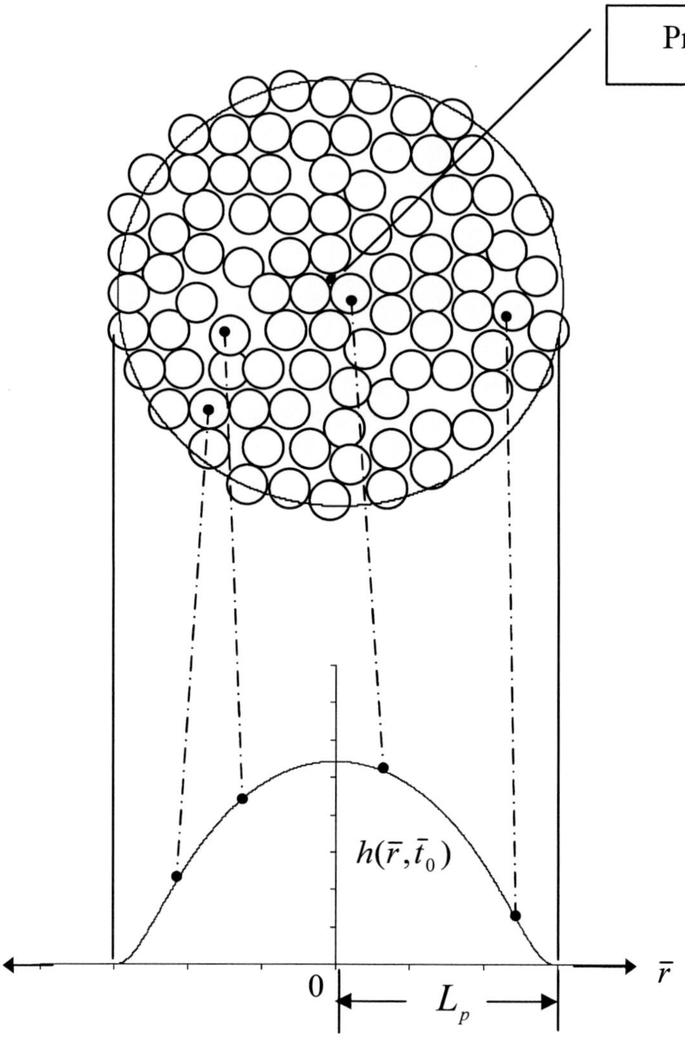

Figure 4. Concept of weighting for space averaging (the same idea applied to time averaging).

According to the averaging method, the maroscopic properties obtained conform with those in the common balance equations in the continuum approach. Moreover, the method is in principle applicable to all flow regimes. These advantages are particularly useful to the analysis of complex granular flows such as those in hoppers and blast funace (Zhu & Yu, 2005).

3. Application Examples

DEM has been extensively used to study the dynamics of various particulate systems. The main results of these studies have been reviewed elsewhere (Zhu et al., 2007a and b). Here, we only discuss a few typtical case studies to illustrate the effectiveness of DEM simulation, including packing of cohesive particles, hopper flow and bladed mixer.

3.1 Packing of Cohesive Particles

As discussed above, cohesive particles are associated with various non-contact forces including, for example, the van der Waals and electrostatic forces associated with fine particles and/or capillary force as a result of liquid bridge among particles. These forces affect the packing of particles significantly, as highlighted by the variation of packing density with particle size and liquid content (German, 1989).

The packing of fine particles can be simulated by DEM if the van der Waals force is introduced. The applicability of DEM to such systems can be illustrated by the quanlitative agreements between the simulated and measured results under compraiable conditions, as shown in Figure 5. It can be seen that packing structures become looser as particle size decreases, and the simulated relationship between porosity and particle size is comparable with the measured data (Feng, 1998). Clearly, the packing structure varies with particle size or porosity, and the coordination number decreases with particle size. Quantitatively, the structural variation can be identified from the RDF (radial distribution function). As shown in Figure 6, as particle size decreases, the first peak in RDF becomes narrower, with a sharp decrease to the first minimum. The first component of the second peak vanishes when particle size is less than 100 μm and the peaks beyond the second one gradually vanish. When particle size is less than 100 μm, the gravity is not the dominant force, and the van der Waals becomes more important. The cohesive force provides a resistant force to the relative motion between particles, leading to relatively loose packing.

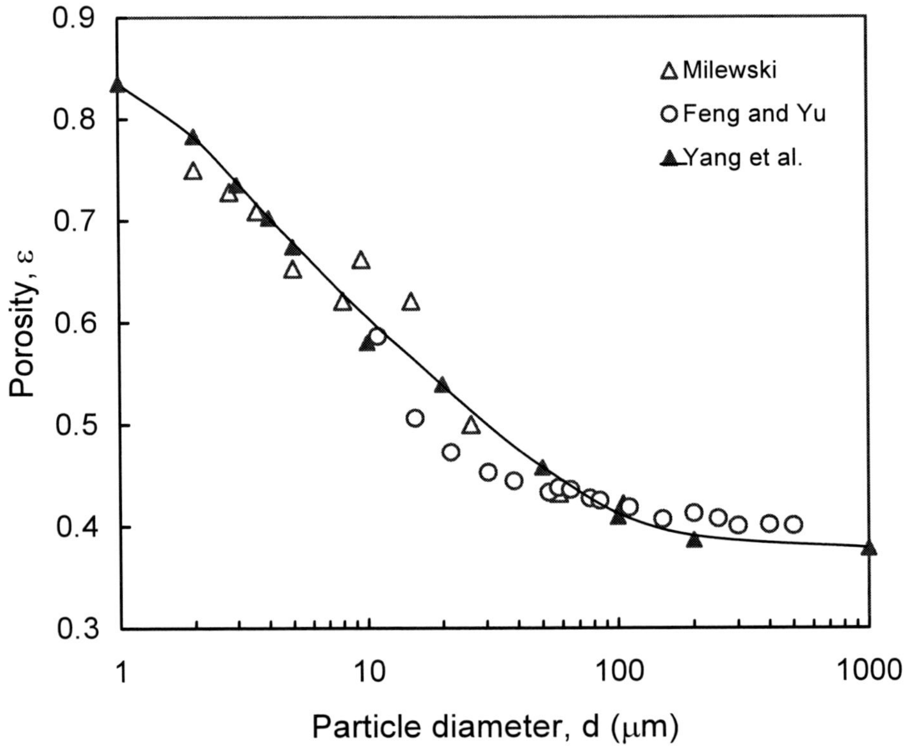

(a)

Figure 5. Continued on next page..

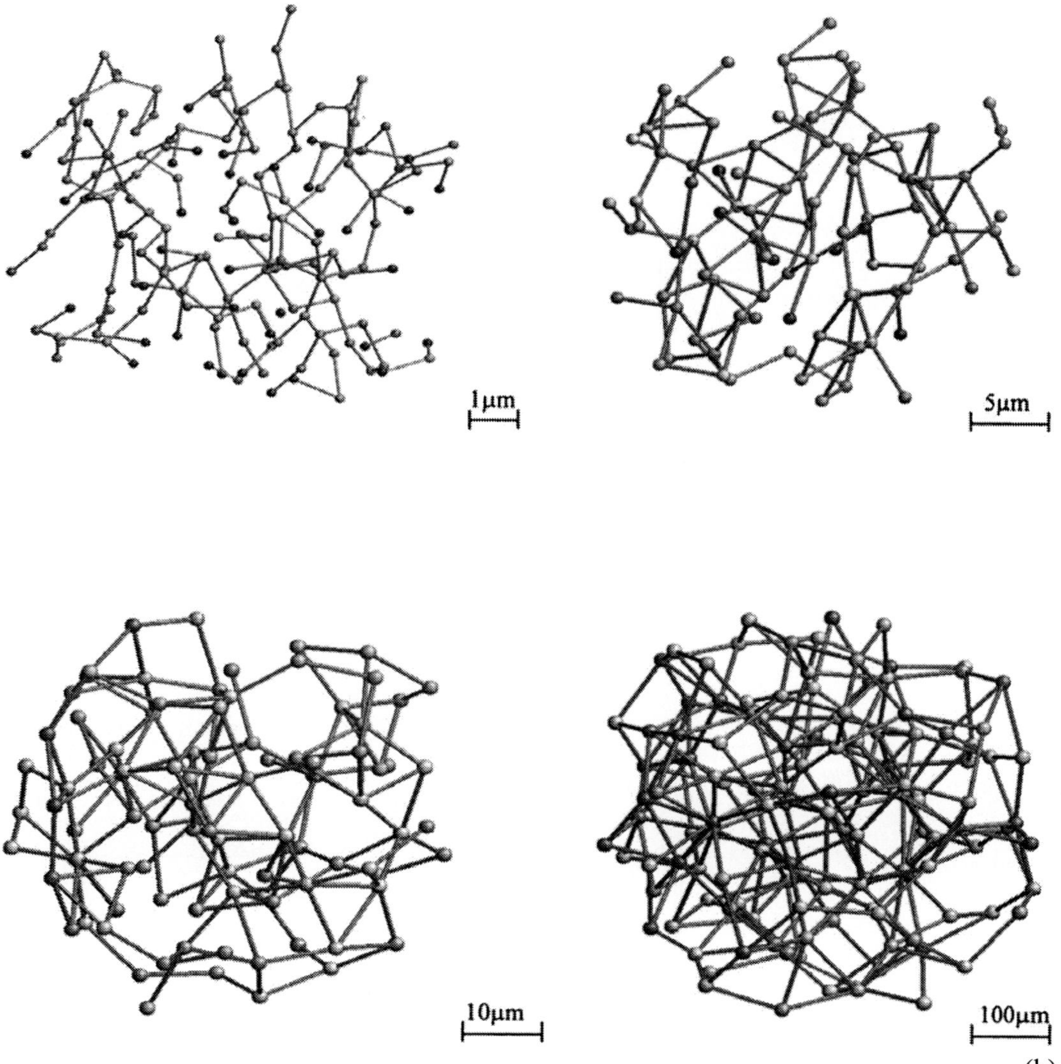

Figure 5 (a), Porosity as a function of particle size: simulated vs. measured; and (b) the contact networks in a spherical sample taken from the packing of different sized particles, where small spheres represent the centres of particles, and sticks represent the contacts between particles (Yang et al., 2000).

The capillary force plays a similar role to the van der Waals force. By explicitly taking into account the capillary force, the packing of wet spheres with different moisture contents can be simulated. It was shown that the simulated relationship between dry-based porosity and moisture content is reasonably consistent with that measured (Yang et al.; 2003b). Further, it was found that the packing structures, if expressed as a function of porosity, are qualitatively comparable to those for fine particles.

Cohesive forces may have different forms, e.g. van der Waals force or capillary force, but function similarly. It is therefore very helpful to formulate a general equation to describe the relationship between porosity and cohesive forces. For the packing of fine particles in air or vacuum, it was found that porosity can be related to the ratio of van der Waals force to the gravity force (Yang et al., 2000). For the packing of particles in liquid, if the concept of effective gravity defined as the difference between the gravity and buoyancy forces is used, then their relationship is also applicable (Dong et al., 2006). It has been reported that this relationship also applies to other cohesive force such as capillary force (Yang et al., 2003b). The results further support the argument that there is a general relationship between porosity and inter-particle force for commonly used particles, as illustrated in Figure 7 (Yu et al., 2003). Since the relationship concerns only the final state of particle packing, it can be treated as an equation of state highlighting the importance of inter-particle forces governing the state of particle packing.

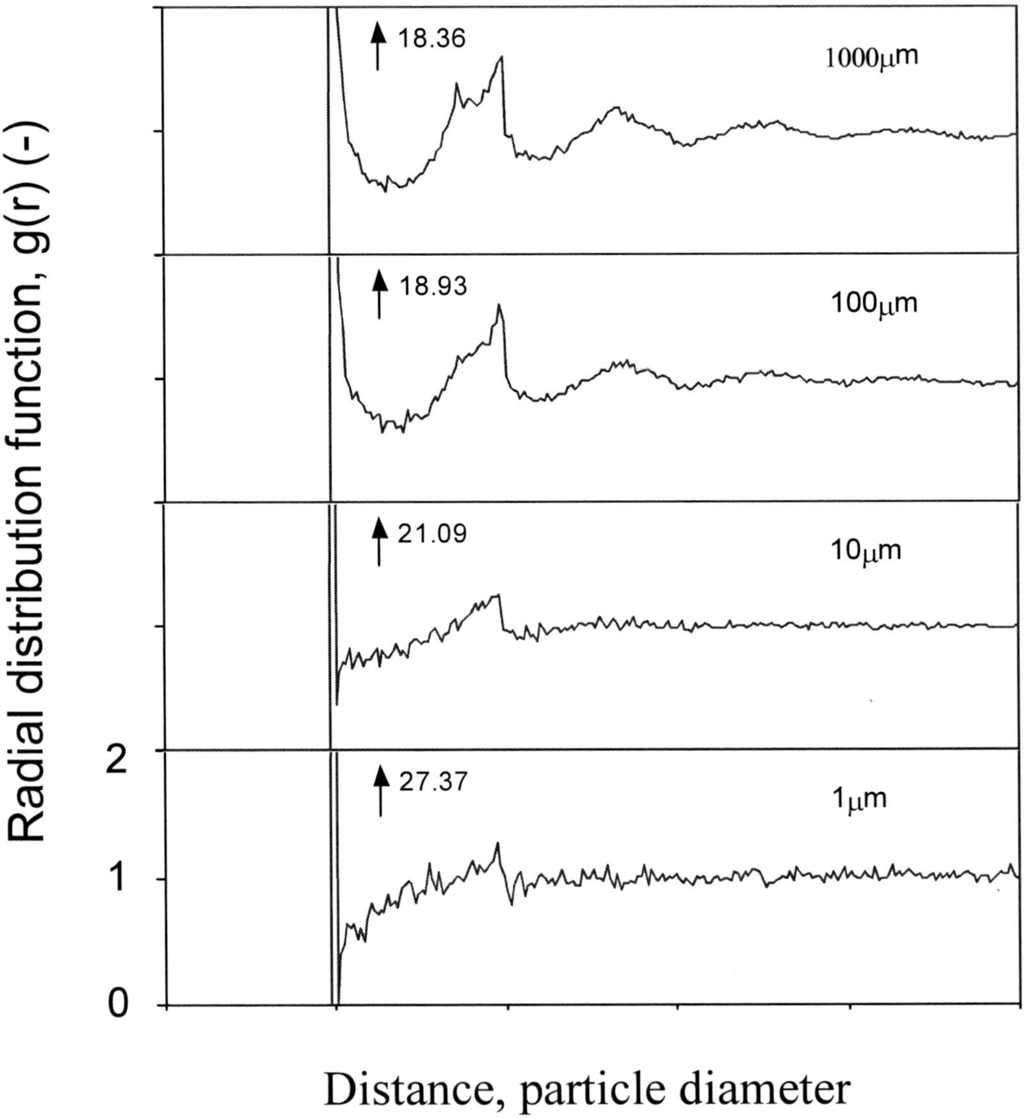

Figure 6. Radial distribution function as a function of particle size (Yang et al., 2000).

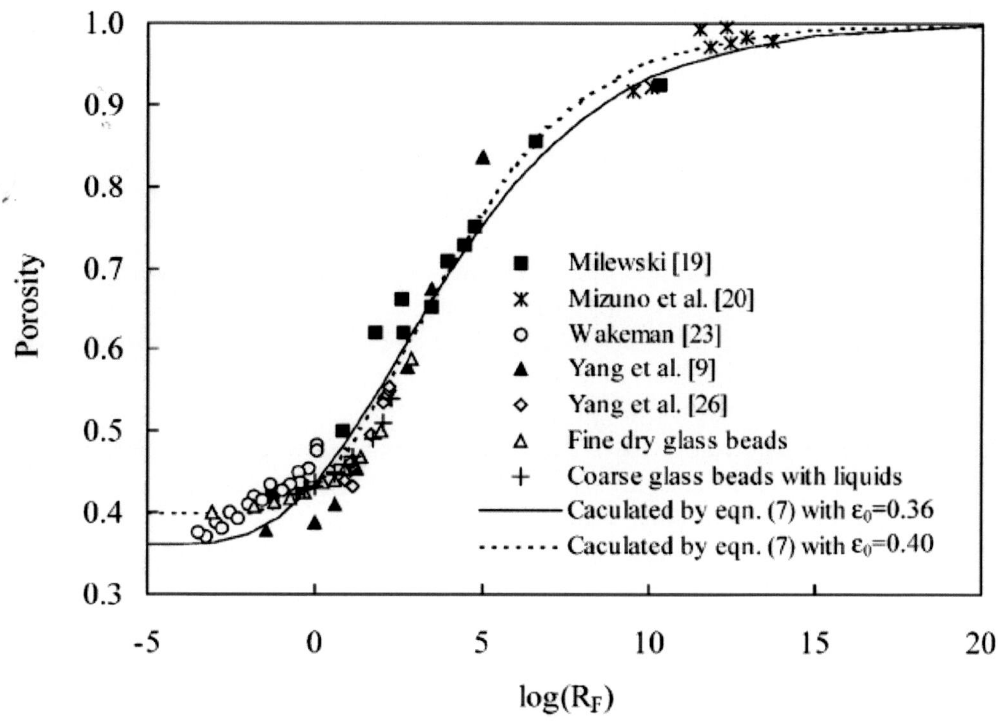

Figure 7. Porosity as a function of interparticle force, data from various sources and generated using various method of analysis (Yu et al., 2003).

3.2. Hopper Flow

Hoppers are widely applied in engineering practice mainly due to their capacity to enhance flow conditions for granular materials. They must be properly designed for reliable control. Therefore, a comprehensive understanding of the dynamic behaviour of granular flow in a hopper is essential. Previous work in this area is mainly at a macroscopic or bulk scale (Janssen, 1895, Jenike et al., 1973), which is helpful in developing a broad understanding of hopper flow of particular interest. However, the methods developed thus far rely on the macroscopic assumptions and ignore the effect of the microstructure of flow, which makes it difficult to generate a reliable method for the general purpose of design and control without substantial empirical input. In fact, it has been illustrated that the continuum models developed based on the plasticity theory or kinetic theory can not satisfactorily be applied to hopper flow (Potapov and Campbell, 1996). In recent years, to overcome this problem, numerical simulations have been developed to study the dynamics of hopper flow at a microscopic or particle scale (Zhu and Yu, 2005).

It is important to understand the microscopic structure and its relations to the mechanisms governing the hopper flow. DEM simulation takes into account the discrete nature of granular materials, and therefore is very effective for this purpose. The particle velocity or kinetic energy, flow structure and force structure of granular materials in hoppers can be investigated based on the information from DEM simulation. As an example, Figures 8 and 9 show the microscopic structure of the granular flow in a cylindrical hopper with flat bottom. It can be seen from Figure 8 that DEM simulations confirm that there are several flow zones in a hopper flow. In different zones, the flow pattern, flow structure and force structures are different. These studies indicate that the hopper flow can be qualitatively described in terms of these spacial distributions. The results shown in Figure 9 indicate that the statistical distributions of the properties such as velocity, porosity and force can be used to describe the hopper flow quantitatively. It is suggested that variables such as velocity and force vary in a large range, and geometry of hopper and particle material properties affect the flow.

As discussed above, through a proper averaging method, macroscopic quantities such as density, velocity, stress and couple stress can be obtained based on the DEM results. Figure 10 gives an example of the application of such averaging method, where the velocity, density, stress and couple stress distributions of a hopper flow are shown. It can be observed that these distributions are largely related to the corresponding microscopic structures. In particular, the results about velocity and mass density are qualitatively in agreement with the experimental results and the microdynamic analysis. The consistency between the micro- and macro-scopic variables confirms the applicability of the average method to the granular flow in hoppers. The results on stresses indicate that the distributions of the stresses are related to the force structures, controlled by the magnitude and direction of the interaction forces and the contact direction between particles. Only near the walls of hopper, couple stress is significant; when far from the walls, it can be ignored.

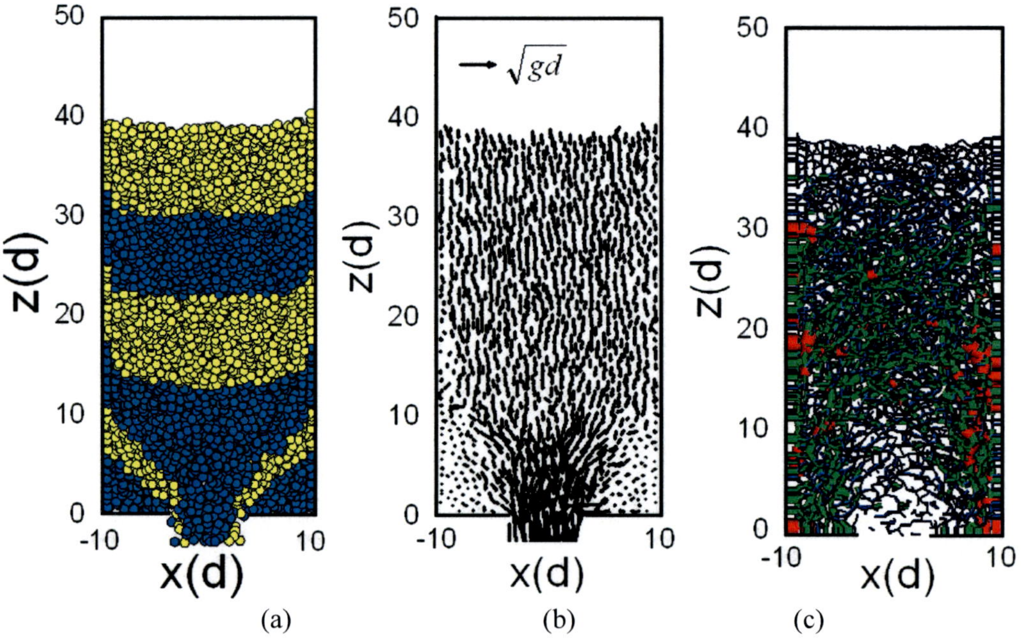

Figure 8. Snapshots of: (a) flow pattern; (b) velocity field; and (c) force structure of the unsteady state granular flow in a cylindrical hopper with flat bottom (Zhu and Yu, 2005).

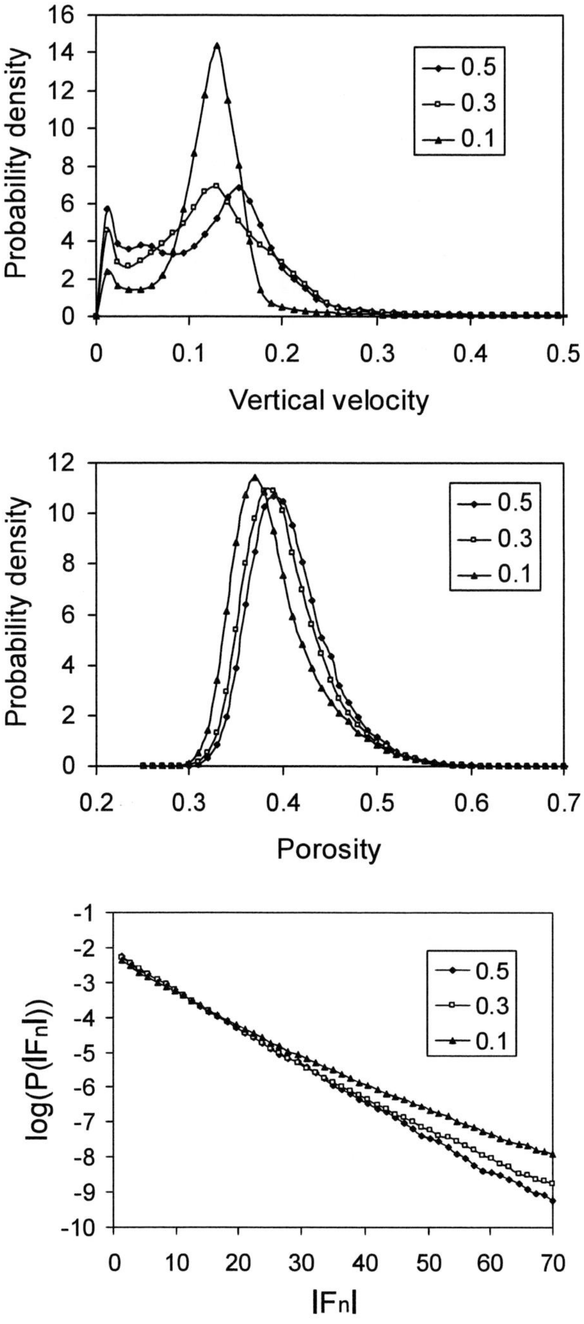

Figure 9. Statistical distributions of vertical velocity, porosity and normal forces of the steady state granular flow in a cylindrical hopper with flat bottom for different wall roughness (Zhu and Yu, 2004).

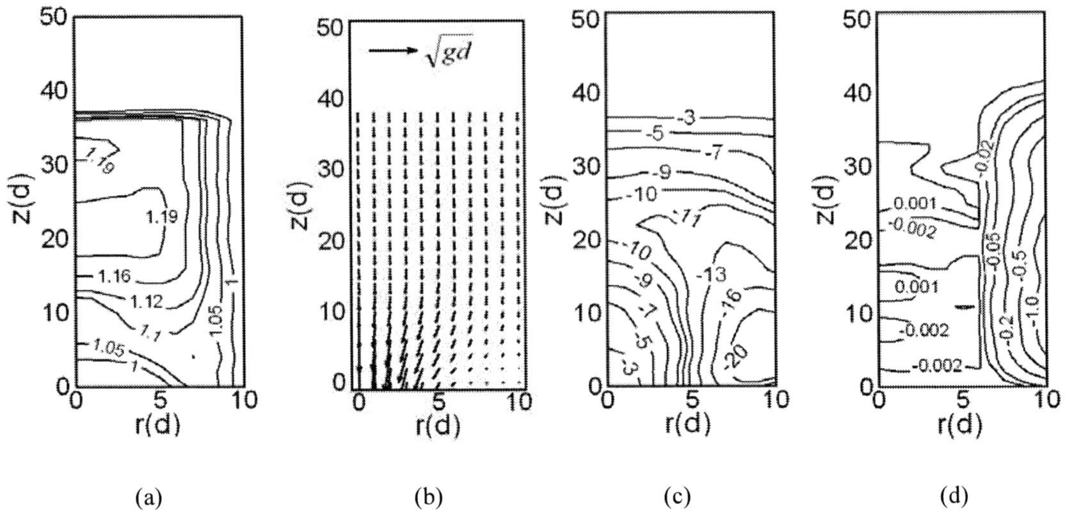

Figure 10. Distributions of: (a) mass density; (b) velocity; (c) stress; and (d) couple stress of the unsteady state granular flow in a cylindrical hopper with flat bottom (Zhu and Yu, 2005).

3.3 Bladed Mixer

The mixing of particular materials in mechanically stirred vessels is an important process widely used in various industries, and has been a subject of research for years (see, for example, reviews by McCarthy et al., 1996). Extensive work has been carried out in this area. While significant progress has been made, previous experimental studies have been largely focusing on the phenomenological description of the process, and therefore, to date, reliable scale-up, design and control/optimization of processes are largely ad hoc.

Research at an individual particle level would improve our knowledge of the flow behavior since the bulk behavior of flow depends on the collected outcome of the interactions among particles and between particle and devices. Earlier work was mainly conducted by photographing experiments through a transparent wall, or by freezing bed to reveal the inner flow patterns, which were laborious and suffered a great deal of uncertainty (Bridgwater and Bagster, 1968; Malhotra et al, 1988). Recently, some non-invasive experimental techniques have been developed, such as Positron Emission Topography (PET) (Broadbent et al., 1995), Positron Emission Particle Tracking (PEPT) Parker et al., 1997) and Magnetic Resonance Imaging (MRI) (Nakagawa et al., 1993). While these experiment techniques are useful, they all have their limitations and particularly in information generation. As an illustration, Table 1 shows the advantages and disadvantages of PEPT relative to DEM, or vice versa.

DEM simulation can be validated by compared quantitatively with experimental results at particle scale. Figure 11 shows the comparison of the flow of particles in vertical bleaded mixer, where the experimental results were obtained by PEPT. The results show that there is a good agreement between the simulated and measured flow fields of particles in the vertical bladed mixer under comparable experimental conditions (Stewart et al., 2001).

Table 1. PEPT vs DEM for particle scale research (Zhou et al., 2004)

Technique	PEPT	DEM
Feature	An experimental technique for tracing the motion of a radioactively labeled particle within a particle system	A numerical technique for simulating the dynamics of a particle system
Information generated	Trajectory of a labeled particle from which flow field can be derived	Trajectories and velocities of all particles Particle/particle and particle/device interactions
Advantages	Real physical measurement Tracing for long time	Low cost Full dynamic information
Disadvantages	Expensive No force information and hence incomplete dynamic information Indirect link between behaviours of a single particle and particles in a system	Limited by computing capability (relatively small number of particles and short time simulation) Only applied to situations where equations for calculating inter-particle forces are well established

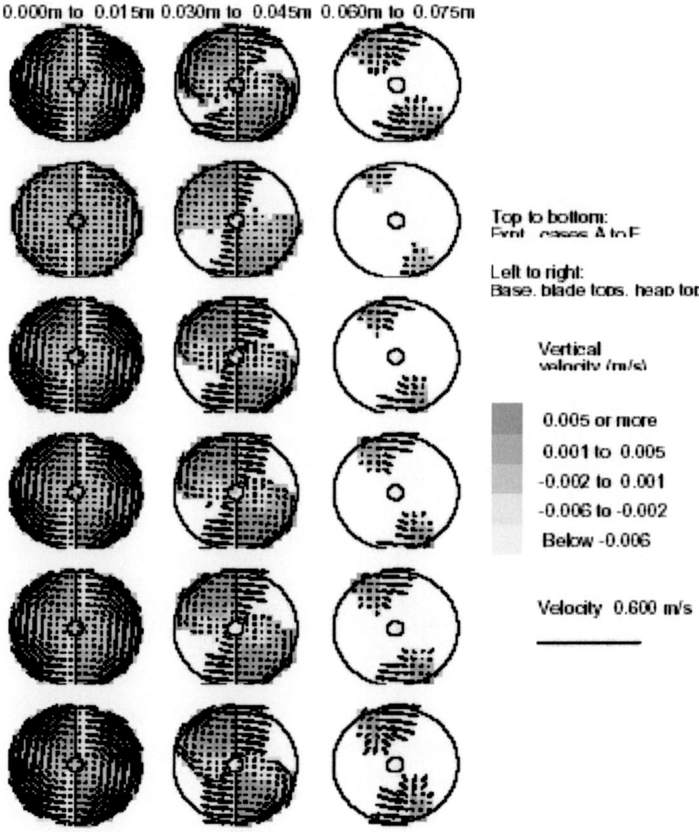

Figure 11. Velocity fields of a bladed mixer from experiment and DEM simulations. The cell shade indicates the vertical velocity while the vectors show horizontal velocity (Stewart et al., 2001).

Simulations using DEM can generate dynamic information such as transient forces acting on individual particles which is extremely difficult to obtain by physical experimentation. Such information is very important to understanding the underlying physics governiong the dynamics of the particulate systems. Figure 12 shows an example, where the force networks of particles in different horizontal sections at different depths of a cylindrical bladed mixer are shown. It can be seen that contact forces propagate mainly in the angular direction, starting from the particles in front of the blades where they are driven by the blades and the wall and then forced to move in the system. This is the reason why particles at the lower part mainly flow in an angular direction (Zhou et al, 2004). Moreover, the wall generates strong forces to restrict particles flowing outside of the mixer, which causes that particles in force of the blade and those at the top part in particular move toward to the centre of the mixer. Therefore, the flow and force structures are related.

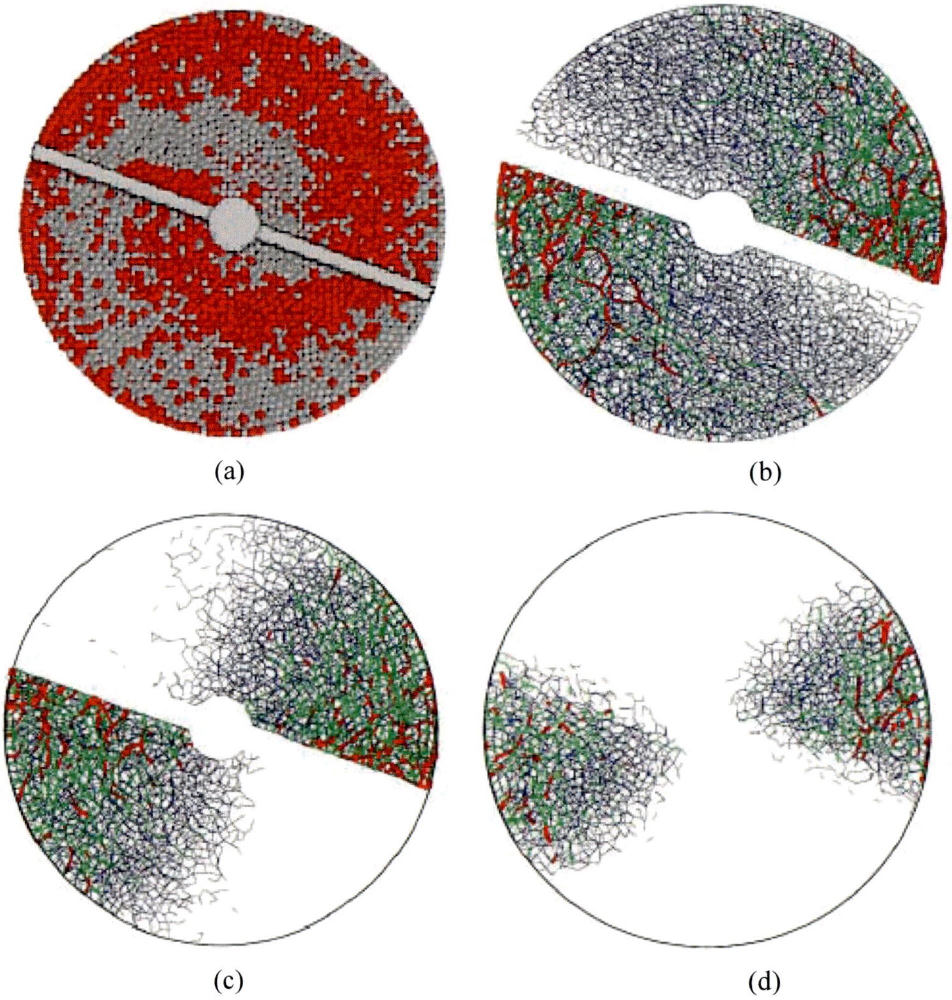

Figure 12. Flow pattern and normal force network in different horizontal sections at different heights of a bladed mixer: (a) flow pattern; (b) between 0 and 20 mm; (c) between 20 and 40 mm; and (d) between 40 and 60 mm (Zhou et al, 2004).

4. CONCLUSION

Since the bulk behaviour of particulate matter depends on the collective interactions among individual particles, it is the particle scale modelling and analysis that plays a crucial role in elucidating the underlying fundamentals and linking fundamental to applied research. DEM-based discrete particle simulation is an effective way to achieve this goal. While DEM, as a state-of-the-art simulation technique, is increasingly used to study the physics of particulate and/or multiphase flow related to various industrial problems, as pointed out by Yu (2004) and Zhu et al. (2007a), future effort will be made to meet the following research needs.

- Microscale: to develop a more comprehensive theory and experimental techniques to study and quantify the interaction forces between particles, and between particle and fluid under various conditions, generating a more concrete basis for particle scale simulation.
- Macroscale: to develop a general theory to link the discrete and continuum approaches, so that particle scale information, generated from DEM or DEM-based simulation, can be quantified in terms of (macroscopic) governing equations, constitutive relations and boundary conditions that can be implemented in continuum-based process modeling.
- Application: to develop more robust models and efficient computer codes so that the capability of particle scale simulation can be extended, say, from two-phase to multi-phase and/or from simple spherical to complicated non-spherical particle system, which is important to transfer the present phenomenon simulation to process simulation and hence meet real engineering needs.

REFERENCES

[1] Bertrand, F., Leclaire, L.A., & Levecque, G.. (2005). DEM-based models for the mixing of granular materials. *Chemical Engineering Science, 60*, 2517-2531.

[2] Bridgwater, J., & Bagster, D.F. (1968). Geometric and dynamic similarity in particle mixing. *Powder Technology, 2*, 198–206.

[3] Brillianton, N.V., & Pöschel, T. (1998). Rolling friction of a viscous sphere on a hard plane. *Europhysics Letters, 42*, 511.

[4] Broadbent, C.J., Bridgwater, J., & Parker, D.J. (1995). The effect of fill level on powder mixer performance using a positron camera. *Chemical Engineering Journal and the Biochemical Engineering Journal, 56*, 119-125.

[5] Cundall, P.A., & Strack, O.D.L. (1979). A discrete numerical model for granular assemblies. *Geotechnique, 29*, 47-65.

[6] Di Renzo, A., & Di Maio, F.P. (2004). Comparison of contact-force models for the simulation of collisions in DEM-based granular flow codes. *Chemical Engineering Science, 59*, 525-541.

[7] Dong, K.J., Yang, R.Y., Zou, R.P., & Yu, A.B. (2006). Role of Interparticle Forces in the Formation of Random Loose Packing. *Physical Review Letters, 96*, 145505.

[8] Dziugys, A., & Peters, B. (2001). An approach to simulate the motion of spherical and non-spherical fuel particles in combustion chambers. *Granular Matter, 3*, 231–266.

[9] Favier, J.F., Abbaspour-Fard, M.H., & Kremmer, M. (2001). Modeling nonspherical particles using multisphere discrete elements. *Journal of Engineering Mechanics-ASCE, 127*, 971-977.

[10] Fisher, R.A. (1926). On the capillary forces in an ideal soil, correction of formulate given by W.B. Haines. *Journal of Agricultural Science, 16*, 492-505.

[11] Gallas, J.A.C., & Sokolowski, T. (1993). Grain non-sphericity effects on the angle of repose of granular material. *International Journal of Modern Physics B, 7*, 2037–2046.

[12] German, R.M. (1989). *Particle Packing Characteristics*. Princeton, USA: Metal Powder Industries Federation.

[13] Hamaker, H.C. (1937). The london-van der waals attraction between spherical particles. *Physica (Amsterdam), 4*, 1058-1072.

[14] Hertz, H. (1882). Über die Berührung fester elastischer Körper. *Journal fur die reine und angewandte Mathematik, 92*, 156-171.

[15] Israelachvili, J.N. (1991). *Intermolecular and surface forces*. London, UK: Academic Press.

[16] Iwashita, K., & Oda, M. (1998). Rolling resistance at contacts in simulation of shear band development by DEM. *Journal of Engineering Mechanics-ASCE, 124,* 285-292.

[17] Janssen, H. A. (1895). Texts on grain pressure in silos. *Zeitschrift des Vereines Deutscher Ingenieure, 39*, 1045-1049.

[18] Jenike, A. W., Johanson, J. R., & Carson, J. W. (1973). Bin Loads-part 2, concepts, Part 3, mass flow bin. *Journal of Engineering for Industry-Transaction of the ASME, 95*, 1-12.

[19] Johnson, K.L. (1985). *Contact Mechanics*. Cambridge, UK: Cambridge University Press.

[20] Kohring, G.A., Melin, S., Puhl, H., Tillemans, H.J., & Vermohlen, W. (1995). Computer-simulations of critical, nonstationary granular flow - through a hopper. *Computer Methods in Applied Mechanics and Engineering, 124*, 273-281.

[21] Krupp, H. (1967). Particle adhesion theory and experiment. *Advances in Colloid and Interface Science, 1*, 111-239.

[22] Langston, P.A., Tuzun, U., & Heyes, D.M. (1994). Continuous potential discrete particle simulations of stress and velocity-fields in hoppers – transition from fluid to granular flow. *Chemical Engineering Science, 49*, 1259-1275.

[23] Langston, P.A., Tuzun, U., & Heyes, D.M. (1995). Discrete element simulation of granular flow in 2Dand 3D hoppers - dependence of discharge rate and wall stress on particle interactions. *Chemical Engineering Science, 50*, 967-987.

[24] Malhotra, K., Mujumdar, A.S., Imakoma, H., & Okazaki, M. (1988). Fundamental particle mixing studies in an agitated bed of granular materials in a cylindrical vessel. *Powder Technology, 55*, 107–114.

[25] McDowell, G.R., Harireche, O., Konietzky, H., Brown, S.F., & Thom, N.H. (2006). Discrete element modelling of geogrid-reinforced aggregates. *Proceedings of the Institution of Civil Engineers-Geotechnical Engineering, 159*, 35-48.

[26] Mikami, T., Kamiya, H., & Horio, M. (1998). Numerical simulation of cohesive powder behavior in a fluidized bed. *Chemical Engineering Science, 53*, 1927-1940.

[27] Mindlin, R.D., & Deresiewicz, H. (1953). Elastic spheres in contact under varying oblique forces. *Journal of Applied Mechanics, 20*, 327-344.

[28] Muguruma, Y., Tanaka, T., Kawatake, S., & Tsuji, Y. (2000). Numerical simulation of particulate flow with liquid bridge between particles (simulation of a centrifugal tumbling granulator). *Powder Technology, 109*, 49-57.

[29] Munjiza, A., Latham, J.P., & John, N.W.M. (2003). 3D dynamics of discrete element systems comprising irregular discrete elements—integration solution for finite rotations in 3D. *International Journal for Numerical Methods in Engineering, 56*, 35–55.

[30] Nakagawa, M., Altobelli, S.A., Capihan, A., Fukushima, E., & Jeong, E.K. (1993). Noninvasive measurements of granular flows by magnitic-response-imaging. *Experiments in Fluids, 12*, 54-60.

[31] Parker, D.J., Dijkstra, A.E., Martin, T.W., & Seville, J.P.K. (1997). Positron emission particle tracking studies of spherical particle motion in rotating drums. *Chemical Engineering Science, 52*, 2011-2022.

[32] Potapov, A.V., & Campbell, C.S. (1996). Computer simulation of hopper flow. *Physics of Fluids, 8*, 2884-2894.

[33] Princen, H.M. (1968). The effect of capillary liquid on the force of adhesion between spherical solid particles-comments. *Journal of Colloid and Interface Science, 26*, 249-253.

[34] Rumpf, H. (1962). The strength of granules and agglomerates, In: Knepper, W.A. (ed.), *Agglomeration* (edition, pp. 379), New York, USA: Wiley Interscience.

[35] Stewart, R.L., Bridgwater, J., Zhou, Y.C., & Yu, A.B. (2001). Simulated and measured flow of granules in a bladed mixer - a detailed comparison. *Chemical Engineering Science, 56*, 5457-5471.

[36] Yamane, K. (2004). Discrete-element method application to mixing and segregation model in industrial blending system. *Journal of Materials Research, 19*, 623-627.

[37] Yang, R.Y., Zou, R.P., & Yu, A.B. (2000). Computer simulation of the packing of fine particles. *Physical Review E, 62*, 3900-3908.

[38] Yang, R.Y., Zou, R.P., & Yu, A.B. (2003a). Effect of material properties on the packing of fine particles. *Journal of Applied Physics, 94*, 3025-3034.

[39] Yang, R.Y., Zou, R.P., & Yu, A.B. (2003b). Numerical study of the packing of wet coarse uniform spheres. *AIChE Journal, 49*, 1656-1666.

[40] Yu, A.B. (2004). Discrete element method-an effective method for particle scale research of particulate matter. *Engineering Computations, 21*, 205-214.

[41] Yu, A.B., Liu, L.F., Zhang, Z.P., Yang, R.Y., & Zou, R.P. (2003). Computer simulation of the packing of particles. *International Journal of Materials and Product Technology, 19,* 324-336.

[42] Zhou, Y. C., Wright, B.D., Yang, R.Y., Xu, B.H., & Yu, A.B. (1999). Rolling friction in the dynamic simulation of sandpile formation. *Physica A, 269*, 536-553.

[43] Zhou, Y.C., Yu, A.B., Stewart, R.L., & Bridgwater, J. (2004). Microdynamic analysis of the particle flow in a cylindrical bladed mixer. *Chemical Engineering Science, 59*, 1343-1364.

[44] Zhu, H.P., & Yu, A.B. (2002). Averaging method of granular materials. *Physical Review E, 66*, 021302.

[45] Zhu, H.P., & Yu, A.B. (2003). The effects of wall and rolling resistance on the couple stress of granular materials in vertical flow. *Physica A, 325*, 347-360.

[46] Zhu, H.P., & Yu, A.B. (2004). Steady-state granular flow in a three-dimensional cylindrical hopper with flat bottom, microscopic analysis. *Journal of Physics D-Applied Physics, 37*, 1497-1508.

[47] Zhu, H.P., & Yu, A.B. (2005). Steady-state granular flow in a 3D cylindrical hopper with flat bottom, macroscopic analysis. *Granular Matter, 7*, 97-107.

[48] Zhu, H.P., & Yu, A.B. (2006). A theoretical analysis of the force models in discrete element method. *Powder Technology, 161*, 122-129.

[49] Zhu, H.P., Zhou, Z.Y., Yang, R.Y., & Yu, A.B. (2007a). Discrete particle simulation of particulate systems: theoretical treatments. *Chemical Engineering Science, 62,* 3378-3396.

[50] Zhu, H.P., Zhou, Z.Y., Yang, R.Y., & Yu, A.B. (2007b). Discrete particle simulation of particulate systems: applications. to submit.

In: Applied Mathematical Modeling
Editor: Eetu N. Virtanen, pp. 273-296
ISBN: 978-1-60021-975-7

Chapter 7

MATHEMATICAL MODELING AND COMPUTATION OF THREE-DIMENSIONAL, TURBULENT, CONVECTIVE HEAT TRANSFER IN A HEAT EXCHANGER WITH CIRCULAR PIN FINS

Anupam Dewan**, L. Dayanand and Pandaba Patro
Department of Mechanical Engineering
Indian Institute of Technology Guwahati, Guwahati - 781039, India

ABSTRACT

Compact heat exchangers are used in a wide variety of applications. Typical among these are automobile radiators, air-conditioning evaporators and condensers, electronic cooling devices, recuperators and regenerators, and cryogenic exchangers. The need for lightweight, space saving, and economical heat exchangers has driven the development of compact surfaces. A mathematical model to predict the overall performance of such heat exchanges has been presented here. The computational domain considered was a three-dimensional rectangular duct with an array of circular pin fins in the staggered arrangement and mounted on the heated channel bed maintained at a constant temperature. The values of different parameters chosen were based on the practical working conditions. The governing Reynolds-averaged Navier-Stokes and thermal energy equations for the complex turbulent convective heat transfer were numerically solved using the control volume approach and by applying appropriate thermal boundary conditions within the domain. The numerical simulations were performed using two different turbulent models, namely, the standard k-ε model and the RNG k-ε model along with the wall functions to treat the turbulence in the vicinity of the solid walls. The details of the governing equations, numerical procedure, mesh employed, and grid independence study have been presented in this chapter. The predictions have been compared with the experimental data reported in the literature and a good agreement

---

* Corresponding author. Professor. E-mail: adewan@iitg.ernet.in, Fax: 0091-361-2582699, http://www.iitg.ernet.in/adewan

between the two has been observed. The temperature and flow fields have been investigated for different inlet velocities. The present investigation shows that choosing an appropriate turbulence model and near-wall treatment has a significant influence on thermal and fluid dynamics behaviour of the heat exchanger. The results also enable us to present the effects of the various operating parameters on the performance of the heat exchanger.

NOMENCLATURE

A	*Area*
d	Diameter
h	Heat transfer coefficient, height
Nu	Nusselt number
k	Thermal conductivity, turbulent kinetic energy
l	Length
p	Pressure
Q	Heat flux
Re	Reynolds number = $\rho V d_h / \mu$
T	Temperature
Pr	Prandtl number
u, v, w	Cartesian velocity components
R	Radius of the fin

Greek Letters

ε	Effectiveness, heat exchanger efficiency
η	Efficiency
μ	Dynamic viscosity
ν	Kinematic viscosity
ρ	Density
τ	Momentum transport term
ε	Rate of dissipation of turbulent kinetic energy
μ_t	Turbulent viscosity

Subscripts

a	Air
f	Fin, fluid
h	Hydraulic
in	Inlet
out	Outlet

w	Wall
t	Turbulence

1. Introduction

The demand and need for efficient cooling techniques and their performance in electronics is unfortunately accompanied by an ever increasing thermal dissipation. Further, in any complex system design, multiple objectives must be considered in a mathematically rigorous fashion. In addition, in many applications the tradeoffs between the energy efficiency, system size, cost, and potential performance variability need to be considered. Air heat exchangers or heat sinks are the most efficient devices used for the cooling electronic equipment primarily because of their low cost and high reliability. This idea recently prompted fresh attention of the researchers into heat transfer and flow characteristics of such heat exchangers. The flow in the heat exchangers may be laminar or turbulent based on the engineering application in which they are used. There are many ways of increasing the heat and mass transfer using turbulence promoters such as fins, ribs, interrupting plates, cavities, zigzags, etc., as shown by Habib et al. [1]. A review of the insert devices as passive elements was given by Dewan et al. [2]. Of all, extended surfaces (fins) are frequently used in heat exchanging devices for the purpose of increasing the heat transfer between a primary surface and the surrounding fluid. Sahiti et al. [3, 4] demonstrated that pin fins are the devices that offer the most effective way of heat transfer enhancement within a given volume of the heat exchanger. Pin-fins protruding from a heating surface can increase the surface area of dissipation and cause turbulent mixing of flow, thereby enhancing the heat dissipation performance and protecting the reliability and life of devices. The factors affecting the thermal performance of pin-fins include the velocity and thermal properties of the working fluid and the pin-fins, the relative pin-fin height, the cross-sectional shape of the pin-fins, and the arrangement of the pin-fins, etc. Many researchers have considered the effects of the above mentioned parameters on the design of heat exchangers (Yeh [5], Maveety and Jung [6], Jonsson et al. [7], Sahin [8]). The major conclusion that can be drawn from the above-mentioned literature is that the distance between pin fins in their lengthwise and spanwise directions and their height in a heat sink plays a major role in the heat transfer enhancement. By increasing the fin height, it was shown that an increase in the heat transfer coefficient can be achieved without an increase in the pressure drop. Further at equal mean fin height, a high inter fin distance ratio and a high clearance ratio should be preferred for a high thermal performance. Due to the presence of the pin fins, the predicted values of the pressure drop show complex flow patterns with large pressure gradients that generate vorticity, circulation and flow reversal.

Comprehensive numerical investigations were carried out by Larson and Sparrow [9], Behnia and Copeland [10] and Sahiti et al. [11] , who considered pin-fins of different cross-sections. In order to perform a fair comparison of the pin cross-sections and come to conclusions that might be valid for a wide range of applications, two different geometric comparison criteria were applied by Sahiti et al. [11]. Furthermore, both common arrangements, namely, staggered and inline were considered by them. The results of the simulation of six different pin cross-sections show that for both comparison criteria and for the staggered arrangement the elliptic cross section performs better than all other cross sections. In all the cases the staggered geometries performed better than the inline geometries.

The most popular cross-section of the pin-fins is circular because it is widely used in many industrial applications. Some of the researchers who investigated the cylindrical pin fins include Sparrow et al. [12], Metzer et al. [13] and Tahat et al. [14]. These authors concluded that the flow becomes complex due to the presence of pin fins because the flow on the end walls forms horseshoe vortices as the flow approaches the stagnation region of the cylinder. In the mid passage, the flow stagnates on the pins, and then accelerates over the surface before separating. The resulting shedding caused by the unsteady separation is a significant driver of heat transfer on the backside of the pins. In the first few rows high levels of turbulence are produced as spanwise velocity gradients mix out in the wakes of the pins. The intense turbulence levels have a large effect on the transport rates and the flow through the downstream pin rows. The effect of the turbulent flow on the fluid and thermal characteristics of pin fins with different types of arrangements was studied by several researchers (Ames et al. [15], Ames and Dvorak [16], Sutherland [17], Sparrow and Ramsey [18]). The conclusion that can be drawn from their studies is that the relative intensity of shedding increases with Reynolds number. The data indicates that heat transfer is highest in the third row of pins due to the highest effective velocity, while heat transfer augmentation due to turbulence is highest in the fourth row and beyond. Hot wire measurements show higher turbulence intensity and dissipation rates upstream of the fourth row compared to the upstream of third row. Moshfegh and Nyiredy [19] compared five different turbulence models for pin fin heat sinks, namely, the standard k-ε model, RNG k-ε model, the realizable k-ε model, the k-ω model, and the Reynolds stress transport model. Different near-wall treatments were employed by these investigators. The RNG k-ε model provided the best agreement with the experimental data. The non-equilibrium wall approach resulted in a poor agreement with the measured data, while the standard wall function and enhanced wall functions provided good agreement with the experimental data. On the other hand Lu and Jiang [20] experimentally and numerically investigated the heat transfer in a rectangular channel with angled ribs. The ribs were taken at various angles and the comparison between the experimental and numerical results showed that the SST k- ω turbulence model was more suitable for the convection heat transfer in such channels than the RNG k-ε turbulence model. The numerical results indicated that the heat transfer coefficients were the largest with 60° ribs, but the channel with 20° ribs had the best overall thermal and hydraulic performance considering the heat transfer and pressure drop when the spacing between the ribs was 4.0 mm.

The complex turbulent flow regimes are encountered in many thermal-fluid engineering applications, e.g., in the cooling of electronics components, cooling of gas turbine blades, hot water boilers, etc. The presence of turbulence has proven resistant to the design of such devices including compact heat exchangers. This is because it is computationally expensive to use CFD tools including a suitable turbulence model, particularly for the later stages of the design where accurate solutions are required. Thus the application of iterative optimization algorithms becomes extremely time consuming. The problem is further complicated by the fact that no single turbulence model is appropriate to model inherently complex reqimes. Thus significant modeling simplifications and assumptions need to be made in their simulations, resulting in approximate solutions only.

1.1. Choosing a Turbulence Model

It is unfortunate that no single turbulence model is universally accepted as being good for all classes of problems. The choice of turbulence model for a particular flow depends on several considerations such as the physics encompassed in the flow, the established practice for a specific class of problem, the level of accuracy required, the available computational resources and the amount of time available for the simulation. The simplest complete models of turbulence are the two-equation models in which the solution of two separate transport equations allows the turbulent velocity and length scales to be independently determined. The standard k-ε model falls within this class of turbulence model and has become the workhorse of practical engineering flow calculations. Robustness, economy, and reasonable accuracy for a wide range of turbulent flows explain its popularity in industrial flow and heat transfer simulations.

1.2. Aim of the Present Work

During the past few decades, several turbulence models have been developed to investigate the heat transfer, pressure drop and other flow characteristics in pin fin arrays. However, it is not clear which turbulence model is appropriate for a paticular flow configuration. Therefore, it is a rather challenging task to choose the right turbulence model for a specific application involving heat exchangers. The objective of the present study is to investigate the performance of two well-known turbulence models, namely, the standard k-ε model and RNG k-ε model along with the wall functions to predict the air flow and heat transfer in a three-dimensional rectangular duct with circular pin fins. The numerical simulations are performed using two turbulence models with different wall functions for a range of Reynolds numbers from 100 to 1000. An assessment of two turbulence models was made so as to conclude which model is appropriate for the present flow configuration. The flow characteristics such as the temperature, pressure, and turbulence parameters such as turbulence kinetic energy, turbulence dissipation rate are discussed in detail in the subsequent sections.

2. Flow Characteristics and Geometric Parameters

The important parameters that need to be considered in the present flow configuration are the distance between the pin fins both in the transverse and longitudinal directions, height of the pin fin and its radius. Although studies on infinite tube banks (Zukauskas [21]) are usually taken as a reference in order to characterize the flow regime through pin fins, thermal and fluid dynamic characteristics of the flow through the pins may show significant differences from that of the tube banks. This difference is particularly critical for the flow over short pins with a small length to diameter ratio (l/d), as a larger influence of the side walls on the flow over pin arrays is to be expected. According to Zukauskas and Ulinskas [22], at low Re ($< 10^3$), the mainstream within the bank is laminar with regions of circulating macro scale eddies whose effect on the boundary layer over the front part of tubes is attained by the viscous forces and the favorable pressure gradient. This flow pattern over a tube bank which occurs at $Re < 10^3$ can be regarded as predominantly laminar with large scale vortices in the recirculation regions. Further, Bergelin et al. [23] have shown that the flow regimes in the range of $200 < Re < 5000$ could be considered in transition regime and below $Re = 200$ laminar. They also indicated that the transition process is not the same for the staggered and in-line arrangements. Fluid flow across the first row of pins resembles the flow across a singular fin but further downstream the flow in the domain with staggered pin fins is similar to the flow in a channel with periodically converging diverging cross section. Hwang and Lui [24] observed a 4-5 times higher pressure drops in an staggered arrangement than that in an in-line arrangement. Fluid flow across the first row of pins resembles the flow across a singular fin for both in-line and staggered pin arrangements. For the second and subsequent rows, the flow patterns are specific for each arrangement. For the staggered arrangement, the nature of the flow over the inner rows differs only slightly from that over the first row. However, for the in-line arrangement a lesser influence of the vortex shedding is observed, as these are positioned in the vortex region of the upstream pins and a substantial part of the fluid flows through the longitudinal passage between the pins. The intensity of occurrence of such effects depends mainly on the distance between the pins, their hydraulic diameter and the flow regime.

Due to the complexity of the flow, it is necessary to make some assumptions to simplify the analysis. They are

1. Each pin is of uniform diameter (d) and height (h) with circular cross section.
2. The pins are uniformly spaced on the base plate.
3. The fin tips are adiabatic.
4. There is no airflow bypass, i.e., the heat sink is fully ducted.
5. The airflow is normal to the pin-axis.
6. Radiation heat transfer is negligible.
7. Body forces are negligible. The effects of gravitational or buoyancy forces are omitted due to the forced convection. Any effects produced by the buoyancy of the fluid due to temperature differences are assumed to be very small.
8. There is no slip at the base plate, top plate, side plate and the fin surface. The condition of no slip is confirmed by the previous experimental studies and it is acceptable, at least for the fluid flows considered in the present study. The solid surface of the cylinder always contains minute projections and indentations, and under normal conditions of temperature and pressure, some fluid will be trapped within these small cavities. Such trapped fluid, which can be regarded as forming part of the surface does not move with respect to the surface. Also, the nearest high projections in the surface can be regarded on a macroscopic scale as forming the outer limit of the surface and the fluid is also stationary with respect to the surface. It is therefore a satisfactory approach on physical grounds to accept the no slip condition at the wall.

Two types of arrangements are most commonly used in practical applications: staggered and inline. As we have seen from the previous study by Isachenko et al. [25], the staggered arrangement provides more heat transfer rate than the inline arrangement, and therefore the staggered arrangement is considered in the present work The geometric parameters taken in the present work are similar to those used by Kays [26] in their experiments. The staggered pin fin surface which was manufactured by Kays [26] with a good and clean pin-to-plate bond is close to the present smooth surface pins. The geometric parameters of the circular pin fin cross section are shown in Figure 1.

3. MATHEMATICAL FORMULATION

To solve the convective heat transfer problems, a mathematical model need to be expressed in the form of the governing equations with appropriate boundary conditions. The governing equations comprise the continuity, momentum and thermal energy equations. To numerically solve the governing equations, the geometry of the problem domain is divided into discrete components. The present computations are performed for a three-dimensional, turbulent, steady state flow over a heated pin fin array, considering the fluid to be incompressible. The governing equations are

Continuity equation:

$$\frac{\partial(\rho u_i)}{\partial x_i} = 0 \tag{1}$$

Momentum equation:

$$\rho u_i \frac{\partial u_j}{\partial x_i} = -\frac{\partial p}{\partial x_j} - \frac{\partial \tau_{ij}}{\partial x_i} \tag{2}$$

Where for Newtonian fluids, the molecular dependent momentum transport term is given by

$$\tau_{ij} = -\mu_t \left(\frac{\partial u_i}{\partial x_j} + \frac{\partial u_j}{\partial x_i} \right) + \frac{2}{3}\delta_{ij} k \tag{3}$$

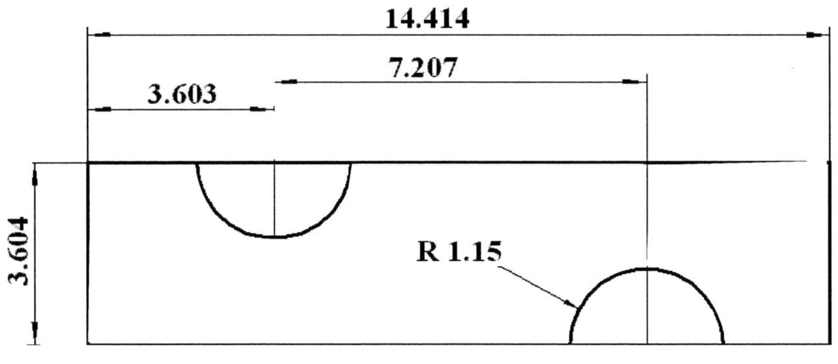

Figure 1. Dimensions of unit cells (in mm) of staggered pin fin arrangement.

Thermal energy equation:

$$\rho c_p u_i \frac{\partial T}{\partial x_i} = k_f \frac{\partial^2 T}{\partial x_i^2} - \frac{\partial}{\partial x_i}(\rho c_p \overline{u_i' T'}) \tag{4}$$

Where u_i denotes the velocity components in the cartesian coordinate system with its coordinates x_i, T denotes fluid temperature, p the pressure and k_f the thermal conductivity of the fluid. Turbulent heat fluxes in the thermal energy equations are modeled as $-\overline{u_i' T'} = -\frac{\nu_t}{\Pr_t}\frac{\partial T}{dx_i}$

The law of the wall for temperature used in the present work is of the following form:

$$T^+ = \Pr_t[\ln(EY^+) + P]$$

Where $P = 9.24\left[\left(\frac{\Pr}{\Pr_t}\right)^{3/2} - 1\right][1 + 0.28e^{-0.007\Pr/\Pr_t}]$

Here $\Pr_t$ is the turbulent Prandtl number and is taken as 0.85 and Pr = 0.71 for air. The physical properties of air considered are ρ = 1.225 kg.m^{-3}, μ = 1.789x10^{-5} kg/m.s , and c_p = 1006.43 Jkg^{-1} K^{-1}.

The conjugate heat transfer from pin fin arrays implies the simultaneous solution of the Eqs. (1) to (4), and the energy equation through the solid, which reads

$$\frac{\partial^2 T_s}{\partial x_i^2} = 0 \tag{5}$$

Standard *k*-ε Model

The following modeled transport equations for turbulence kinetic energy (*k*) and its rate of dissipation (ε) are used

$$\frac{\partial}{\partial t}(\rho k) + \frac{\partial}{\partial x_i}(\rho k u_i) = \frac{\partial}{\partial x_j}\left[\left(\mu + \frac{\mu_t}{\sigma_k}\right)\frac{\partial k}{\partial x_j}\right] + G_k - \rho\varepsilon \tag{6}$$

$$\frac{\partial}{\partial t}(\rho \varepsilon) + \frac{\partial}{\partial x_i}(\rho \varepsilon u_i) = \frac{\partial}{\partial x_j}\left[\left(\mu + \frac{\mu_t}{\sigma_\varepsilon}\right)\frac{\partial \varepsilon}{\partial x_j}\right] + C_{1\varepsilon}\frac{\varepsilon}{k}G_k - C_{2\varepsilon}\rho\frac{\varepsilon^2}{k} \tag{7}$$

In these equations, G_k represents the generation of turbulence kinetic energy due to the mean velocity gradients. $C_{1\varepsilon}$ and $C_{2\varepsilon}$ are the model constants. σ_k and σ_ε are the turbulent Prandtl numbers for k and ε respectively.

The turbulent (or eddy) viscosity μ_t is computed by combining k and ε as follows

$$\mu_t = \rho C_\mu \frac{k^2}{\varepsilon} \tag{8}$$

Where C_μ is a constant (= 0.09) and the other model constants are

$C_{1\varepsilon} = 1.44, C_{2\varepsilon} = 1.92, C_\mu = 0.09, \sigma_k = 1.0, \sigma_\varepsilon = 1.3$.

RNG *k-ε* Model

The following transport equations are used in the RNG *k-ε* model:

$$\frac{\partial}{\partial t}(\rho k) + \frac{\partial}{\partial x_i}(\rho k u_i) = \frac{\partial}{\partial x_j}\left(\frac{\mu_{eff}}{\sigma_k}\frac{\partial k}{\partial x_j}\right) + G_k - \rho\varepsilon \tag{9}$$

$$\frac{\partial}{\partial t}(\rho \varepsilon) + \frac{\partial}{\partial x_i}(\rho \varepsilon u_i) = \frac{\partial}{\partial x_j}\left(\frac{\mu_{eff}}{\sigma_\varepsilon}\frac{\partial \varepsilon}{\partial x_j}\right) + C_{1\varepsilon}\frac{\varepsilon}{k}G_k - C_{2\epsilon}\rho\frac{\varepsilon^2}{k} - R_\varepsilon \tag{10}$$

Where $R_\varepsilon = \dfrac{c_\mu \rho \eta^3 (1 - \eta/\eta_0)}{1 + \beta\eta^3}\dfrac{\varepsilon^2}{k}$

Here G_k denotes the generation of turbulence kinetic energy due to the mean velocity gradients. The quantities σ_k and σ_ε are the Prandtl numbers for k and ε respectively.

The scale elimination procedure in RNG theory results in a differential equation for the turbulent viscosity:

$$d\left(\frac{\rho^2 k}{\sqrt{\varepsilon\mu}}\right) = 1.72 \frac{\hat{v}}{\sqrt{\hat{v}^3 - 1 + C_v}} d\hat{v} \tag{11}$$

Where $C_v = 100, \hat{v} = \mu_{eff} / \mu$

4. Computational Domain and Boundary Conditions

The computational domain in the present work as shown in Figure 2 is a three-dimensional rectangular duct with pin fins mounted on the base plate. It has a width *a*, flow length *l* and height *h* which is equal to the pin height. The pins, which are considered as solid are placed in the computational domain and surrounded by the fluid. The flow developing inlet block was taken to be $5\,d_h$ whereas the outlet block length was set equal to $15\,d_h$ in order to avoid any influence of the eventual back flow streams on the final results. Here d_h denotes the hydraulic diameter.

Flow over tube banks with more than sixteen rows is considered to be fully developed (Zukauskas [27]), where there is no change in the flow and the temperature. Thus heat transfer and pressure drop characteristics derived from the array of sixteen rows are valid also for pin fin arrays with a large number of pins in the flow direction. According to the above consideration the computational domain in the present work consists of sixteen rows of pin fins in the flow direction.

The inlet, outlet, wall and symmetry boundary conditions were applied in the computational domain. The boundary conditions are as follows:

A unidirectional uniform flow field, with a constant temperature, was assumed for the inlet section 1-8-16-9:

$$u(0, y, z) = u_{in}, v(0, y, z) = 0,\ w(0, y, z) = 0 \text{ and } T(0, y, z) = T_{in} = 293K \quad (12)$$

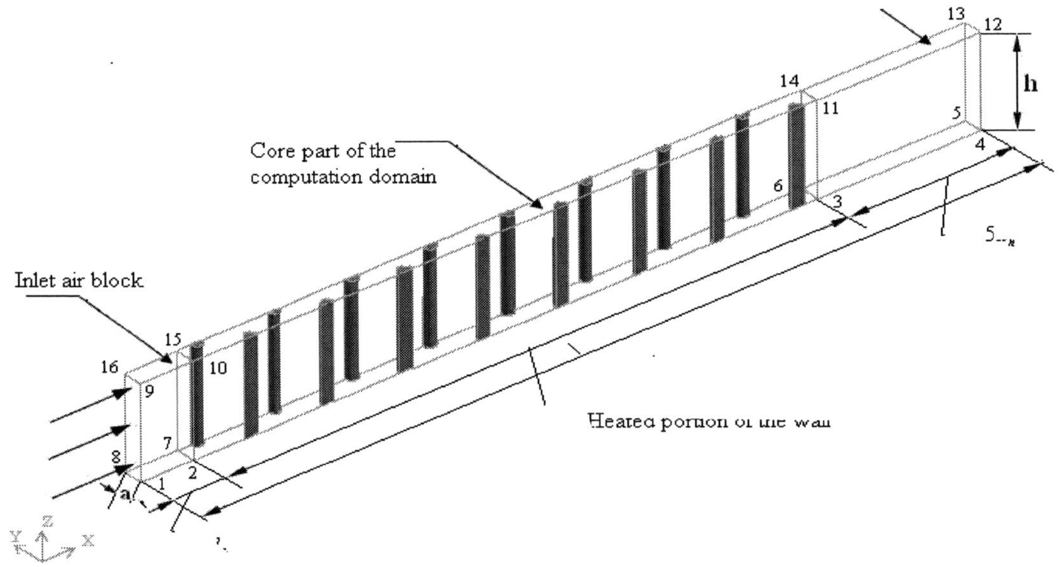

Figure 2. Computational domain with boundary conditions employed in the present work.

The no slip condition and constant temperature were applied for the wall 1-4-5-8 except for the inlet part 1-2-7-8 and outlet part 3-4-5-6, which were assumed to be adiabatic:

For section 2-3-6-7:

$$u(x,y,0) = v(x,y,0) = w(x,y,0) = 0 \text{ and } T(x,y,0) = T_w = 343K \quad (13)$$

For sections 1-2-7-8 and 3-4-5-6:

$$u(x,y,0) = v(x,y,0) = w(x,y,0) = 0 \text{ and } \left(\frac{\partial T}{\partial z}\right)_{x,y,0} = 0 \quad (14)$$

The top wall 9-12-13-16 was considered to be adiabatic (Figure 2), where the no slip condition for the velocity components was applied:

$$u(x,y,h) = v(x,y,h) = w(x,y,h) = 0 \text{ and } \left(\frac{\partial T}{\partial z}\right)_{x,y,h} = 0 \quad (15)$$

For sections 1-4-12-9 and 8-5-13-16, the symmetry boundary condition was applied:

$$\frac{\partial u}{\partial y} = \left(\frac{\partial w}{\partial y}\right)_{x,o,z} = 0\,,\, v(x,0,z) = 0\,, \text{ and } \left(\frac{\partial T}{\partial y}\right)_{x,o,z} = 0 \text{ (section 1-4-12-9)} \quad (16)$$

$$\frac{\partial u}{\partial y} = \frac{\partial w}{\partial y}_{x,a,z} = 0\,,\, v(x,a,z) = 0, \text{ and } \left(\frac{\partial T}{\partial y}\right)_{x,a,z} = 0 \quad \text{(section 8-5-13-16)} \quad (17)$$

In the outlet section 4-12-13-5, the variables were calculated by the interpolation from the upstream and the assumption of zero gradients in the flow direction:

$$\frac{\partial u}{\partial x} = \frac{\partial v}{\partial x} = \left(\frac{\partial w}{\partial x}\right)_{l,y,z} = 0\,, \text{ and } \left(\frac{\partial T}{\partial x}\right)_{l,y,z} = 0 \quad (18)$$

In the present work a hydraulic diameter (d_h) of 2 mm was chosen, as this pin diameter is commonly encountered in pin fin assemblies for cooling of electronic circuits and for the cooling of gas turbine blades. Here Reynolds number is defined as $Re = \rho V d_h/\mu$ and Nusselt number is defined as $Nu = (h.d_h)/k$, where V denotes the inlet velocity, ρ the fluid density, μ the dynamic viscosity, *h* the overall heat transfer coefficient, and *k* the thermal conductivity of the fluid.

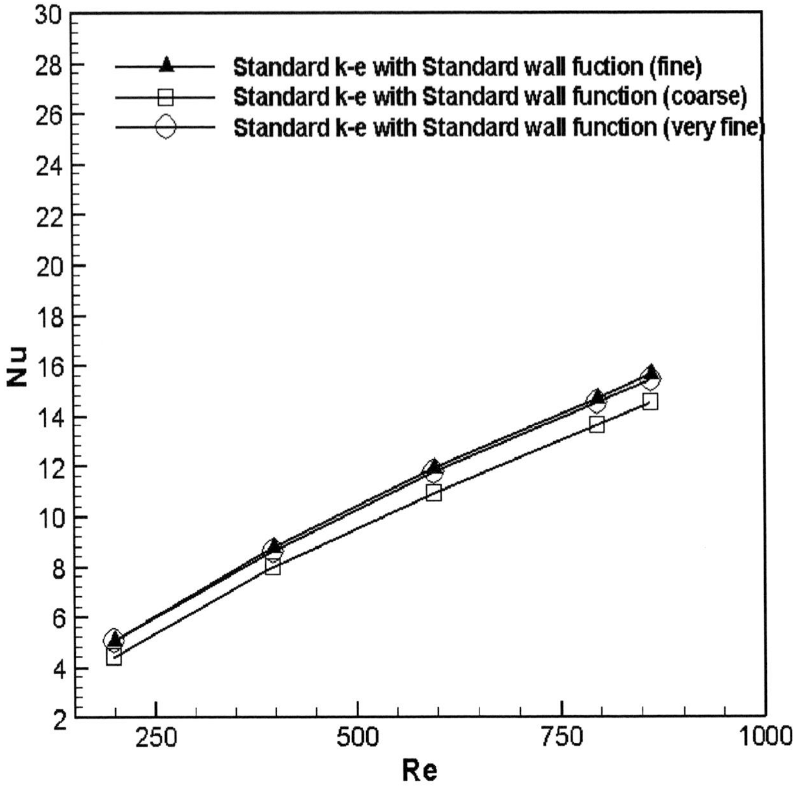

Figure 3. Nusselt number versus Reynolds number for the grid independence study.

5. Near-Wall Treatment

Three different near-wall treatments were used in the present investigation, namely, (a) the standard wall function proposed by Launder and Spalding [28], (b) the non-equilibrium wall function and (c) the enhanced wall function. In the standard wall approach the first grid point near the wall should lie in the logarithmic layer [29]. The enhanced wall function subdivides the near wall region into a viscous sub-layer and a fully turbulent flow. In the viscous sub-layer region, only the k equation is solved, while in the fully turbulent flow equations for both k and ε are computed. The term wall distance is introduced to compute the interface between the two layers. The non-equilibrium wall function not only considers the equilibrium assumption (production = dissipation) for computing the budget for the kinetic energy adopted in the standard wall function approach, but it also takes into account the budget of kinetic energy, which can vary widely in highly non-equilibrium flows involving strong pressure gradients. This effect can be implemented by introducing the pressure gradient term in the log law in order to sensitise the mean velocity.

6. NUMERICAL METHOD

The flow inside a 3D channel is characterized by a complex flow field, which is affected by the blockage and recirculation zones. The pre-processor GAMBIT was used to create the geometry and a computational mesh and the solutions of the governing equations presented in the previous section were achieved using the commercial CFD software FLUENT 6.2.16. The governing equations were solved numerically by using the finite volume technique. The fluid flow inside the heat exchanger channel was considered to be incompressible and steady. The standard k-ε model and the RNG k-ε model were used for modeling the turbulence while the conjugate heat transfer model was used for solving the energy equation. The turbulence in the vicinity of solid walls was treated using the standard, non-equilibrium and enhanced wall functions [29]. The second order upwind scheme was used for the discretization of all the governing equations. The SIMPLE algorithm was used for the pressure-velocity coupling. The SIMPLE algorithm uses the relationship between the velocity and pressure corrections to enforce mass conservation and to obtain the pressure field. The fluid was chosen to be air with constant specific heat and thermal conductivity. The solid part of the computational domain (pins) was considered to be of aluminum with constant isotropic properties (ρ = 2719 kg.m^{-3}, c_p = 871 Jkg^{-1} K^{-1}, and k = 202.4Wm^{-1}K^{-1}).

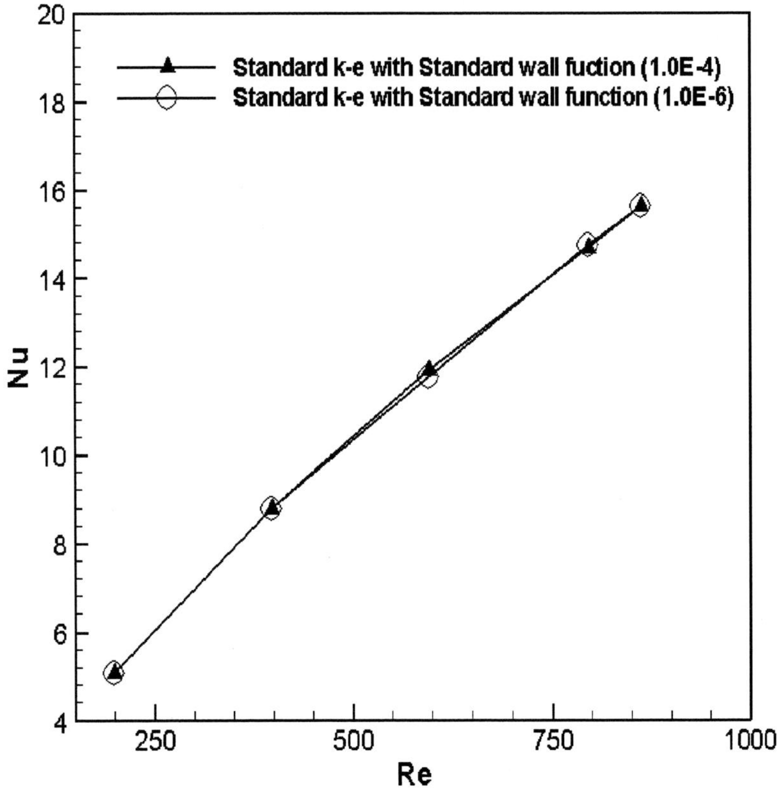

Figure 4. Nusselt number versus Reynolds number for the comparison of convergence criteria.

7. Grid Independence Study

The grid independence test was performed by taking the standard *k-ε* model with the standard wall function with the inlet velocities from the range of 1.0 m/s to 6.0 m/s. Three meshes of different sizes, namely, coarse mesh with 520000 cells, fine mesh with 810000 cells and very fine mesh with 1020000 cells were used for the grid independence study. In all cases, the mesh was very fine in the critical regions, i.e., in the vicinity of solid walls and both upstream and downstream of the pin fins. As can be seen in Figure 3, the difference between the fine mesh and very fine mesh was less than 1.5 %. Hence the fine mesh was used for all the subsequent computations. The convergence criterion also plays an important role in the solution of the problem and therefore an appropriate convergence criterion must be selected so as to make sure that the solution is converged. In the present work two convergence criteria of 10^{-4} and 10^{-6} were used and it was observed that the numerical values of the flow parameters taken at the two convergence criteria were almost identical (Figure 4). Therefore a convergence criterion of 10^{-4} was used for all subsequent computations. The solution was said to be converged only after all the residuals satisfied this criterion.

8. Validation of the Code

Validation of the present computational model was performed by comparison of the computations with the experimental data of Kays [26] who investigated pin fins with different fin geometries in both inline and staggered arrangements. The staggered pin fin surface which was manufactured with a good and clean pin to plate bond is close to the present smooth surface and hence used for the validation. A comparison between the predictions by two turbulence models, namely, the standard *k-ε* model and the RNG *k-ε* model along with the wall functions for different inlet velocities from 1.5 m/s to 6.5 m/s (corresponding Re from 200 to 900) is shown in Figure 5. The predictions by both the models with the standard wall functions are close to the experimental data while the two models with enhanced wall function under-predict and with the non-equilibrium wall treatment overpredict. At low Reynolds numbers, i.e., in the range of 200 to 500 the predictions by both the models show a slight deviation from the experimental data but as Re is increased, the RNG *k-ε* model predictions are in good agreement with the experimental data. However, the predictions by standard k-ε model show a deviation of approximately 7% from the experimental data. Hence it can be concluded that the two models with the standard wall functions are good for the type of the present flow configuration. The RNG *k-ε* model was found to be more accurate compared to the standard *k-ε* model.

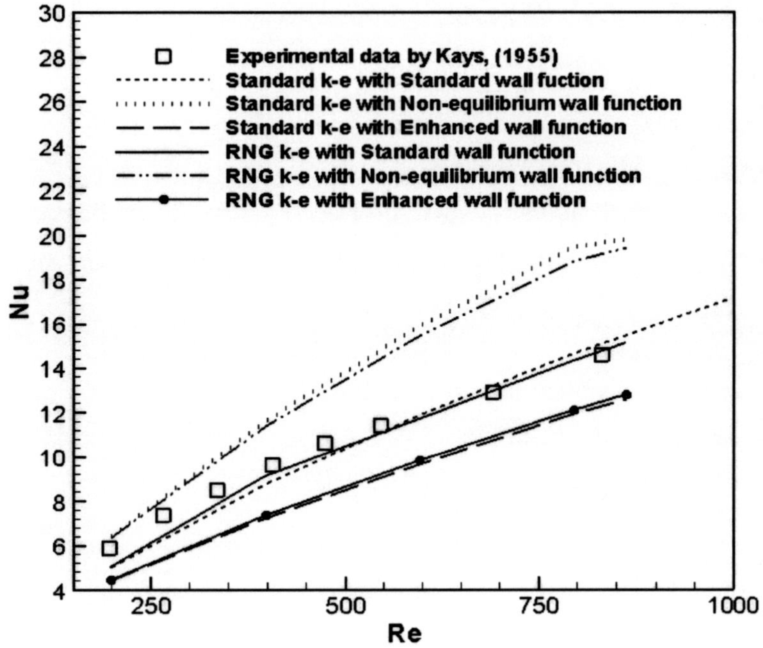

Figure 5. Comparison between the numerical data obtained in the present work and the experimental data by Kays (1955).

9. Results and Discussion

The flow physics over circular pins with different inlet velocities has been assessed in the present work. An understanding of the flow behaviour can be obtained by analyzing the velocity field around the pins. In addition, the temperature field in the air and solid (pins) offers detailed information about the locations where high heat transfer rates occur and the locations where a heat hold-up takes place. For the present computations, three-dimensional velocity and temperature flow fields are considered. The presentation of these parameters was done for the planes normal to directions of largest changes (y- and z- axes) by considering the RNG k-ε model with the standard wall functions.

The characteristics of the convective heat transfer from pin arrays can be understood from the temperature field in fluid and solid parts of the computational domain. The complete temperature changes in the computational domain, for the plane which passes through half of pin length (z = 11.5 mm), and for the plane which coincides with the section 1-4-12-9 (Figure 2) for different inlet velocities are presented in Figure 6. It can be seen that the fluid adjacent to the fin attains the maximum temperature. Further the fluid passing through the mid plane of the computational domain takes thermal energy away from the fluid that is in the vicinity of the fin (Figure 6). For an inlet velocity the temperature difference between the fin and the fluid decreases as we move in the downstream direction of the computational domain. Since the fluid takes heat from the fin and surface, the bulk temperature of the fluid increases due to the turbulent convective heat transfer and consequently the temperature of the fin increases in the downstream direction.

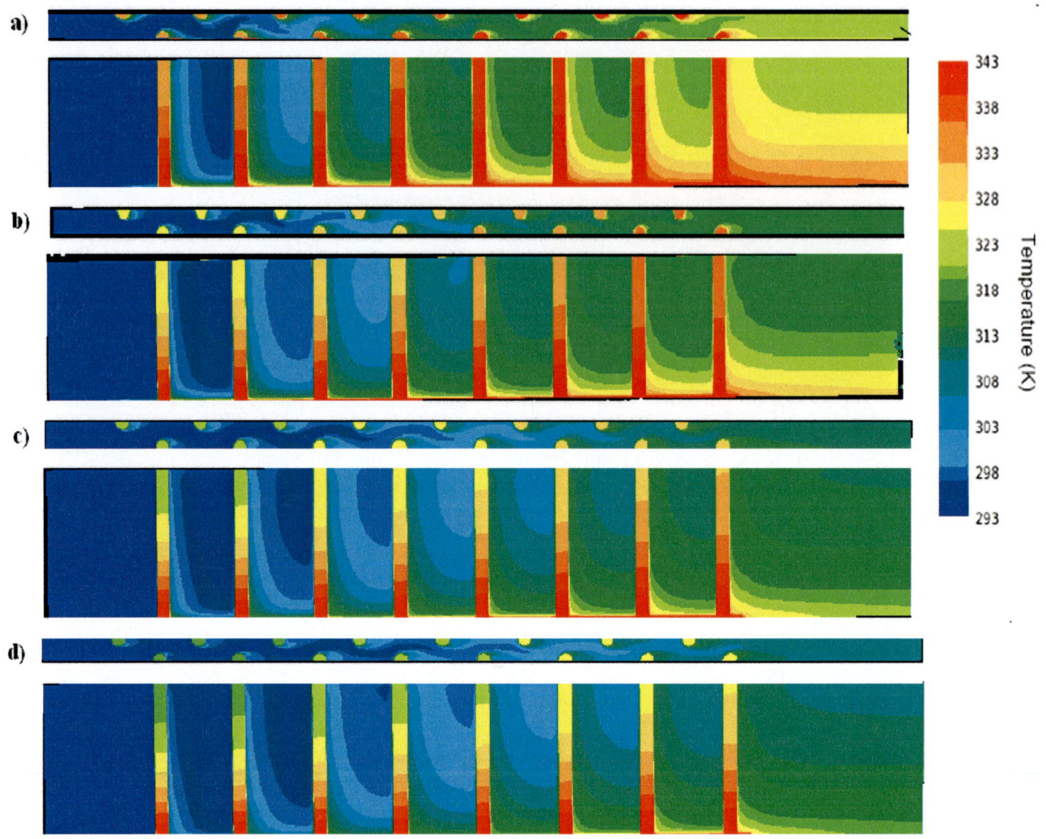

Figure 6. Temperature profiles in air and pins in the computational domain for the RNG k-ε model for (a) 1.5 m/s, (b) 3.0 m/s, (c) for 4.5 m/s, (d) 6.0 m/s.

In the computational domain the intensive reduction of the temperature difference between the pin wall and surrounding air depends on the air velocity in addition to the material of the fin chosen. Since the base plate thickness is negligible, the material of the base plate does not affect the solution. However, in the present work all the pins chosen are of the same material aluminum (it has the highest thermal conductivity among metals used in fin). In the present work we have considered the same pin length for all pins and same cross sections, and therefore the temperature changes of the fluid are mainly influenced by the inlet velocities. It was observed that as the inlet velocity increases the outlet bulk temperature decreases (Figure 7) due to larger convective heat transfer at higher inlet velocity compared to that at lower velocity. This trend is clearly seen in the above mentioned temperature counters (Figure 6). It was also observed in Figure 7 that between any two fin rows in the spanwise direction there is small drop and sudden increase in the temperature of the fluid. The reason for this behaviour is that as the fluid enters from the inlet, the fluid first comes into contact with the first row of fins and the fluid starts taking heat away from the fin and gets heated up. This heated fluid upon its motion to the next pin rows moves into the gap between this fin rows and pushes the colder fluid on to the next fin rows. This colder fluid takes heat from the fin which is at a higher temperature than the first row of fins because this fin row has been taking heat continuously from the base plate (same as that of the first fin row) and the heat is taken by the fluid only after it takes the heat from the first fin row. Due to this behaviour, there is increase and decrease in the temperature in the downstream direction which follows a periodic trend after certain fin rows depending upon the inlet velocity. Further the temperature of the fin at a particular location decreases with an increase in the inlet velocity. This is due to the fact that a higher fluid inlet velocity leads to larger convective heat transfer resulting in an larger increase in the fluid temperature compared to that at low inlet velocity.

The pressure behaves completely differently from temperature. At the beginning of the pin array a pressure drop takes place as a consequence of the sudden flow contraction. Further downstream the flow path resembles a converging-diverging channel. Hence in each subsequent pin row a pressure loss due to the contraction and a pressure gain due to expansion of the flow occurs. Such a kind of pressure behaviour is demonstrated by the evaluation of the pressure variation in the vicinity of the seventh pin row (shown by means of two vertical lines in Figure 8). The total pressure drop over the entire computational domain can be easily obtained from the pressure at the inlet and outlet sections.

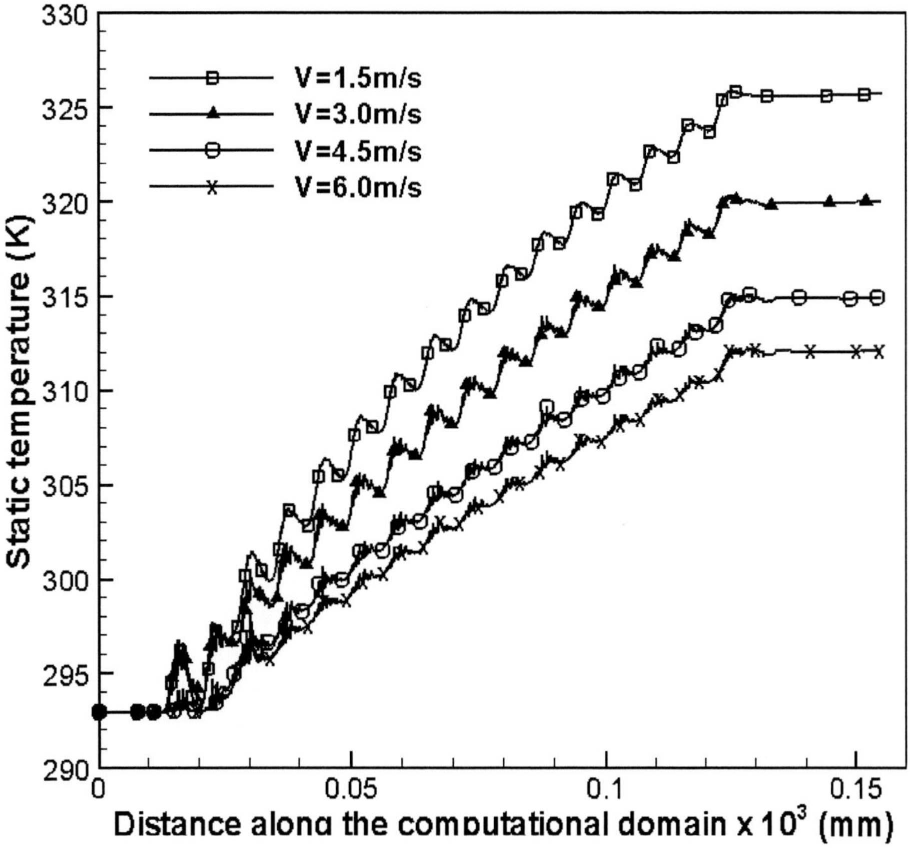

Figure 7. Comparison of the static temperature for different inlet velocities (V).

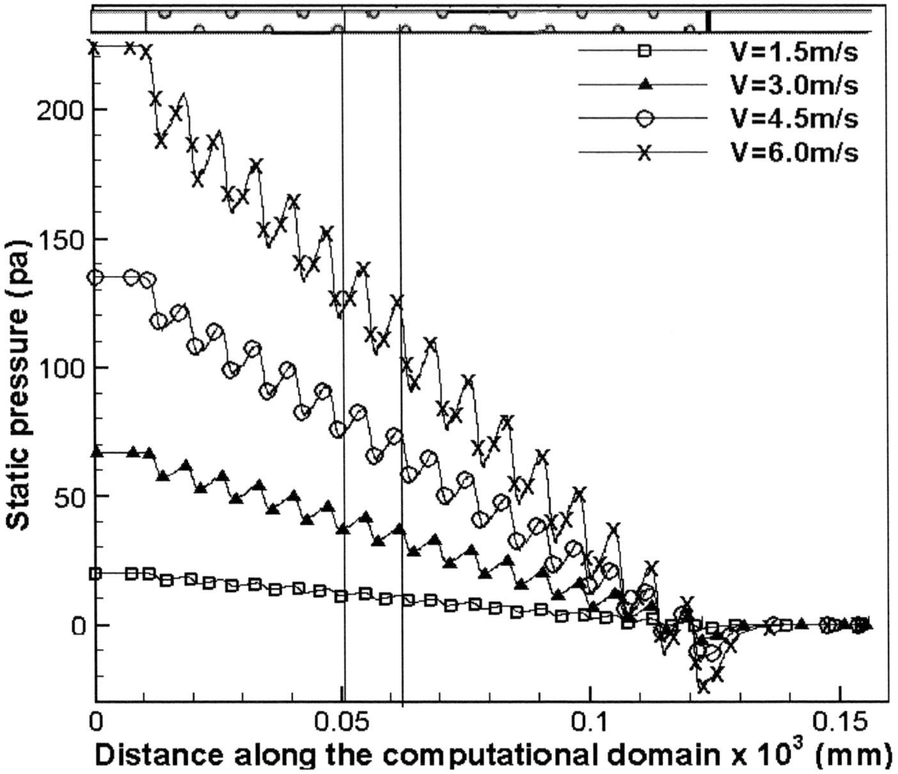

Figure 8. Comparison of static pressure for different inlet velocities (V).

As shown by Zukauskas [27], the temperature and flow fields after the first four to five pin rows start to stabilize. This means that for subsequent rows no substantial change in the fluid dynamic parameters can occur. A similar behaviour was also observed in the present computations. This is confirmed by the contour plots presented in Figure 9. The periodic nature is observed both in the temperature and the flow parameters in the computational domain and this periodicity is attained depending upon the given inlet velocity. From the variation of the X- velocity component along the computational domain (Figure 10), we can notice that the velocities at the inlet and the outlet blocks are almost constant but in the computational domain velocity fluctuations occur due to the presence of the fins which causes the converging and diverging paths. When the velocity is 1.5 m/s the periodicity is obtained after two pin rows and when the velocity is 3.0 m/s it has shifted downstream to three to four pin rows and the same trend is followed with an increase in the velocity. High velocities were observed at the converging portion of the computational domain shown in Figure 9 (the red colored portion shows the maximum velocity) due to which there is a sudden rise in the dynamic pressure ($p_d = 1/2\rho v^2$, where v is the magnitude of the local fluid velocity). The recirculation zones formed in the computational domain are shown in Figure 11. As the domain consists of converging and diverging sections there is a sudden rise and fall in the dynamic pressure as we move along the computational domain (Figure 12).

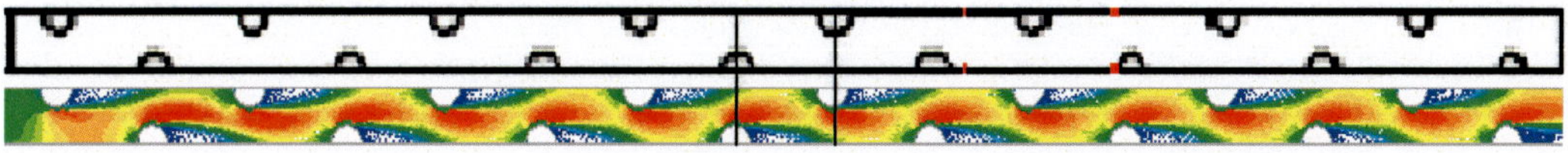

Figure 9. Velocity contour for an inlet velocity of 3.0 m/s, showing the periodic nature in the computational domain.

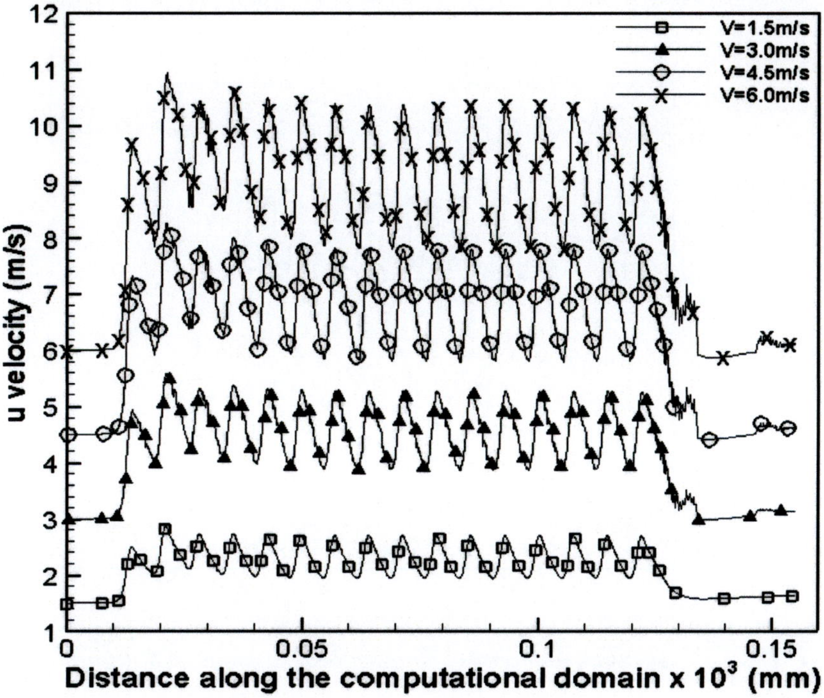

Figure 10. Variation of the X-component velocity along the computational domain for different inlet velocities.

When we consider the turbulent kinetic energy distribution (Figure 13) the maximum turbulent kinetic energy was observed approximately at the centre of the fins. This is probably because in general the Reynolds stresses are large near the walls which cause the maximum production of turbulent kinetic energy at similar locations. Since the fin causes the fluid to deflect upward, the turbulent kinetic energy is found to be the maximum in the vicinity of the fins. Depending upon the inlet velocity similar periodic trend after the flow stablizes was observed both in the dynamic pressure variation and also in the turbulent kinetic energy variation. The local Nusselt number variation is shown in Figure 14. Since the bulk temperature of the fluid increases in the downstream direction, the local Nusselt number decreases. The local Nusselt number is zero at the inlet and outlet locations, since the walls are considered to be adiabatic. Figure 15 shows the skin friction coefficient in the downstream direction. It is observed that the skin friction coefficient increases with the velocity and also noticed that the amplitude of variation between any two fin rows also increases with an increase in inlet velocity.

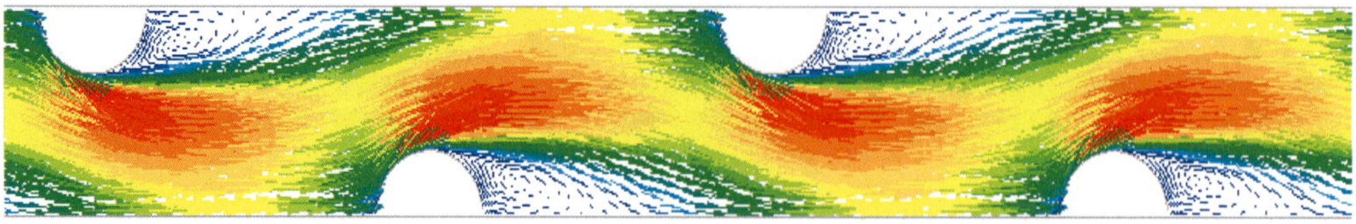

Figure 11. Velocity vectors for an inlet velocity of 3.0 m/s showing the recirculation zone at the rear portion of the cylinder.

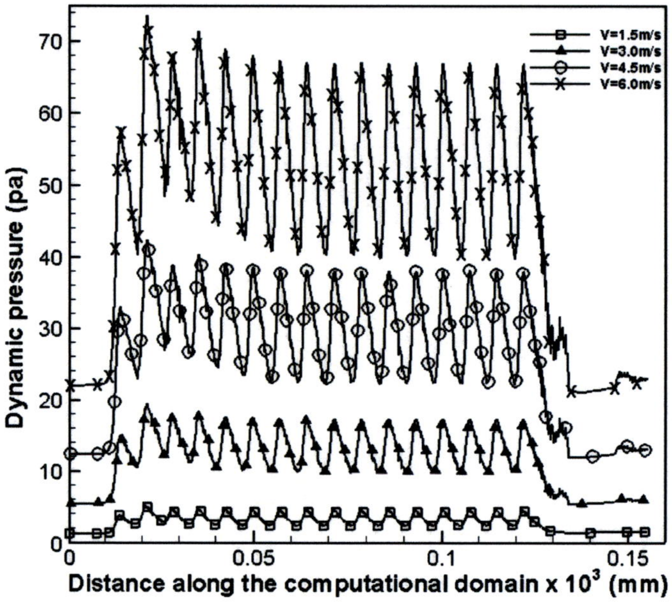

Figure 12. Comparison of the dynamic pressure variation along the computational domain for different inlet velocities.

10. Conclusion

In the present work, heat transfer phenomenon is studied in a rectangular duct with circular pin fins in staggered arrangement, for a turbulent flow using two different turbulence models. Both the standard *k-ε* model and the RNG *k-ε* model with the standard wall functions predict results better than those with the enhanced and non-equilibrium wall functions. It is observed that the pressure variation is different at the beginning of pin fin array compared to that further downstream. In the beginning the pressure drop is as a consequence of the sudden contraction in flow area, but downstream, the flow area resembles a converging-diverging channel, hence a different pressure variation is observed. Temperature profiles are obtained for different inlet velocities. For low inlet velocities, the profiles suggest that the fin temperatures are high, compared to those with high inlet velocities. The bulk temperature of the fluid at the outlet decreases with an increase in inlet velocity. For any inlet velocity, it is observed that the flow gets separated at the rear portion of the pin fin. For low velocities, the flow gets separated early before reaching the rear portion of the pin fin and vice versa. The flow was observed to be stabilized and a periodic behaviour was observed in the flow parameters in the downstream of the computational domain depending upon the value of the inlet velocity.

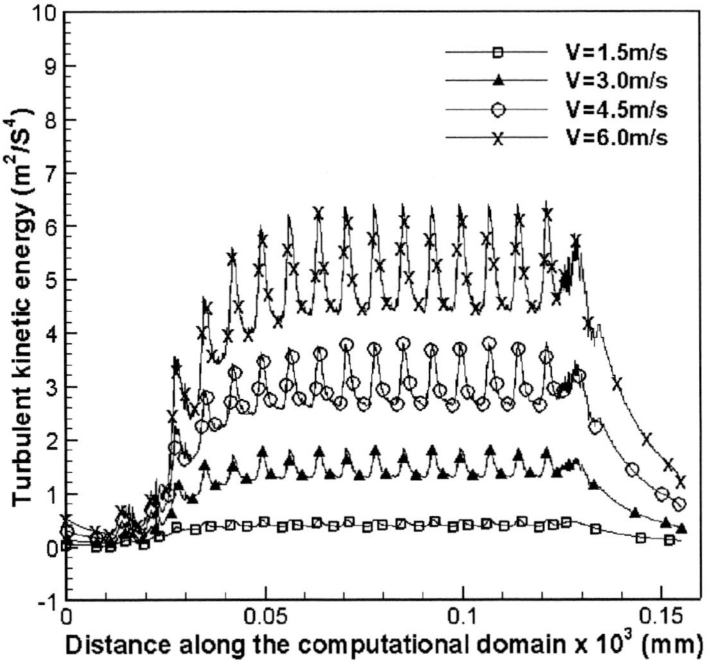

Figure 13. Variation of turbulent kinetic energy along the computational domain for different inlet velocities.

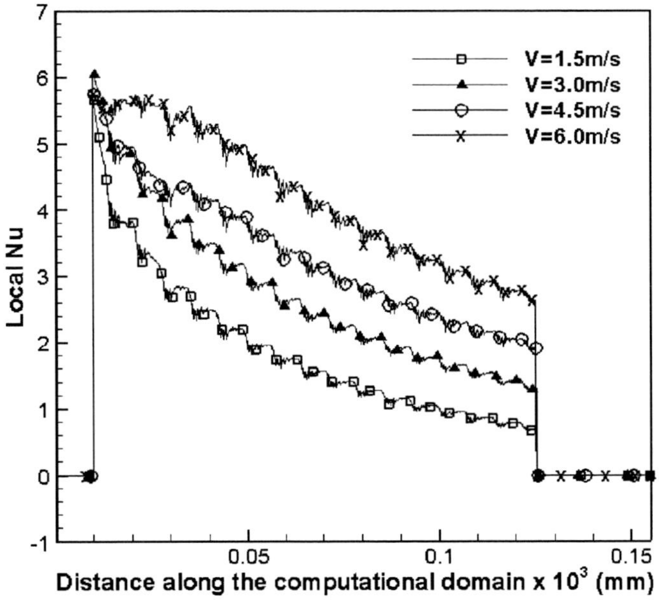

Figure 14. Local Nusselt number variation along the computational domain for different inlet velocities at the centre line at the bottom of the wall (x from 0 to 155.312 mm, y = 1.802 mm, and z = 0 mm).

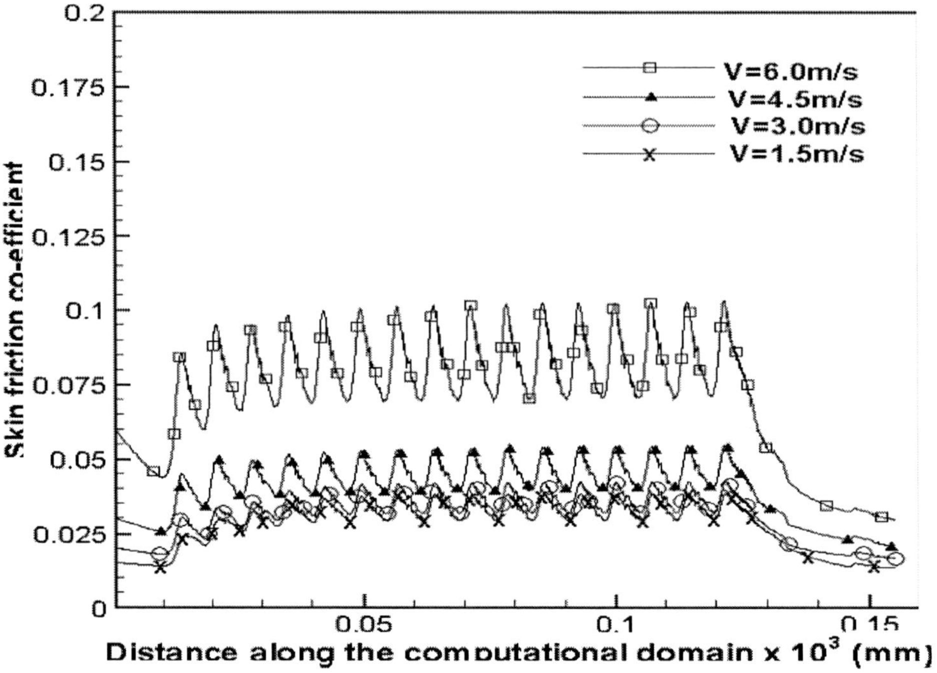

Figure 15. Variation of skin friction co-efficient along the computational domain for different inlet velocities at the centre line at the bottom of the wall (x from 0 to 155.312 mm, y = 1.802 mm, and z = 0 mm).

REFERENCES

[1] M.A. Habib, I.U. Haq, H.M. Badr, and S.A.M. Said, "Calculation of Turbulent Flow and Heat Transfer in Periodically Converging and Diverging Channels", *Comp. Fluids* 27 (1998) 95-120.

[2] Dewan, P. Mahanta, K.S. Raju, and P. S. Kumar, "A Review of Passive Heat Transfer Augmentation Techniques", Proc. Institution of Mechanical Engineers, *Part A J. of Power and Energy* 218 (2004) 509-527.

[3] N. Sahiti, F. Durst, and A. Dewan, "Heat Transfer Enhancement by Pin Elements", *Int. J. Heat Mass Transfer* 48 (2005) 4738-4747.

[4] N. Sahiti, F. Durst, and A. Dewan, "Strategy for Selection of Elements for Heat Transfer Enhancement", *Int. J. Heat Mass Transfer* 49 (2006) 3392-3400.

[5] R.H. Yeh, "An Analytical Study of the Optimum Dimensions of Rectangular Fins and Cylindrical Pin Fins", *Int. J. Heat Mass Transfer* 40 (1997) 3607-3615.

[6] J.G. Maveety and H.H. Jung, "Design of an Optimal Pin Fin Heat Sink with Air Impingement Cooling", *Int. Comm. Heat Mass Transfer* 27 (2000) 229-240.

[7] H. Jonsson and B.Moshfegh, "Enhancement of the Cooling Performance of Circular Pin Fin Heat Sinks under Flow Bypass Conditions", *Inter Society Conference of Thermal Phenomena* (2002).

[8] Sahin, "A Taguchi Approach for Determination of Optimum Design Parameters for a Heat Exchanger Having Circular-cross Sectional Pin Fins", *Int. J. Heat Mass Transfer* 43 (2007) 493-502.

[9] E.D. Larson and E.M. Sparrow, "Performance Comparison Among Geometrically Different Pin Fin Arrays Situated in an Oncoming Longitudinal Flow", *Int. J. Heat Mass Transfer* 25 (1982) 723-725.

[10] M. Behnia and D. Copeland, "A Comparison of Heat Sink Geometries for a Laminar Forced Convection: Numerical Simulation of Periodically Developed Flow" The University of New South Wales, Sydney 2002, Australia.

[11] N. Sahiti, A. Lemouedda, D. Stojkovic, F. Durst, and E. Franz, "Performance Comparison of Pin Fin in Ducts Flow Arrays with Various Pin Cross Sections", *Applied Thermal Engineering* 26 (2006) 1176-1192.

[12] E. M. Sparrow, J. W. Ramsey, and C.A.C. Altemani, "Experiments on Inline Pin-fin Arrays and Performance Comparison with Staggered Arrangement", *ASME J. Heat Transfer* 102 (1980) 44-50.

[13] D.E. Metzger, R.A. Berry, and J.P. Bronson, "Developing Heat Transfer in Rectangular Ducts with Staggered Arrays of Short Pin Fin", *ASME J. Heat Transfer* 104 (1982) 700-706.

[14] M. Tahat, Z.H. Kodah, B.A. Jarrah, and S.D. Probert, "Heat Transfers from Pin-fin Arrays Experiencing Forced Convection", *Applied Energy* 67 (2000) 419-442.

[15] F.E. Ames, L.A. Dvorak, and M.J. Morrow, "Turbulent Augmentation of Internal Convection over Pins in Staggered Pin-fin Arrays", *J. Turbomachinery* 127 (2005), 183-190.

[16] F.E. Ames and L.A. Dvorak, "Turbulent Transport in Pin Fin Arrays: Experimental Data and Predictions", *J. Turbomachinery* 128 (2006) 71-81.

[17] W.A. Sutherland, "Improved Heat Transfer Performance with Boundary Layer Turbulence Promoters", *Int. J. Heat Mass Transfer* 10 (1967) 1589-1599.

[18] E.M. Sparrow and J.W. Ramsey, "Heat Transfer and Pressure Drop for a Staggered Wall Attached Array of Cylinders with Tip Clearance", *Int. J. Heat Mass Transfer* 21 (1978) 1369-1377.

[19] Moshfegh and R. Nyiredy, "Comparing RANS Models for Flow and Thermal Analysis of Pin Fin Heat Sinks", *15th Australasian Fluid Mechanics Conference,* University of Sydney, Sydney, December 2004.

[20] B. Lu and P.X. Jiang, "Experimental and Numerical Investigation of Convection Heat Transfer in a Rectangular Channel with Angled Ribs", *Experimental Thermal and Fluid Sciences* 30 (2006) 513-521.

[21] A. Zukauskas, "Heat Transfer from Tubes in Cross Flow", *Advances in Heat Transfer,* Academic Press, New York, 8 (1972) 93-160.

[22] Zukauskas and R. Ulinskas, "Efficiency Parameters for Heat Transfer in Tube Banks", *Heat Transfer Engineering* 6 (1983) 19-25.

[23] Bergelin, G.A. Brown, and S.C. Doberstein "Heat Transfer and Fluid Friction During Across Bank of Tubes", *Trans. ASME* 74 (1952) 953-960.

[24] J.J. Hwang and C.C. Lui, "Measurement of End Wall Heat Transfer and Pressure Drop in a Pin Fin Wedge Duct", *Int. J. Heat Mass Transfer* 45 (2002) 877-889.

[25] V. Isachenko, V. Osipova, and A. Sukomel, *"Heat Transfer",* Mir (1969) Moscow.

[26] W. M. Kays "Pin Fin Heat Exchanger Surfaces", *Trans. ASME* 77 (1955) 471-483.

[27] A. Zukauskas, "Heat Transfer from Tubes in Cross Flow", in: J.P. Hartnett, T.F. Irvine (Eds.), *Advances in Heat Transfer, A*cademic Press 18 (1987).

[28] B.E Launder and D.B. Spalding, "The Numerical Computation of Turbulent Flow", *Computer Methods in Applied Mechanics and Engineering* 3 (1974) 269-289.

[29] Fluent 6.2.16 *User Manual,* "Modeling of Turbulence", Documentation Report, Chapter 11, 2006, Fluent Inc, Lebanon, USA.

In: Applied Mathematical Modeling
Editor: Eetu N. Virtanen, pp. 297-320

ISBN: 978-1-60021-975-7

Chapter 8

AN ASSESSMENT OF SIX TURBULENCE MODELS FOR PREDICTION OF SINGLE-PHASE FLOW IN A VESSEL STIRRED BY A GRID DISC IMPELLER

***Anupam Dewan*[*, 1], Rahul Swarnkar*[*], Mithilendu K. Jha*[*], P. K. Jha*[*,2] and Vivek V. Buwa*[#]**

*Department of Mechanical Engineering, Indian Institute of Technology Guwahati, Guwahati - 781039, India

[#]Department of Chemical Engineering, Indian Institute of Technology Delhi, Hauz Khas, New Delhi - 110016, India

ABSTRACT

Fluid turbulence is one of the biggest challenges to scientists and engineers, who need to understand turbulent flows in order to control them and design for their adverse effects or utilize them for their best effects. In all cases, prediction is an important element in all engineering design calculations. The performance of six different turbulence models, namely, the standard k-ε model, realizable k-ε model, renormalization group k-ε model, standard k-ω model, shear stress transport k-ω model, and Reynolds stress transport model, is assessed in this chapter for the simulation of a turbulent single-phase flow generated by a grid disc impeller in a stirred vessel. The measurements of three components of the mean velocity and turbulent kinetic energy at different locations within the stirred vessel reported in the literature are used to validate the numerical predictions. The numerical simulations were carried out with a commercial flow solver FLUENT 6.1.22 using the multiple reference frames (MRF) approach. A fairly good agreement was observed between the predicted and experimental velocity components along axial, radial and tangential directions. The predictions of the standard k-ε model were better than the other turbulence models considered in the present work. The performances of other turbulence models were found to be approximately similar.

[1] Corresponding author. Professor. Ph: +91-361-2582656, FAX: +91-361-2582699, Email: adewan@iitg.ernet.in

1. INTRODUCTION

The phenomenon of fluid mixing in stirred vessels is important to several industries such as chemical processing, wastewater treatment, mineral processing, blending of miscible liquids, dispersion of gases or immiscible liquids into a liquid-phase, suspension of solids, etc. Several types of impeller have been investigated and are widely used (Mavros et al., 2001; Li et al., 2004; Derksen et al., 1999; Lee and Yianneskis, 1998; Schafer et al., 1998; and Aubin et al., 2001). The shape of an impeller and the vessel configuration are the major factors contributing to the complexity of a highly turbulent flow inside the vessel and these significantly influence the mixing process in a stirred vessel. The flow generated in a stirred vessel is three-dimensional, unsteady, and rotating turbulent flow. CFD has emerged as a powerful tool to investigate the performance of a vessel stirred by an impeller (see for example, Bujalski et al., 2006; Singh et al., 2007). However, because of the complexity of the flow involved and choice of different mathematical approaches that are available, it is essential to arrive at a reliable prediction technique, which can help in the design of an impeller and stirred vessel.

The numerical simulations of turbulent flows can be performed in two broad ways by solving (1) Three-dimensional, time-dependent, Navier-Stokes equations for the instantaneous flow and then time averaging the solution to obtain the mean flow properties and (2) Reynolds-Averaged Navier-Stokes (RANS) equations. The first approach is used in the large eddy simulation (LES) and direct numerical simulation (DNS). In LES only large scales are resolved and the smallest scales are modelled (see for example, Derksen and Akker, 1999) and in DNS all scales ranging from the largest to the smallest Kolmogorov scale are resolved. The computational requirements for performing LES are somewhat less than those for DNS. Therefore, LES can be performed for a prediction of the turbulent flow generated by impellers (for example, Yeoh et al., 2004 and Hartmann et al., 2006). It may be noted that Reynolds numbers encountered in practical applications are quite large and therefore, DNS and LES cannot be used as a tool for the design of such stirred vessels (Montante et al., 2001). The RANS approach is the most widely used approach for practical engineering flows (see, for example, Menter, 1994; Bakker et al. 1996; Micale et al., 1999; Oshinowo et al., 2000; Li et al., 2004; and Khopkar et al., 2004), mainly because it is computationally much less expensive compared to DNS and LES. The objective of the present investigation is to compare the performance of six different turbulence models covering a wide spectrum to close the system of the RANS governing equations. It has been shown that though the value of Reynolds number considered in the present work is relatively small, the overall predicted trends are independent of this value as several researchers have observed similar trends at different values of Reynolds number considered by them (for example, Bakker et al., 1996;, Oshinawa et al., 2000; Jones et al., 2001; Aubin et al., 2004; and Bujalski et al., 2006).

A large number of RANS equations based turbulence models that have been developed, include the simplest zero-equation models to the most complex Reynolds stress transport models. Of these, the zero-equation model and one-equation model are not being used these days because they are incomplete models. A two-equation model is the simplest complete model of turbulence. Of several two-equation models, the standard k-ε model has been widely applied to study practical flows. In a Reynolds stress transport (RST) model, six additional

[2] Present address: Department of Mechanical & Industrial Engineering, IIT Roorkee, Roorkee-247667, India

transport equations for each component of Reynolds stress are solved. It may be noted that several types of Reynolds stress transport models and two-equation models exist in the literature, and each turbulence model has its advantages and drawbacks. The present chapter compares the performance of six different turbulence models, namely, the standard k-ε model, realizable k-ε model, renormalization group k-ε model, standard k-ω model, shear stress transport k-ω model and Reynolds stress transport model, for the investigation of flow field generated by a grid disc impeller in a baffled vessel. The grid disc impeller developed by Buwa et al. (2006) has been considered for the present study. The predicted profiles of the three components of mean velocity and turbulent kinetic energy at three different radial locations within the vessel have been compared with the LDA measurements of Buwa et al. (2006). Aubin et al. (2004) investigated the effect of the approach to model rotational motion of impeller, discretization scheme and two turbulence models, namely, the standard k-ε model and RNG k-ε model, on the single-phase fluid flow in a tank stirred by a down- and up-pumping pitched blade turbine. They compared the predictions with their LDV measurements. Earlier, Jones et al. (2001) had investigated the performance of six different two-equation turbulence models in predicting the flow field in an unbaffled stirred tank reactor with a single-paddle impeller. The presence of four baffles in the present case makes the flow more complex and, further, in addition to several two-equation turbulence models, a Reynolds stress transport model has also been included to cover a wide spectrum of turbulence models that are available in the literature.

2. Numerical Simulations

2.1. Modelling of Rotational Motion of Impeller

The approaches available to simulate the rotational motion of the impeller are: (a) Black box; (b) Multiple reference frames (MRF); and (c) Sliding mesh. Detailed information on these approaches and their advantages and disadvantages can be found in the literature (for example, Brucato et al., 1998; and Ranade, 2002). The black box approach does not allow *a priori* simulations of the flow field (Ranade and Joshi, 1990). The sliding mesh approach, though quite accurate, requires high computational resources (Singh et al., 2007). See Aubin et al. (2004) for a comparison of the predictions using the sliding grid and multiple frames of reference approach. We have used the MRF approach for modelling the rotational motion of the impeller with respect to the stationary vessel and baffles. In the MRF approach (Luo et al., 1994) the flow domain is divided into subdomains, each of which may be rotating/ translating with respect to the laboratory (inertial) frame. The governing equations in each subdomain are written with respect to a particular subdomain's reference frame. At the boundary between the two subdomains, the diffusion and the other terms in the governing equations in one subdomain require values for the velocities in the adjacent subdomain.

2.2. Governing Equations

The flow simulations were carried out by solving the Reynolds-averaged Navier-Stokes (RANS) equations in the tank and impeller regions using the stationary and rotating frames of reference, respectively. For the stationary frame of reference, the governing equations can be written as

$$\frac{\partial}{\partial t}(\rho)+\nabla\cdot\left(\rho\vec{U}\right)=0 \tag{1}$$

$$\frac{\partial}{\partial t}\left(\rho\vec{U}\right)+\nabla\cdot\left(\rho\vec{U}\otimes\vec{U}\right)=-\nabla P+\nabla\cdot\overline{\overline{\tau}}+\vec{F} \tag{2}$$

where ρ denotes the fluid density, $\vec{U}$ the mean velocity, P the mean pressure, $\vec{F}$ the body force and $\overline{\overline{\tau}}=-\rho\overline{u_i'u_j'}$ the Reynolds stress tensor, which can be written on the basis of the Boussinesq approximation as

$$\overline{\overline{\tau}}=\mu_t\left[\left(\nabla\vec{U}\right)+\left(\nabla\vec{U}\right)^T\right]-\frac{2}{3}\left(\rho k+\mu_t\nabla\cdot\vec{U}\right)\mathbf{I} \tag{3}$$

where $\mathbf{I}$ is the identity matrix. In the stationary frame of reference $\vec{F}=-\rho g\hat{z}$. In the rotating frame of reference attached to the rotor, the body force $\vec{F}$ has the additional components due to centrifugal and coriolis forces, which are

$$\left.\begin{aligned} f_r &= \rho\left(\omega^2 r+2\omega u_\theta\right)\\ f_\theta &= \rho(-2\omega u_r)\\ f_z &= -\rho g \end{aligned}\right\}\quad \vec{F}=f_r\hat{r}+f_\theta\hat{\theta}+f_z\hat{z} \tag{4}$$

where f_r, f_θ and f_z are the components of $\vec{F}$ in the radial, tangential and axial directions, respectively, and ω denotes the rotational speed of the coordinate system. The effect of the gravity was not considered in the present simulations.

2.3. Turbulence Models

As already mentioned in the introduction, a wide variety of turbulence models exist in the literature and the choice of a particular turbulence model for a certain application depends on several factors, such as, the flow physics, accuracy required, available computational resources and time for the simulation. In the present work, we have considered six different turbulence models, which are briefly described in the following sections. Further details on these models may be found in Fluent (2006).

2.3.1. Standard k-ε Model

It is one of the most widely used models and employs transport equations for turbulent kinetic energy (k) and its dissipation (ε) to compute the eddy viscosity (Launder and Spalding, 1974).

$$\frac{\partial}{\partial t}(\rho k)+\frac{\partial}{\partial x_i}(\rho k U_i)=\frac{\partial}{\partial x_j}\left[\left(\mu+\frac{\mu_t}{\sigma_k}\right)\frac{\partial k}{\partial x_j}\right]+\tau_{ij}\frac{\partial U_i}{\partial x_j}-\rho\varepsilon \tag{5}$$

$$\frac{\partial}{\partial t}(\rho \varepsilon)+\frac{\partial}{\partial x_i}(\rho \varepsilon U_i)=\frac{\partial}{\partial x_j}\left[\left(\mu+\frac{\mu_t}{\sigma_\varepsilon}\right)\frac{\partial \varepsilon}{\partial x_j}\right]+C_{1\varepsilon}\frac{\varepsilon}{k}\tau_{ij}\frac{\partial U_i}{\partial x_j}-C_{2\varepsilon}\rho\frac{\varepsilon^2}{k} \tag{6}$$

The turbulent (or eddy) viscosity μ_t is given by

$$\mu_t=\rho C_\mu\frac{k^2}{\varepsilon} \tag{7}$$

This model assumes isotropy for the eddy viscosity and the model constants are given standard values as $C_{1\varepsilon}=1.44, C_{2\varepsilon}=1.92, C_\mu=0.09, \sigma_k=1.0, \sigma_\varepsilon=1.3$.

2.3.2. Renormalization Group (RNG) k-ε Model

The RNG *k-ε* model (Yakhot and Orszag, 1986) has similar form as other two equation models but uses the RNG theory (see Choudhury, 1993), which is a statistical technique. The additional terms in the governing equations improve the accuracy of the predictions by this model for flows with rapid strain and streamline curvature (FLUENT 6.1.22).

2.3.3. Realizable k-ε Model

The realizable *k-ε* model was proposed by Shih et al. (1995) and differs from the standard *k-ε* model in two ways: (a) it contains a different formulation for the eddy viscosity and (b) the transport equation for the dissipation rate *ε* is derived from an exact equation for the transport of the mean-square vorticity fluctuation. The term "realizable" is used as the model satisfies some mathematical constraints on the Reynolds stresses based on the physics of turbulent flows. The model is likely to provide good performance for flows involving rotation, boundary-layers under strong adverse pressure gradients, separation and recirculation.

2.3.4. Standard k-ω Model

The standard *k-ω* model proposed by Wilcox (1998) is based on the modelled transport equations for the turbulence kinetic energy (*k*) and the specific dissipation rate (*ω*), which can also be thought of as the ratio of *ε* to *k*. The spreading rates predicted by this model are in close agreement with measurements for far wakes, mixing layers, and plane, round, and radial jets compared to those by the standard k-ε model,

2.3.5. Shear Stress Transport (SST) $k-\omega$ Model

The shear stress transport (SST) $k-\omega$ model proposed by Menter (1994) uses the standard $k-\omega$ model in the near wall region and the standard $k-\varepsilon$ model in the far field. Therefore this model blends the robust formulation of the k-ω model in the near-wall region with the free-stream independence of the k-ε model in the far field. Further, this model incorporates a damped cross-diffusion derivative term in the ω equation. The definition of the turbulent viscosity is modified to account for the transport of the turbulent shear stress.

2.3.6. Reynolds-Stress Transport Model (RSM)

The exact transport equation for Reynolds stress $\overline{\overline{\tau}} = -\rho\overline{u_i' u_j'}$ can be found in several literatures (for example Wilcox, 1998). Reynolds stress transport model is an anisotropic model and is physically one of the most rigorous models based on RANS equations. However the literature shows (for example, Bakker et al., 1996 and Sheng et al., 1998) that the flow fields predicted by RSM might not always be superior to the simpler models in several types of flows. Therefore an additional computational expense involved with this model may be unwarranted. Of the various terms in the exact transport equation for Reynolds stress, convection, molecular diffusion, stress production and production by system rotation do not require any modelling. The turbulent transport was modelled based on the gradient diffusion hypothesis. The pressure strain ϕ_{ij} was decomposed as

$$\phi_{ij} = \phi_{ij,1} + \phi_{ij,2} + \phi_{ij,w} \tag{8}$$

The slow pressure-strain term $\phi_{ij,1}$ was modelled as

$$\phi_{\mathrm{ij},1} = C_1 \frac{\varepsilon}{k}\left[\tau_{ij} + \frac{2}{3}\rho\delta_{ij}k\right] \tag{9}$$

where C_1 = 1.8. The rapid pressure-strain term, $\phi_{ij,2}$ was modelled as

$$\phi_{ij,2} = -C_2\left[\left(P_{ij} + F_{ij} + G_{ij} - C_{ij}\right) - \frac{2}{3}\delta_{ij}\left(P + G - C\right)\right] \tag{10}$$

where $P = \frac{1}{2}P_{kk}, G = \frac{1}{2}G_{kk}$ and $C = \frac{1}{2}C_{kk}, C_2$ = 0.60. The wall-reflection term $\phi_{ij,w}$ is responsible for the redistribution of the normal stress near the wall and was modelled as

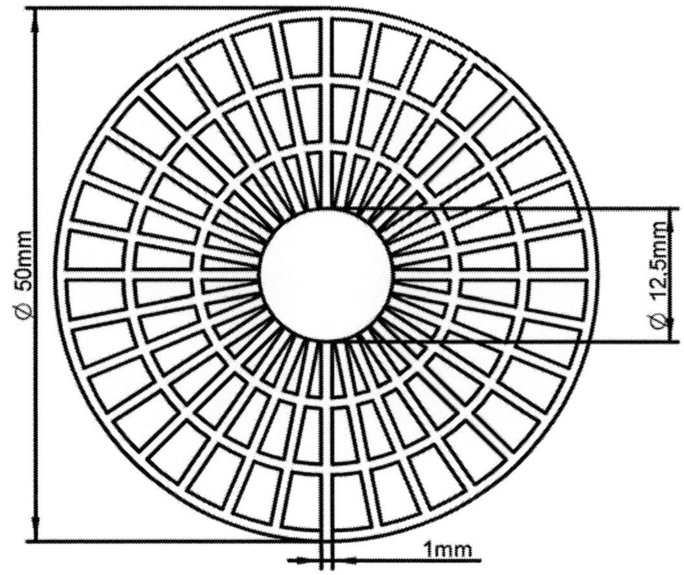

Figure 1(a). Grid disc impeller developed by Buwa et al. (2006).

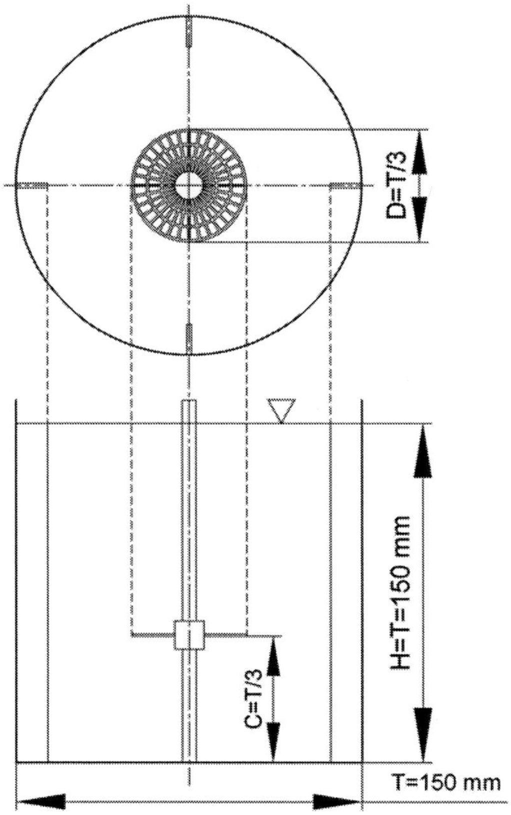

Figure 1(b). Stirred vessel configuration used in the present investigation.

$$\phi_{ij,w} = C_1' \frac{\varepsilon}{k} m \left(-\overline{u_k' u_m'} n_n n_m \delta_{ij} - \frac{3}{2} \overline{u_i' u_k'} n_j n_k - \frac{3}{2} \overline{u_j' u_k'} n_i n_k \right) \frac{k^{3/2}}{C_l \varepsilon d}$$
$$+ C_2' \left(\phi_{km,2} n_k n_m \delta_{ij} - \frac{3}{2} \phi_{ik,2} n_j n_k - \frac{3}{2} \phi_{jk,2} n_i n_k \right) \frac{k^{3/2}}{C_l \varepsilon d} \qquad (11)$$

where n_k denotes the x_k component of the unit normal to the wall, d the normal distance to the wall and $C_l = C_\mu^{3/4} / \kappa$, where C_μ = 0.09 and κ the von Karman constant (= 0.4187). The dissipation tensor ε_{ij} was modelled as $\varepsilon_{ij} = \frac{2}{3} \delta_{ij} \rho \varepsilon$, where the scalar dissipation rate ε was modelled similar to that in the case of standard $k - \varepsilon$ model. To obtain the boundary conditions for the Reynolds stresses, a transport equation for turbulent kinetic energy was solved (Wilcox, 1998).

2.4. Vessel Geometry

Figure 1a shows the grid disc impeller developed by Buwa et al. (2006) which was used in the present investigations. This grid disc contains 32 radial rods along with three radial rings, both having a square cross-section with each side of 1.0 mm. A schematic of the stirred vessel geometry used in the present work is shown in Figure 1b. The central part (i.e., the hub) had a diameter of 12.5 mm and was connected to a shaft of 6 mm diameter. There were four baffles in the tank each with a thickness of 1.0 mm and 13.5 mm width. The impeller was located at a height of 50 mm from the bottom of the tank. The impeller diameter (*D*) was 50 mm and the tank diameter (*T* = tank height *H*) was 150 mm (Figure 1b). The vessel geometry was divided into two major subdomains: (a) the rotating impeller zone and (b) the stationary tank zone, which included the baffles and tank walls. The impeller region extended from $0 < r < 42.5$ mm and 40 mm $< z <$ 65 mm. Due to the symmetical features of the vessel configuration, we considered only one fourth of the entire physical domain and the two isolating vertical planes were specified as a pair of periodic boundary. While computing the flow through the periodic boundary adjacent to a fluid cell, the flow conditions at the fluid cell adjacent to the opposite periodic plane were used.

Table 1. Grid size used for different turbulence models

	Impeller region	**Tank region**	**Total elements**
Very fine	98813	91654	190467
Fine	55483	39841	95324
Medium	26162	32181	58343
Coarse	12414	22991	35405

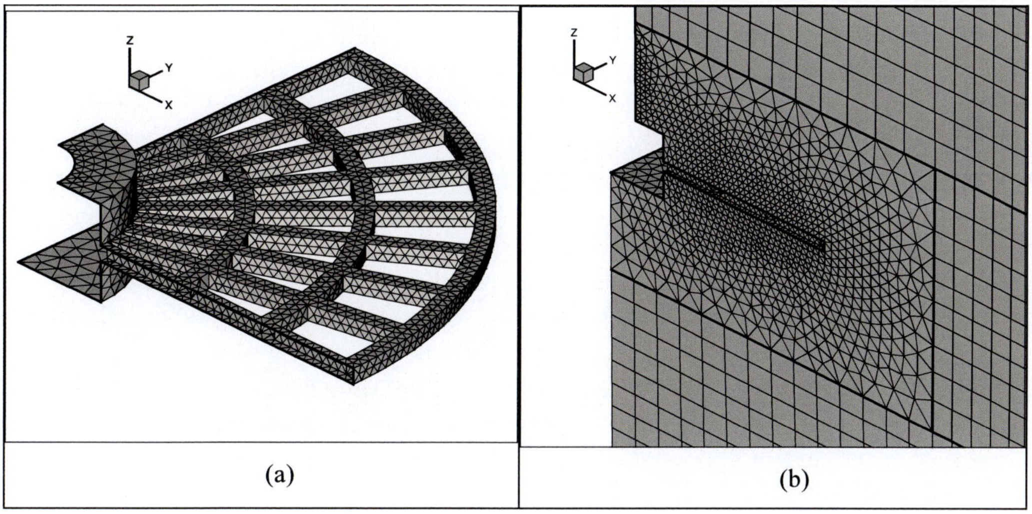

Figure 2 (a) Grid on impeller surface and (b) Tetrahedral mesh generated in the impeller zone.

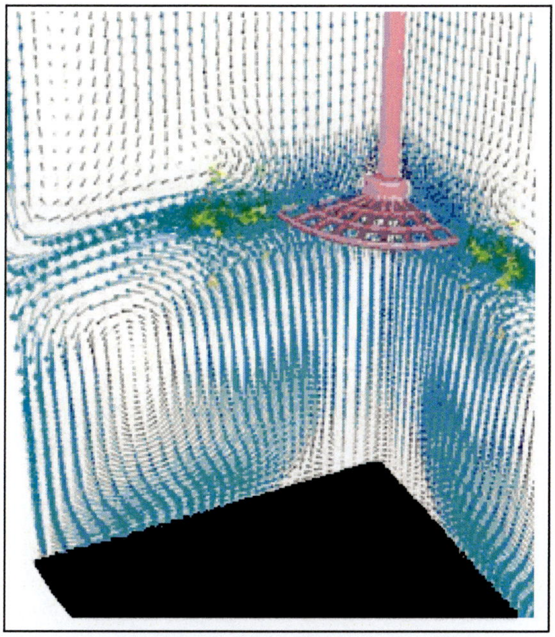

Figure 3. Recirculation zones below the grid disc impeller.

2.6. Grid Independence Study

First, a grid independence study was performed with the standard *k*-ε model using the second order scheme for momentum and first order scheme for turbulence (hereafter termed as FOST) using four grid sizes as shown in Table 1. The rotational speed of the impeller was 1500 rpm based on the conditions in the experiments performed by Buwa et al. (2006). The corresponding Reynolds number (Re) with the kinematic viscosity $\nu = 1.5 \times 10^{-5}$ m^2/s of the white paraffin oil considered in the present work was 4167 (Re = ND^2/ν where N denotes the angular velocity of the impeller in revolutions/sec and ν the kinematic viscosity of the liquid in m^2/s). The comparison of the predicted and measured mean velocity components and turbulent kinetic energy was carried out at three different axial positions corresponding to lines in the *x-z* plane at three axial locations of z = 25 mm, 52 mm and 100 mm. The lines were located behind the baffle at an angle of 22.5° in the counter clockwise direction. These lines have been named as L2, L4 and L6, respectively (see Figure 4).

The results of the grid independence study are shown in Figure 5 for line L4. For axial positions of L2 and L6, the predictions of the mean velocity components and turbulent kinetic energy were not influenced by the different grid resolutions (not shown here). The profiles of the turbulent kinetic energy, radial velocity and tangential velocity along line L4 (see Figures 5a, 5c and 5d) indicated that the grid independent solution is achieved for the the fine mesh and no further refinement was needed. Based on this conclusion we carried out all further investigations using the fine mesh.

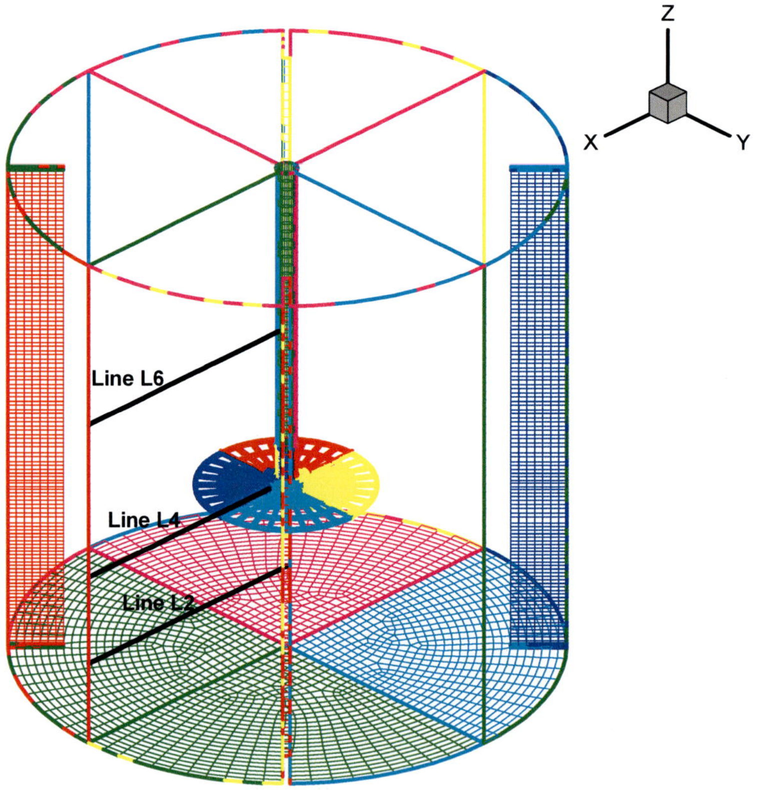

Figure 4. Locations of three radial lines along which different profiles were studied.

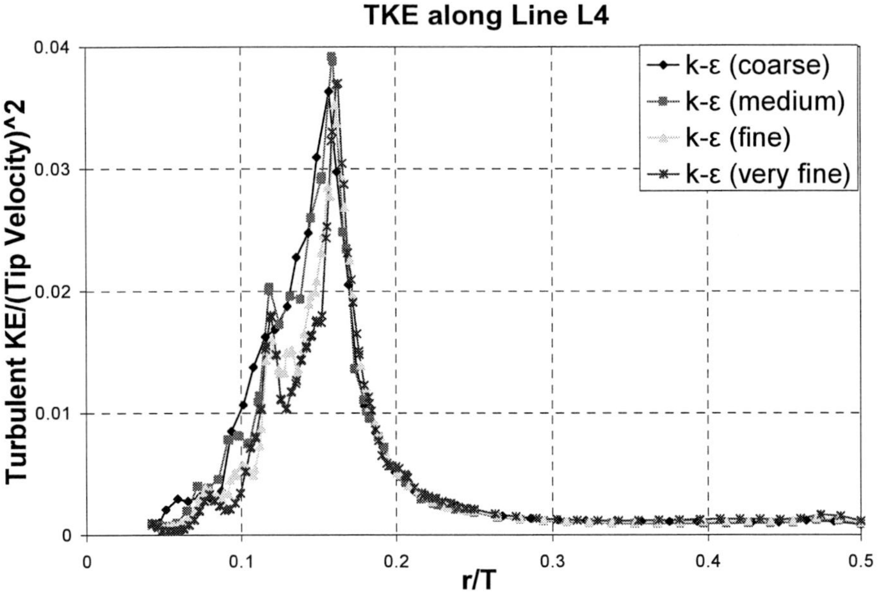

(a)

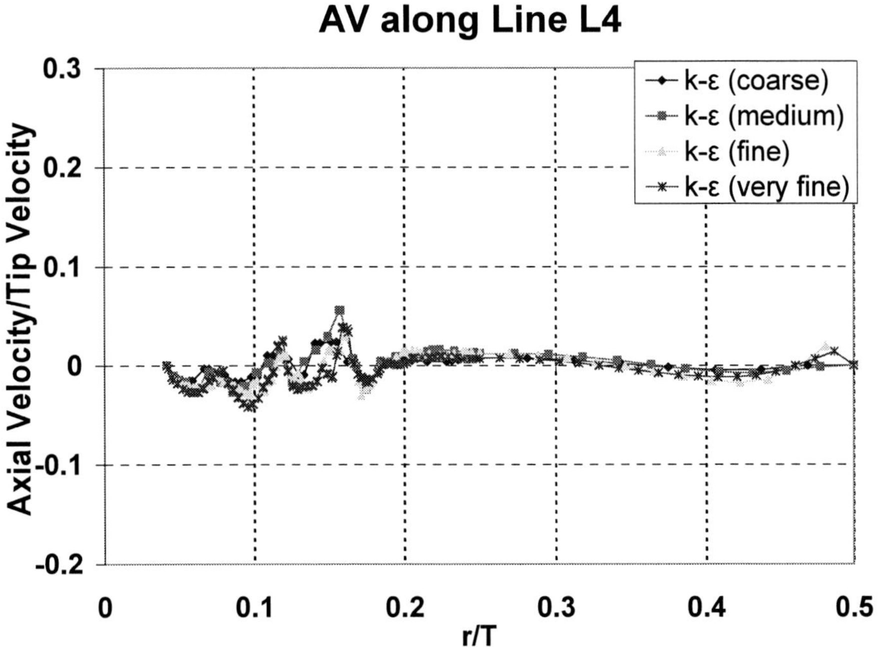

(b)

Figure 5. (Continued on next page)

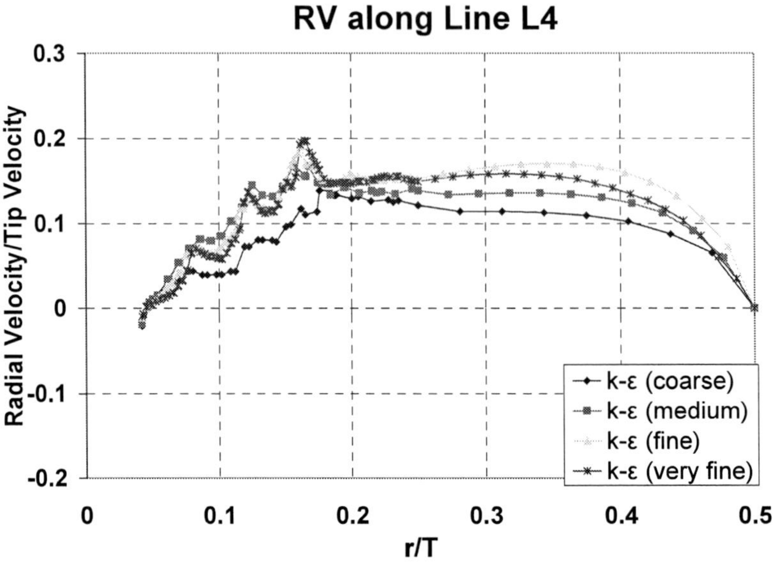

(c)

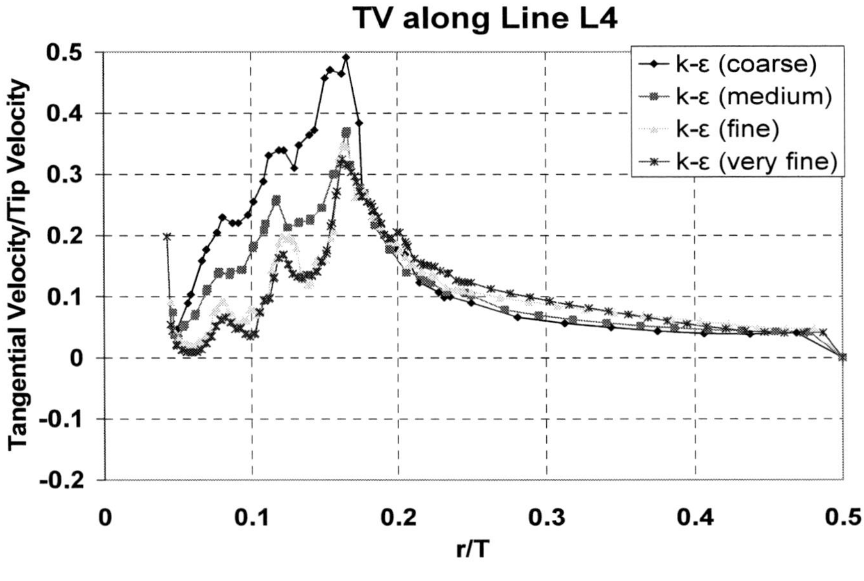

(d)

Figure 5. Grid independence study. (a) Turbulent kinetic energy, (b) Axial velocity, (c) Radial velocity and (d) Tangential velocity profiles using four grid sizes along line L4.

3. Results and Discussion

3.1. Comparison of First and Second Order Turbulence Results

We compared results obtained by the first and second order schemes for the discretization of turbulence equations and with second order discretization for the momentum equations using the standard k-ε model. Figure 6a shows the improvement in the predictions by SOST (i.e., second order scheme for both turbulence and momentum equations) over FOST results. The SOST computations of the peak in turbulent kinetic energy are under-predicted by 22.22% whereas the FOST under-predicts by 66.67% compared to the LDA measurements of Buwa et al. (2006). The corresponding first order turbulence predictions made by Buwa et al. (2006) were 73.33% less than their corresponding experimental data. The present observations of the effects of the discretization schemes on the prediction of the mean and turbulent quantities are in accordance with the observations of Aubin et al. (2004), who compared the prediction of down- and up-pumping pitched blade turbine using the standard k-ε model. They emphasised the need for using higher order scheme for an accurate prediction of turbulent quantities. They also showed that the predictions of the mean quantities were unaffected by the order of the discretization scheme. A small deviation between the FOST predictions by Buwa et al. (2006) and those in the present work can be attributed to the different meshing style and different flow solvers, CFX and FLUENT, respectively, employed in the two works. For the axial velocity plot (Figure 6b), both FOST and SOST results are identical in the inner region (r/T = 0 to 0.3) with a slight deviation from the corresponding LDA measurement. However beyond r/T = 0.3, the FOST results show some deviation whereas the SOST results continue to maintain a fair agreement throughout the line. Similarly in Figures 6c and 6d the SOST results for the radial and tangential velocity components continue to show somewhat better predictions than those by using the FOST. The predictions of the four quantities along lines L2 and L6 by the SOST scheme were somewhat better than those by the FOST (not shown here). This case study leads us to the conclusion that the second order discretized turbulence equations give better prediction than the corresponding first order discretized turbulence equations and this observation formed the basis of our further detailed comparison of six different turbulence models.

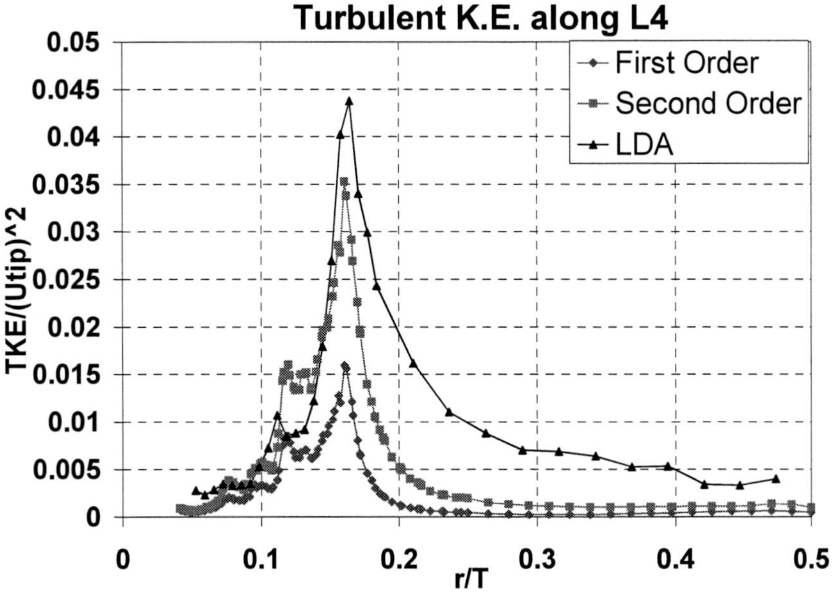

(a)

Figure 6. (Continued on next page)

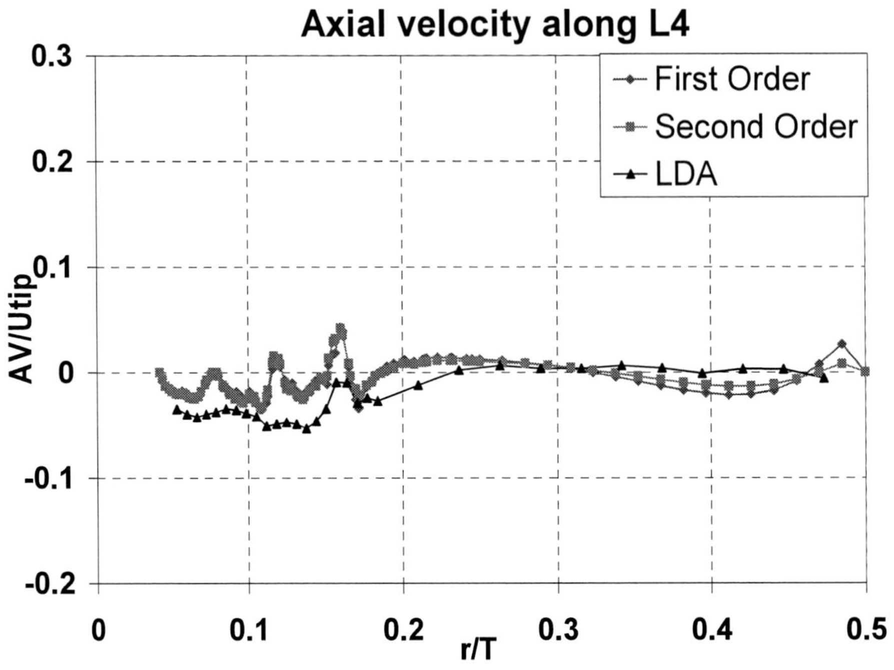

(b)

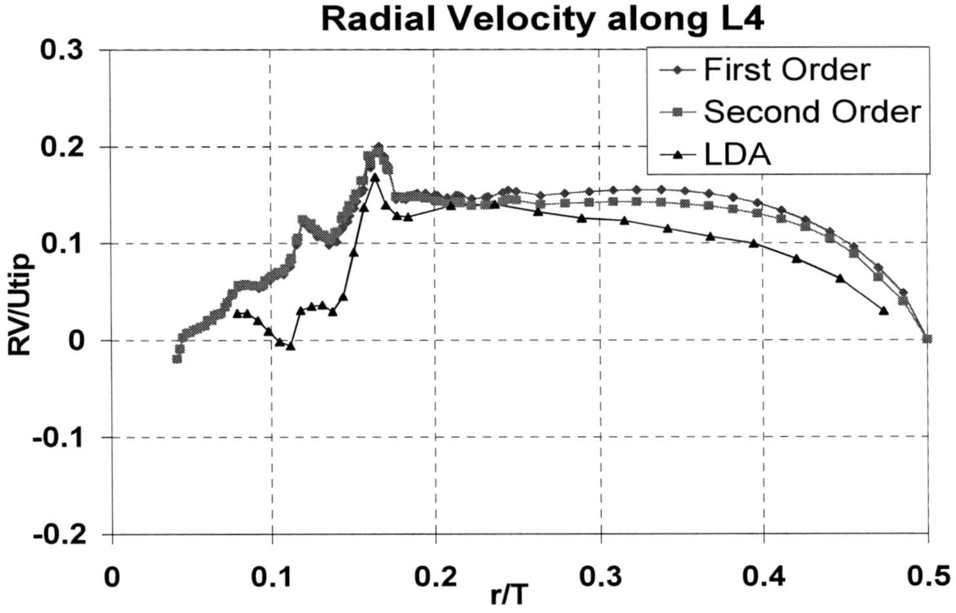

(c)

Figure 6. (Continued on next page)

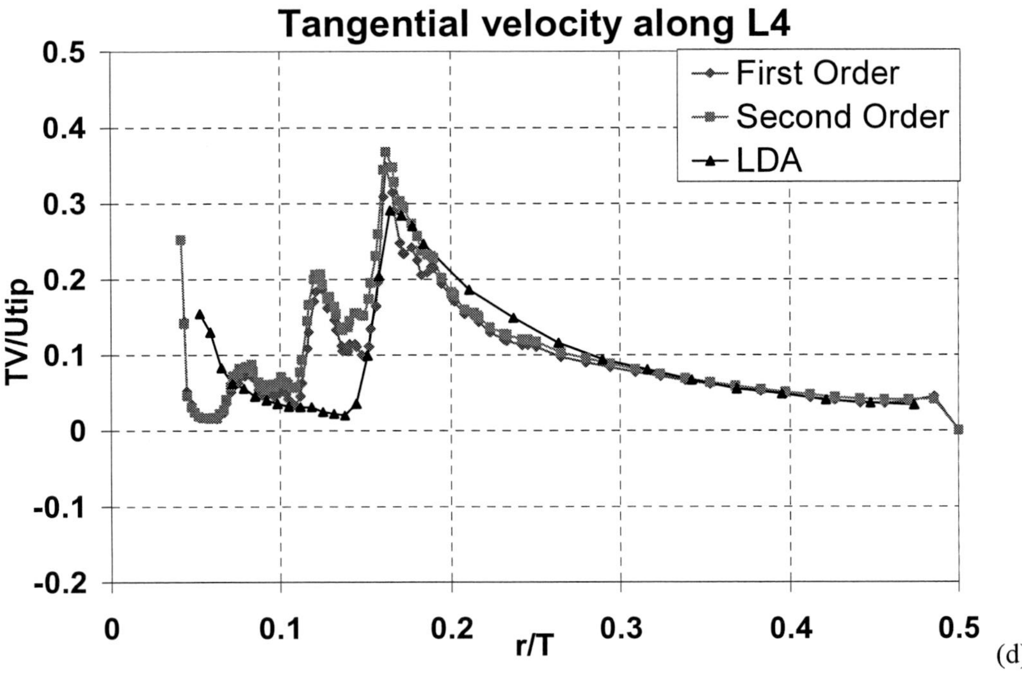

Figure 6. Comparison of first and second order turbulence results along line L4 (just above the impeller) for fine grid and standard k-ε model, (a) Turbulent kinetic energy, (b) Axial velocity, (c) Radial velocity and (d) Tangential velocity.

3.2. Assessment of Six Turbulence Models

We assessed the performance of six different turbulence models, namely, (a) Standard k-ε (SKE) model; (b) Realizable k-ε (RKE) model; (c) Renormalization group k-ε (RNG) model; (d) Standard k-ω (SKW) model; (e) Shear stress transport k-ω (SST) model; and (f) Reynolds stress transport model (RSM). The second order scheme both for turbulence and momentum equations (SOST) for all the above models was used. Figure 7 shows the computed axial and radial velocity profiles along all three radial lines. Along L2 (located below the impeller) the axial velocity is well predicted in the region $0.25 < r/T < 0.44$. However, a slight overprediction by all models is observed up to some distance from the tank shaft [Figure 7 i(a)]. The computation effectively predicts the direction of the primary vortex along L4 as the sign of axial velocity is almost accurately predicted by all the models [Figure 7 ii(a)]. The experimental data along L4 indicates that impeller discharge is projected slightly downwards and all the models seem to predict this downward projection. Significant overpredictions in the axial velocity components near the impeller were also observed by Jones et al. (2001) and were attributed by them to the failure of the two equation models to account for the relaxational effects and stress convection. While predicting flow patterns of a down-pumping four blade pitch blade turbine using three turbulence models, namely, standard k-ε model, RNG k-ε model and Reynolds-stress model Bakker et al. (1996) found that the predictions by three models were similar and provided good agreement with the experimental data. The flow condition in the impeller stream close to the rotor shaft changes rapidly downstream and therefore these effects may be important in the downstream region. A small value of the axial velocity near the impeller justifies that the impeller is of a non axial type [Figure 7 ii(a)]. Along L6, above the impeller, the models under-predict the axial velocity [Figure 7 iii(a)]. Along this line (L6) the velocity predicted is mostly negative and shows the primary vortex to be downward directed for larger distance along L6.

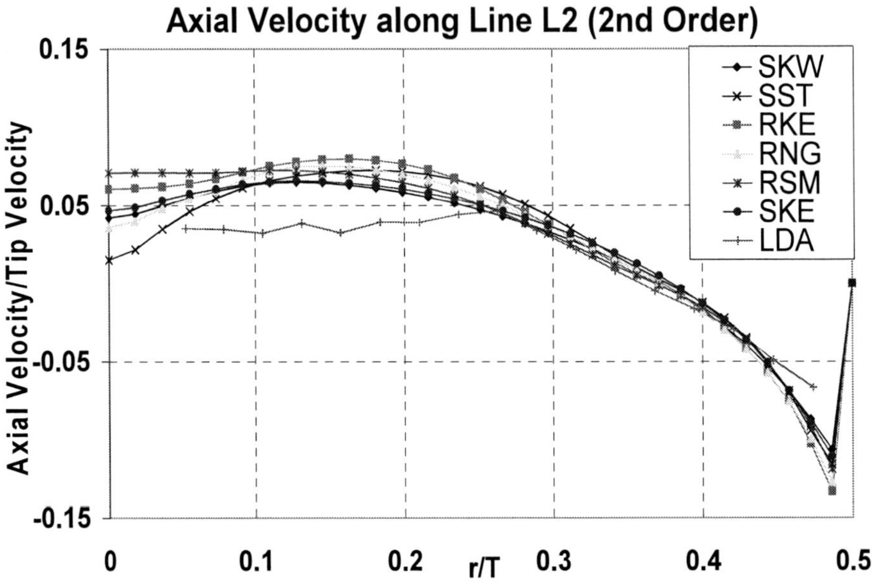

i (a)

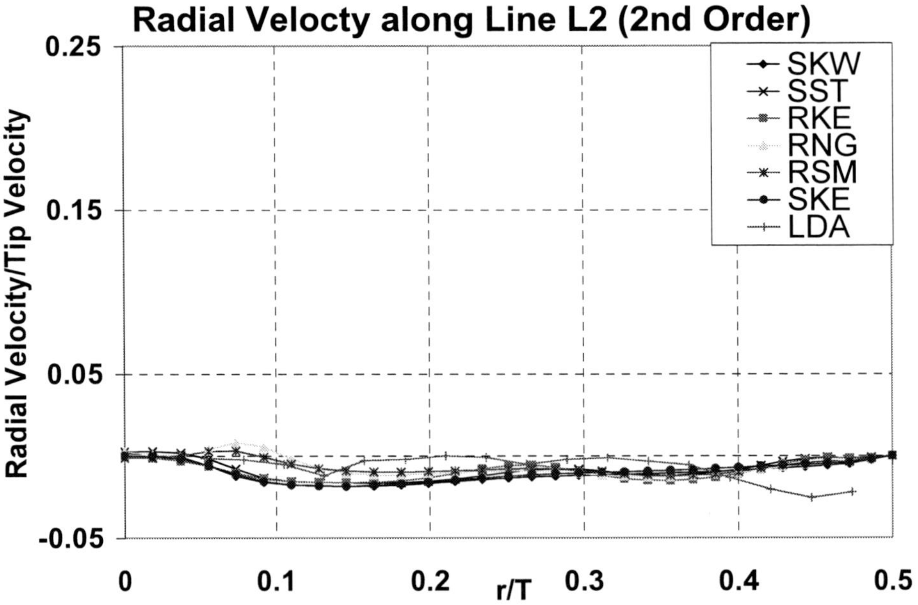

i (b)

Figure 7. (Continued on next page)

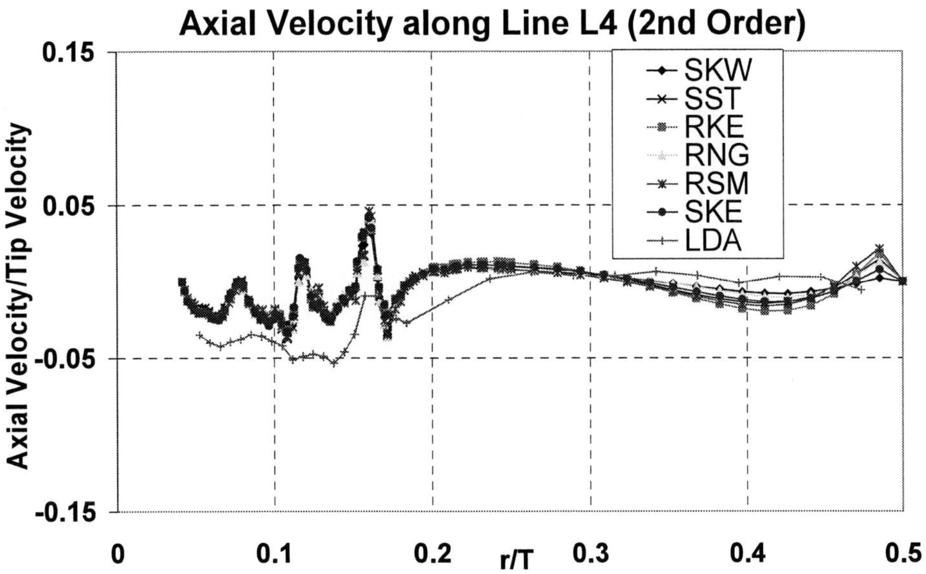

ii (a)

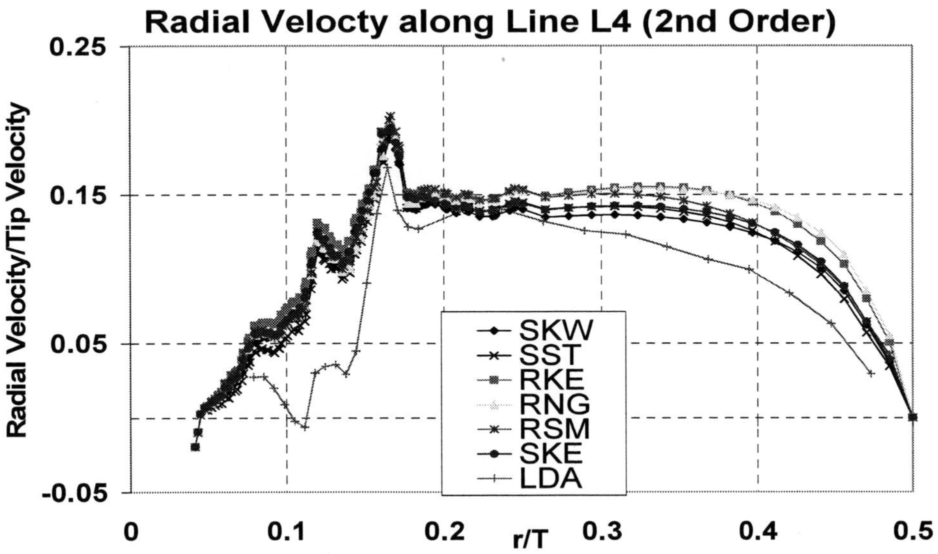

ii (b)

Figure 7. (Continued on next page)

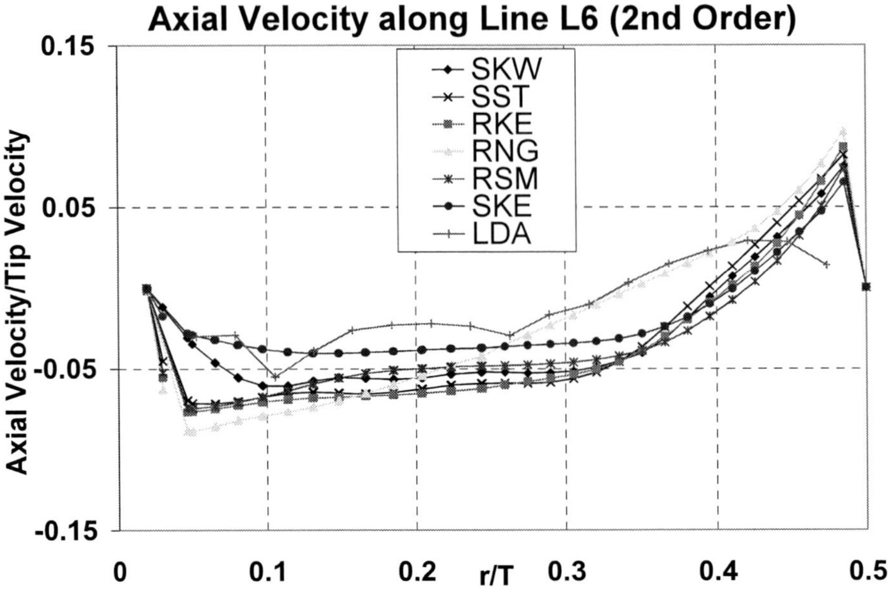

iii (a)

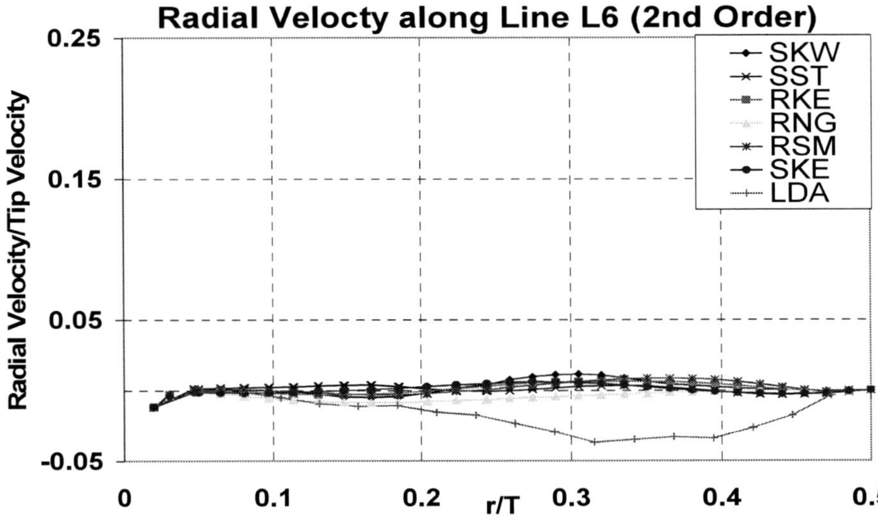

iii (b)

Figure 7. Comparison of experimental and predicted (a) axial velocity and (b) radial velocity profiles at axial locations of (i) z = 25 mm, (ii) z = 52 mm and (iii) z = 100 mm from the bottom of the tank.

The radial velocities along L2 and L6 are nearly zero as predicted almost equally by all the models [Figures 7 i(b) and 7 iii(b)]. The experimental data shows the inward radial motion along L6 whereas the computational results fail to reflect it [Figure 7 iii(b)]. Along L4, all the models over-predict the radial velocity [Figure 7 ii(b)]. The position of the peak near the impeller tip is almost accurately captured by all the models [Figure 7 ii(b)]. Along L4 away from the impeller tip it can be noticed that the renormalization group k-ε model and realizable k-ε model have the highest prediction while the k-ω model has the lowest prediction. The computed tangential velocity along L2 as shown in Figure 8 i(a) is nearly zero as compared to a small positive value obtained from the experiment. Similarly along L6 the tangential velocity almost closely follows the zero velocity line and the predictions are relatively close to experimental values [Figure 8 iii(a)]. The tangential velocity has significant values only in the region near the impeller plane as shown in Figure 8 ii(a). The predictions closely follow the experimental data in the region after the impeller tip. The peak near the tip is slightly overpredicted by all the models.

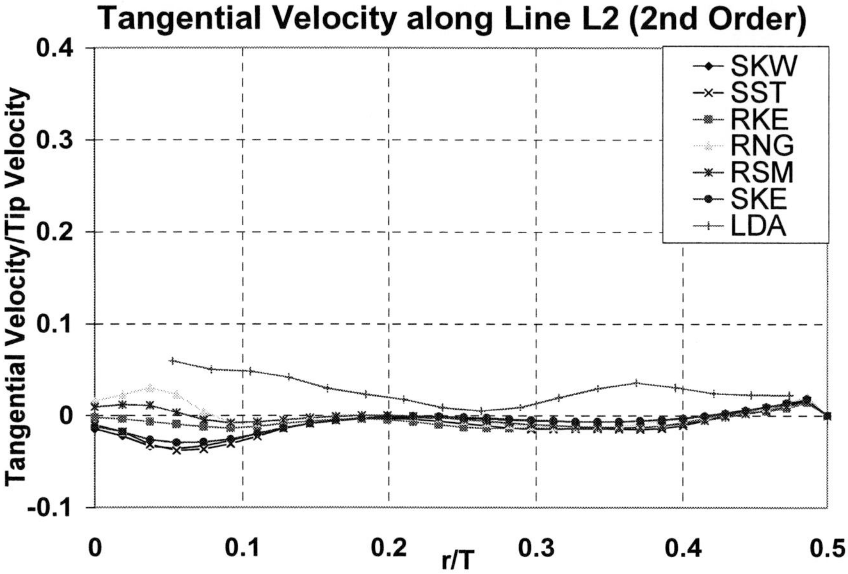

i (a)

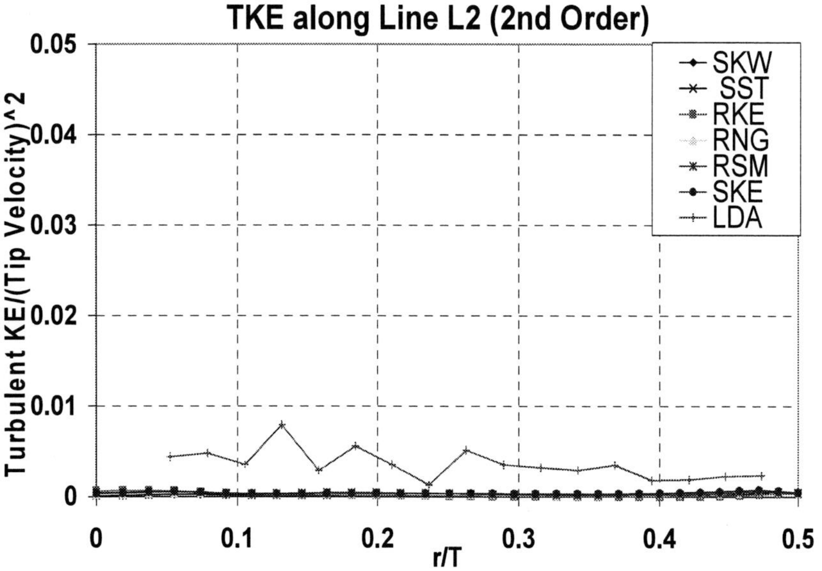

i (b)

Figure 8. (Continued on next page)

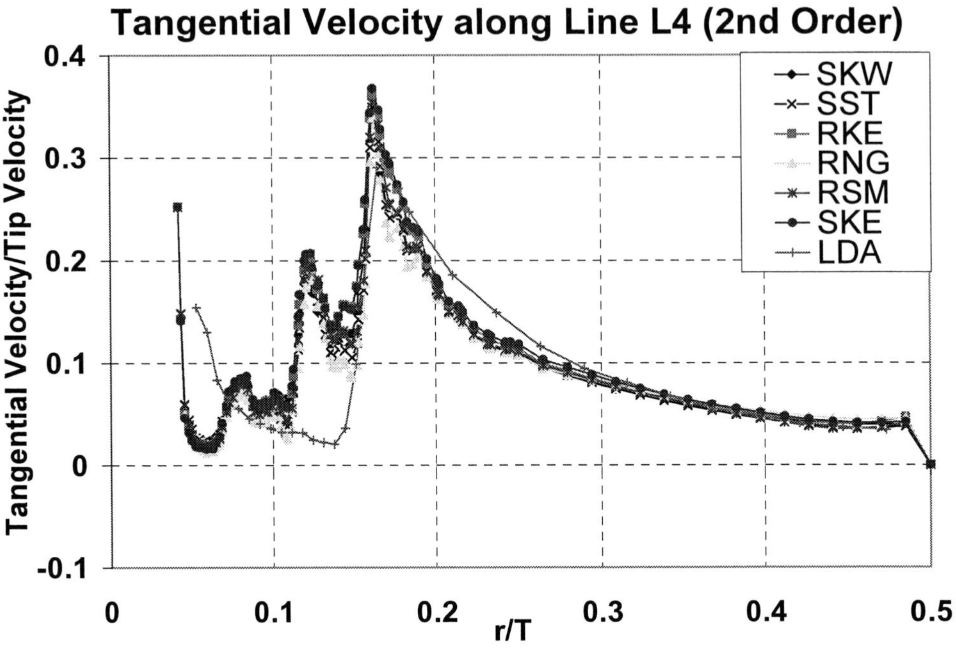

ii (a)

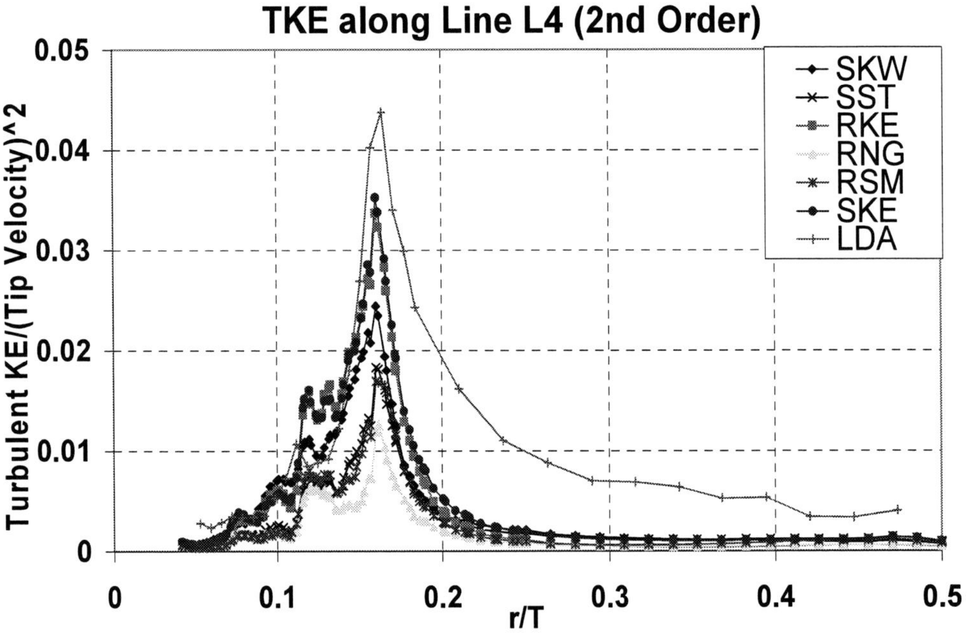

ii (b)

Figure 8. (Continued on next page)

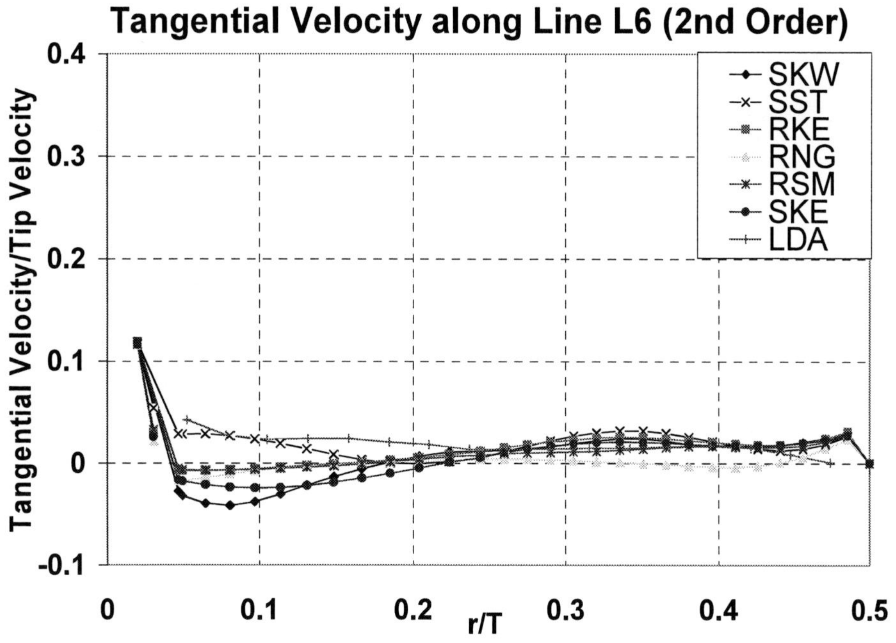

iii (a)

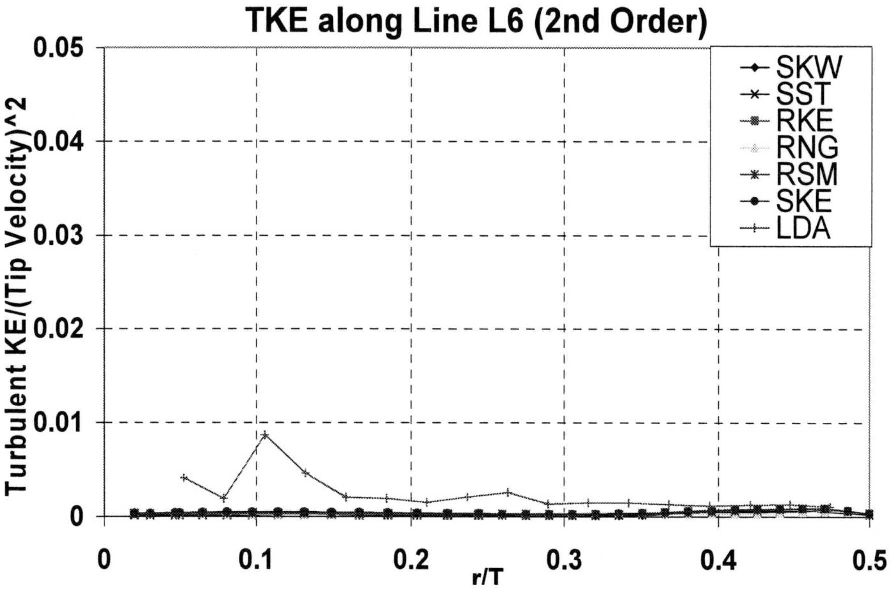

iii (b)

Figure 8. Comparison of experimental and predicted (a) Tangential velocity and (b) Turbulent kinetic energy profiles at axial locations of (i) z = 25 mm, (ii) z = 52 mm and (iii) z = 100 mm from the bottom of the tank.

The turbulence kinetic energy profiles are shown in Figures 8 i(b), 8 ii(b) and 8 iii(b) for lines L2, L4 and L6, respectively. It can be seen that the turbulent kinetic energy along lines L2 and L6 are quite low and closely follow the zero turbulent kinetic energy line while the experimental data along these lines have somewhat higher values [Figures 8 i(b) and 8 iii(b)]. A clear difference in the predictions of all six models can be easily seen in the turbulent kinetic energy plot along L4 as shown in Figure 8 ii(b). Although near the shaft the predictions are close to the LDA results, the difference in the numerical and experimental results grow significantly as we move away from the shaft along L4. The computations accurately capture the location of the peak at the impeller tip but the magnitude of the maximum turbulent kinetic energy is quite underpredicted by all the models. The standard k-ε and realizable k-ε models are most satisfactory while the RNG k-ε model has the poorest performance [Figure 8 ii(b)]. It may be noted that Bujalski et al. (2006) reported a similar amount of underprediction of turbulent kinetic energy by different k-ε based models. They observed that the predicted mixing times were 2-3 times larger than the measured values due to an underprediction of turbulence even with the sliding mesh approach within the framework of RANS equations.

An overall poor prediction of turbulent kinetic energy could be due to the inability of RANS equations to accurately represent the physics of complex flows such as that encountered in a stirred vessel. It can further be attributed to various modelling issues. It may reflect the inability of the standard wall function approach used in this chapter to properly represent turbulent flows in the vicinity of solid walls. The renormalization group k-ε model has been incorporated with the differential equation for the turbulent viscosity allowing the model to better handle the low Reynolds number and near wall flows compared to other models. However, this model does not seem to provide improved predictions of the turbulent kinetic energy near the impeller.

4. CONCLUSION

The present study compares the predictions by six different turbulence models with the LDA measurements reported in the literature. The investigations were carried out for a baffled stirred vessel having grid disc impeller using commercial CFD code FLUENT-6.1.22. The multiple reference frames approach (MRF) was used to account for the impeller rotation with respect to the stationary baffles and the vessel. Important conclusions that can be drawn from the present study are: (1) the standard k-ε model proved to be the best turbulence model for this geometry, followed by the realizable k-ε model; (2) the second order discretization scheme for momentum equations and those for all turbulence models provides improved results for turbulent kinetic energy and other mean quantities compared to the case with second order scheme for momentum and first order scheme for turbulence equations; (3) the axial velocity is mostly overpredicted above the impeller and underpredicted below the impeller, except for the outer region where the predictions seem to have fairly good agreement with the corresponding experimental results; and (4) the predicted radial and tangential velocity profiles seem to deviate from the experimental results by a small amount, but nevertheless they have an overall good agreement.

REFERENCES

Aubin, J., Mavros, P., Fletcher, D.F., Bertrand, J., Xuereb, C., 2001, Effect of axial agitator configuration (up-pumping, down-pumping, reverse rotation) on flow patterns generated in stirred vessels. *Chemical Engineering Research and Design,* 79, 845-856.

Aubin, J., Fletcher, D.F., Xuereb, C., 2004, Modeling of turbulent flow in stirred tanks with CFD: the influence of modeling approach, turbulence model and numerical scheme. *Experimental Thermal and Fluid Science* 28 (5), 431.

Bakker, A., Meyers, K.J., Ward, R.W., and Lee, C.K., 1996, The laminar and turbulent flow patterns of a pitched blade turbine, *Trans. IChemE* 74A, 485-491.

Brucato, A., Ciofalo, M., Grisafi, F., Micale, G., 1998, Numerical prediction of flow fields in baffled stirred vessels: a comparison of alternative modeling approaches. *Chemical Engineering Science* 53, 3653–3684.

Bujalski, W., Jowraski, Z., and Nienow, A.W., 2002, CFD study of homogenization with dual Rushton turbines – comparison with experimental results. *Chemical Engineering Research and Design,* 80, 97.

Bujalski, J.M., Yang, W., Nikolov, J., Solnordal, C.B., Schwarz, M.P., 2006, Measurement and CFD simulation of single-phase flow in solvent extraction pulsed column, *Chemical Engineering Science,* 61, 2930 - 2938.

Buwa, V., Dewan, A., Nassar, A. F. and Durst, F., 2006, Fluid dynamics and mixing of single phase flow in a stirred vessel with a grid disc impeller: Experimental and numerical investigations. Chemical Engineering Science, 61, 2815-2822.

Choudhury, D., 1993, *Introduction to the Renormalization Group Method and Turbulence Modeling.* Fluent Inc. Technical Memorandum TM-107.

Derksen, J.J., Doelman, M.S., Van den Akker, H.E.A., 1999, Three-dimensional LDA measurements in the impeller region of a turbulently stirred tank. *Experiments in Fluids,* 27, 522-532.

Derksen, J.J., Akker, H.E.A. V.D., 1999, Large eddy simulations on the flow driven by a Rushton turbine, *AIChE Journal,* 45 (2) 209-221.

FLUENT 6.1.22 *User Manual,* 2003, Fluent Inc, Lebanon, USA.

Hartmann, H., Derksen, J.J., and van den Akker, H.E.A., 2006, Mixing times in a turbulent stirred tank by means of LES, *AIChE Journal,* 52 (11), 3696-3706.

Jones, R.M., Harvey, A.D., Acharya, S., 2001, Two-equation turbulence modeling for impeller stirred tanks. *ASME Journal of Fluids Engineering,* 123, 640-648.

Khopkar, A.R., Mavros, P., Ranade, V.V., Bertrand, J., 2004, Simulation of flow generated by an axial-flow impeller: batch and continuous operation. *Chemical Engineering Research and Design,* 82 (A6), 737-751.

Launder, B.E., Spalding, D.E., 1974, The numerical computation of turbulent flows. *Computer Methods in Applied Mechanics and Engineering,* 3, 269-289.

Lee, K.C., Yianneskis, M., 1998, Turbulence properties of impeller stream of a Rushton turbine. *American Institute of Chemical Engineers Journal,* 44, 13-24.

Li, M., White, G., Wilkinson, D., Roberts, K.J., 2004, LDA measurements and CFD modeling of a stirred vessel with a retreat curve impeller. *Industrial Engineering and Chemistry Research,* 43, 6534-6547.

Luo, J.Y., Issa, R.I., Gosman, A.D., 1994, Prediction of impeller-induced flows in vessels using multiple frames of reference. *Institution of Chemical Engineers Symposium Series,* 136, 549-556.

Mavros, P., Mann, R., Vlaev, S.D., Bertrand, J., 2001, Experimental visualization and CFD simulation of flow patterns induced by a novel energy saving dual-configuration impeller in stirred vessels. *Transactions of Institution of Chemical Engineers,* 79, 857-866.

Menter, F. R., 1994, Two-Equation Eddy-Viscosity Turbulence Models for Engineering Applications. *AIAA Journal,* 32(8):1598-1605.

Micale, G., Brucato, Grisafi, F. and Ciofalo, M., 1999, Prediction of flow fields in a dual-impeller stirred vessel. *American Institute of Chemical Engineers Journal,* 45 445-464.

Montante, G., Lee, K.C., Brucato. A., Yianneskis, M., 2001, Numerical simulations of the dependency of flow pattern on impeller clearance in stirred vessels. *Chemical Engineering Science,* 56, 3751-3770.

Oshinowo, L., Jaworski, Z., Dyster, K.N., Marshall, E., Nienow, A.W., 2000, Predicting the tangential velocity field in stirred tanks using the multiple reference frames (MRF) Model with validation by LDA measurements. *Proc. 10th Europ. Conf. Mixing, Delft,* Netherlands, 281–288.

Patankar, S.V., 1980, *Numerical Heat Transfer and Fluid Flow,* Hemisphere, Washington DC.

Ranade, V.V., 2002, *Computational Flow Modeling for Chemical Reactor Engineering,* Academic Press, New York.

Ranade, V.V. and Joshi, J.B., 1990, Flow generated by a disc turbine: Part II. Mathematical modeling and comparison with experimental data. *Transactions of Institution of Chemical Engineers,* 68(A): 34–50.

Schafer, M., Yianneskis, M., Wachter, P., Durst, F., 1998, Trailing vortices around a 45° pitched-blade impeller. *American Institute of Chemical Engineers Journal,* 44, 1233-1246.

Sheng, J., Meng, H., Fox, R.O., 1998, Validation of CFD simulations of a stirred tank using particle image velocity data. *Can. J. Chem. Eng.* 76, 611- 625.

Shih, T.H., Liou, W.W., Shabbir, A., Yang, Z., and Zhu, J., 1995, A New Eddy-Viscosity Model for High Reynolds Number Turbulent Flows- Model Development and Validation. *Computers and Fluids,* 24(3): 227-238.

Singh, K.K., Mahajani, S.M., Shenoy, K.T., Patwardhan, A.W., Ghosh, S.K., 2007, CFD modeling of pilot-scale pump-mixer: Single-phase head and power characteristics. *Chemical Engineering Science* 62 (2007), 1308 -1322.

Wilcox, D.C., 1998, *Turbulence Modeling for CFD.* DCW Industries, Inc., La Canada, California.

Yakhot, V. and Orszag, S.A., 1986, Renormalization Group Analysis of Turbulence: I. Basic Theory. *Journal of Scientific Computing* 1 (1), 1-51.

Yeoh, S.L., Papadakis, G., Lee, K.C., Yianneskis, M., 2004, Large eddy simulation of turbulent flow in a Ruston impeller stirred reactor with sliding-forming mesh methodology. *Chemical Engineering and Technology* 27, 257-263.

In: Applied Mathematical Modeling
Editor: Eetu N. Virtanen, pp. 321-339
ISBN: 978-1-60021-975-7

Chapter 9

ALGORITHMS FOR NONLINEAR DYNAMIC RATIONAL MODEL IDENTIFICATION AND VALIDATION

Quan Min Zhu*

Faculty of Computing, Engineering and Mathematical Sciences (CEMS),
University of the West of England (UWE),
Frenchay Campus, Coldharbour Lane, Bristol, BS16 1QY, UK

ABSTRACT

The Non-linear AutoRegressive Moving Average with eXogenous input (NARMAX) model set has been extensively studied in theory and gradually adopted in applications. There are two main steams of sub-model sets, polynomial and rational. The rational model represents an extension of the polynomial model set, and is defined as a ratio of two polynomial expressions. The introduction of the denominator polynomial terms makes the rational NARMAX model to be non-linear in both the parameters and the regression terms. The justification for using the rational model is that it provides a very concise and parsimonious representation for complex non-linear systems and has excellent interpolatory and extrapolatory properties. However model identification and controller design are much more challenging compared to the polynomial models. This has been a new and fascinating research trend in the area of mathematical modelling and applications, but still within a limited research community. This chapter brings several representative algorithms together, developed by the author and his colleagues, to form an easily referenced archive for promotion of awareness, tutorial, applications, and even further research expansion.

* Tel: +44 (0)117 328 2533; Fax: +44 (0)117 328 3800; Email: quan.zhu@uwe.ac.uk
Website: http://www.ias.uwe.ac.uk/people.htm

1. INTRODUCTION

The Non-linear AutoRegressive Moving Average with eXogenous input (NARMAX) model set has been extensively studied in theory (Sontag 1978, Chen and Billings 1989, Haber and Unbehauen 1990) and gradually adopted in applications (Proll and Karim 1994, Wang 1994). There are two main steams of sub-model sets, polynomial and rational. A polynomial NARMAX model is defined as linear in parameters and non-linear in the regression terms, and can be used to represent a wide range of linear and non-linear systems. The advantage for identification of polynomial models comes from the fact that the model is linear in the parameters. The rational model (Sontag 1979, Billings and Chen 1989, Billings and Zhu 1991) represents an extension of the polynomial model set, and is defined as a ratio of two polynomial expressions. Mathematically a general dynamic non-linear rational model is expressed as

$$y(k) = \hat{y}(k) + e(k) = \frac{a(k)}{b(k)} + e(k) = \frac{a\left[u(k-1),..u(k-n_{nu}), y(k-1),...y(k-n_{ny}), e(k-1),..e(k-n_{ne})\right]}{b\left[u(k-1),..u(k-n_{du}), y(k-1),...y(k-n_{dy}), e(k-1),..e(k-n_{de})\right]} + e(k) \tag{1.1}$$

where $y(k)$ and $\hat{y}(k)$ are the measured and model outputs respectively, $u(k)$ is the input, $e(k)$ is the model error, and k (= 1, 2, ...) is the time index. Generally the numerator $a(k)$ and denominator $b(k)$ are functions of past inputs, outputs, and errors, and can be expressed in terms of polynomials

$$\begin{aligned} a(k) &= \sum_{j=1}^{num} p_{nj}(k)\theta_{nj} \\ b(k) &= \sum_{j=1}^{den} p_{dj}(k)\theta_{dj} \end{aligned} \tag{1.2}$$

The regression terms $p_{nj}(k)$ and $p_{dj}(k)$ are the products of past inputs, outputs, and errors, such as $u(k-1)y(k-3)$, $u(k-1)e(k-2)$, $y^2(k-1)$, and θ_{nj}, θ_{dj} are the associated parameters. The task of model structure detection is to select candidate regression terms from a pre-assigned large candidate base. The task of parameter estimation, for a given or detected model structure, is to estimate the associated parameters from the measured inputs and outputs. Model validation is the final step to diagnose the feasibility of an identified model if it is a real representative of the underlying system. The correlation base model validation is to use date sequences of input, output, and residual (difference between measured system output and model predicted output) to form a series of correlation functions to determine if the residual has been reduced to an uncorrelated sequence.

Several remarks relating to the characteristics of the rational model of (1.1) are noted below:

1) Rational model includes almost all other models as its subsets. For example the polynomial NARMAX model is a special case of rational model (1.1) by setting

denominator polynomial $b(k) = 1$. Many neuro-fuzzy systems have been expressed as non-linear rational models (Wang 1994). For example fuzzy systems with centre defuzzifier, product inference rule, singleton fuzzifier, and Gaussian membership function. The normalised radial basis function network is also a type of rational model. When the centres and widths are estimated this becomes a rational model parameter estimation problem.

2) Rational model can be much more concise than a polynomial expansion, for example

$$y(k) = \frac{u(k-1)}{1+y^2(k-1)} = u(k-1)\left[1-y^2(k-1)+y^4(k-1)\cdots\right] \tag{1.3}$$

3) Rational model can be frequently used to represent complicated system structures with a fairly low degree in both numerator and denominator polynomials. For example, large and quick deviations in a system output, can expressed as

4)

$$y(k) = \frac{1}{1+u(k-1)} \tag{1.4}$$

When $u(k\text{-}1)$ approaches -1, the model output response will be quickly increased. The power to capture quick and large changes is gained by introducing the denominator polynomial.

5) Rational model is non-linear in both the parameters and the regression terms. This is induced by the denominator polynomial. Accordingly the identification and control of rational models are more challenging with the total nonlinear structures.
6) Rational models have been gradually adopted in various applications of non-linear system modelling and control (Ford, Titterington, and Kitsos 1989, Ponton 1993, Kambhampati, Mason, and Warwick 2000), particularly the importance of modelling of chemical kinetics has increased sharply as a consequence of the applicability of modelling of catalytic reactions (Dimitrov and Kamenski 1991, Kamenski and Dimitrov1993). Rational models are not only alternative expressions in approximating a wide range of data set in chemical engineering, but also are class mechanistic models, which most previous experience or theoretical considerations had not put forwarded (Dimitrov and Kamenski 1991).
7) A number of methodologies and algorithms have been developed for rational model structure detection, parameter estimation, and correlation based validity tests. These include prediction error estimator (Billings and Chen 1989), generalised least squares estimators (Billings and Zhu 1991, Zhu and Billings 1991, 1993), recursive least squares parameter estimator (Zhu and Billings 1993), orthogonal model structure detection and parameter estimation algorithm (Billings and Zhu 1994a), genetic model structure detection algorithm (Billings and Mao 1998), back propagation parameter estimator (2003), implicit least squares parameter estimator (Zhu 2005), correlation based model validity tests (Billings and Zhu 1994b, 1995, Zhang, Ashley, and Zhu 2007, Zhu, Zhang, and Ashley 2007).

The rest of the chapter is organised with following sections. In section 2, an extended least squares parameter estimator is presented and its full step by step implementation in computation is listed as a guiding procedure for user's applications. In section 3, back propagation algorithm is tailored for rational model parameter estimation. A step by step implementation of the algorithm is listed as a guiding procedure for user's applications. In section 4, orthogonal computation is introduced to provide a solution for rational model structure detection (term selection) and parameter estimation. A step by step implementation of the algorithm is listed as a guiding procedure for user's applications. In section 5, a latest development of correlation based model validation procedure is presented for validating identified linear and nonlinear models including neural networks and fuzzy logic model. In section 6, a brief conclusion is drawn to summarise the chapter results.

2. Extended Least Squares (ELS) Algorithm for Parameter Estimation (Billings and Zhu 1991)

2.1. Bias Analysis Using Ordinary LS Algorithm

Directly estimating parameters of model (1.1) is difficult in formulation and time consuming in computation even though prediction error algorithm has been developed (Billing and Chen 1989). This is because the model is nonlinear in the parameters. An alternative approach is to multiply the denominator polynomial out (1.1) into a model expressed as the one with linear in the parameters, then least squares algorithms can be applied with proper extension. To this end, multiplying the denominator $b(k)$ on both sides of (1.1) and then moving all the terms except the first term $y(k)p_{d1}(k)\theta_{d1}$ on left hand side to the right-hand side gives

$$\begin{aligned} Y(k) &= a(k) - y(k)\sum_{j=2}^{den} p_{dj}(k)\theta_{dj} + b(k)e(k) \\ &= \sum_{j=1}^{num} p_{nj}(k)\theta_{nj} - y(k)\sum_{j=2}^{den} p_{dj}(k)\theta_{dj} + \zeta(k) \\ &= \phi(k)\Theta + \zeta(k) \end{aligned} \tag{2.1}$$

where

$$\begin{aligned} &Y(k) = y(k)p_{d1}(k)\theta_{d1} \quad \theta_{d1} = 1 \,(\text{without loosing genrality}) \\ &\varphi(k) = [\varphi_n(k) \;\; \varphi_d(k)] = [p_{n1}(k) \;\; \cdots \;\; p_{nnum}(k) \;\; -p_{d2}(k)y(k) \;\; \cdots \;\; -p_{dden}(k)y(k)] \\ &\Theta^T = [\theta_n \;\; \theta_d] = [\theta_{n1} \;\; \cdots \;\; \theta_{nnum} \;\; \theta_{d2} \;\; \cdots \;\; \theta_{dden}] \\ &\zeta(k) = b(k)e(k) \end{aligned} \tag{2.2}$$

To understand the noise effect to the parameter estimation in applying least squares algorithms, Inspect (2.1), in case of noise free ($e(k) = 0$), the parameter estimates are unbiased. If noise (even white noise) existed, a very practical situation, directly using least square algorithms will give biased estimates, simply the noise contaminated output is appeared in both dependent variable and regression terms associated with denominator polynomial $b(k)$. This problem is to be resolved in next section.

2.2. ELS Parameter Estimation

This has been derived into following least squares formulation

$$\hat{\Theta} = \left[\Phi^T \Phi - \sigma_e^2 \Psi\right]^{-1} \left[\Phi^T \mathrm{Y} - \sigma_e^2 \psi\right] \tag{2.3}$$

Where

$\hat{\Theta}$ is the estimate of parameter vector Θ, σ_e^2 is the model error variance,

Φ is the regressor matrix and formulated with dada length N

$$\Phi^T = \begin{bmatrix} \varphi_n^T(1) & \cdots & \varphi_n^T(N) \\ \varphi_d^T(1) & \cdots & \varphi_d^T(N) \end{bmatrix} \tag{2.4}$$

Y is the dependent variable vector and composed of

$$\mathrm{Y} = \left[Y(1) \quad \cdots \quad Y(N)\right]^T \tag{2.5}$$

Ψ is the error matrix associated with $\Phi^T \Phi$ and formulated as

$$\Psi = \begin{bmatrix} 0 & \cdots & 0 \\ 0 & \cdots & \sum_{k=1}^{N} p_d^T(k) p_d(k) \end{bmatrix} \qquad p_d(k) = \left[p_{d2}(k) \quad \cdots \quad p_{dden}(k)\right] \tag{2.6}$$

ψ is the error vector associated with $\Phi^T \mathrm{Y}$ and formulated as

$$\psi = \begin{bmatrix} 0 \\ \sum_{k=1}^{N} p_d^T(k) p_{d1}(k) \end{bmatrix} \tag{2.7}$$

An iterative procedure has been developed to obtain the unbiased parameter estimate, which is listed as below:

Step 1, use an ordinary least squares algorithm (set null matrix Ψ and vector ψ in (2.3)) to computer parameter vector $\hat{\Theta}$. This estimate provides initial values for subsequent iterative computations.

Step 2, Compute the model error sequence by $e(k) = y(k) - \frac{a(\cdots, \hat{\Theta})}{b(\cdots, \hat{\Theta})}$. Then estimate the model error variance by $\sigma_e^2 = \frac{1}{N}\sum e^2(k)$.

Step 3, update matrices Φ and Ψ, vectors $\Phi^T Y$ and ψ through formulations (2.4) to (2.7).

Step 4, use the extended least squares algorithm (2.3) to estimate parameter vector $\hat{\Theta}$.

Step 5, Go back to step 2 and repeat until the parameter estimate vector $\hat{\Theta}$ and model error variance σ_e^2 converge to constant values.

2.3. Statistics of the Estimator

The bias and variance of the estimates have been analysed (Zhu and Billings 1991) in case of model error *e*(*k*) is reduced to an uncorrelated noise sequence, which are given as follows.

$$
\begin{aligned}
&Bias(\hat{\Theta}) = 0 \\
&Cov(\hat{\Theta}) \approx \sigma_e^2 \sigma_b^2 \left[\Phi^T \Phi - \sigma_e^2 \Psi\right]^{-1}, \quad for\ \sigma_e^2 \prec 1, \\
&\sigma_b^2 = E\left[b^2(k)\right] \quad E[*] \text{ is the expectation function}
\end{aligned}
\tag{2.8}
$$

2.4. Summary

Expressing rational model in terms of linear in the parameters make it possible to use least squares algorithms for parameter estimation, on the other hand the expansion induces inherent correlated errors to give biased estimates while existing linear squares algorithms are applied directly. Therefore the extended least squares algorithm has been developed to set up a basis for concise and unbiased rational model parameter estimation from lager batch of data sequence.

3. Back Propagation (BP) Algorithm for Parameter Estimation (Zhu 2003)

3.1. BP Parameter Estimation

The rational model can be treated as a neural network, as shown in Fig. 1, which is structured with an input layer of regression terms $p_{nj}(k)$ and $p_{dj}(k)$, a hidden layer of *a*(*k*) and *b*(*k*) with linear activation functions, an output layer of *y*(*k*) with a ratio activation function. The activation function at the output layer is the division operation of the numerator polynomial divided by the denominator polynomial.

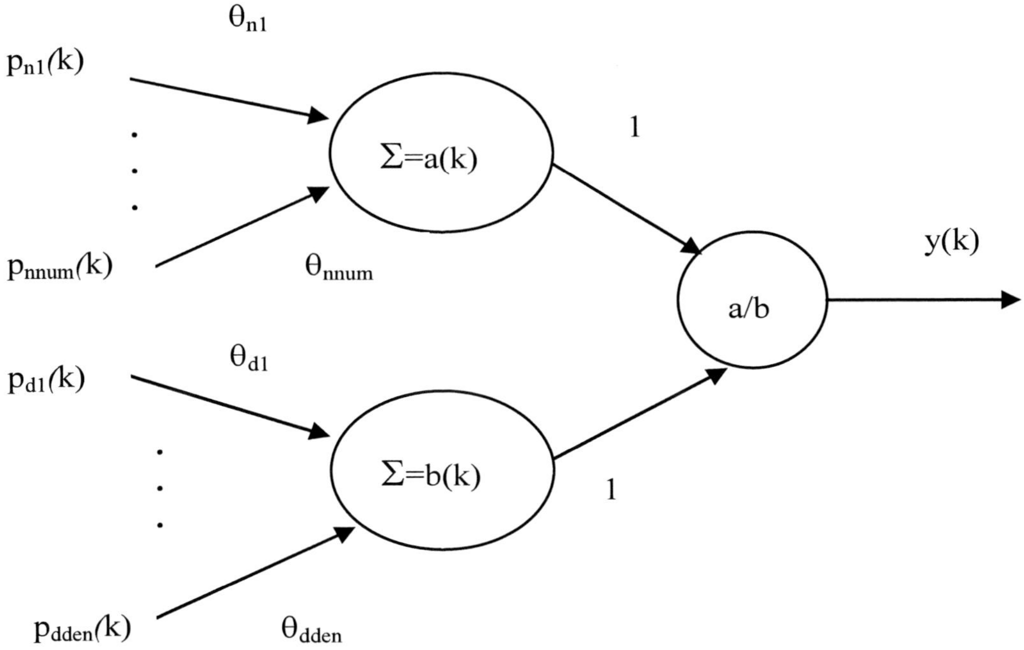

Figure 1. Rational neural network.

Leung and Haykin (1993) presented a rational function neural network. The shortcomings compared with the current study are that Leung and Haykin's network does not have a generalised rational model structure and correlated errors are not accommodated. Hence Leung and Haykin's parameter estimation algorithm cannot provide unbiased estimates with noise corrupted data, and is essentially a special implementation of the procedure of Billings and Zhu (1991) and Zhu and Billings (1991) in the case of noise free data.

To derive the parameter estimation algorithm, with reference to model of (2.1) and (2.2), define the error between the measured output and the model output as

$$e(k) = \frac{1}{2}[y(k) - \hat{y}(k)]^2 \tag{3.1}$$

The task is to determine a set of parameters for the model (2.1) so that the squared error in (3.1) is minimised. The parameter determination can be described as parameter training when the model is interpreted as a neural network structure. To train the parameters θ_{nj} associated with the numerator polynomial $a(k)$, the local gradients are given by

$$\frac{\partial e(k)}{\partial \theta_{nj}} = -[y(k) - \hat{y}(k)]\frac{\partial e(k)}{\partial a(k)}\frac{\partial a(k)}{\partial \theta_{nj}} = -[y(k) - \hat{y}(k)]\frac{1}{b(k)}p_{nj}(k) \tag{3.2}$$

To train the parameters θ_{dj} associated with the denominator polynomial $b(k)$, the local gradients are given by

$$\frac{\partial e(k)}{\partial \theta_{dj}} = -[y(k) - \hat{y}(k)]\frac{\partial e(k)}{\partial b(k)}\frac{\partial b(k)}{\partial \theta_{dj}} = [y(k) - \hat{y}(k)]\frac{a(k)}{b^2(k)} p_{dj}(k) \tag{3.3}$$

To carry out recursive computations, a time index k is introduced to the parameter sets $\theta_{nj}(k)$ and $\theta_{dj}(k)$ as they are updated following the time sequence. Convergence has been proved with $\lim_{k\to\infty} \theta_{nj}(k) \to \theta_{nj}$ $\lim_{k\to\infty} \theta_{dj}(k) \to \theta_{dj}$ (Zhu 2003). Therefore the parameter variation at time k+1 can be defined as follows

$$\begin{aligned} \Delta\theta_{nj}(k+1) = \theta_{nj}(k+1) - \theta_{nj}(k) = -\eta_{nj}\frac{\partial e(k+1)}{\partial \theta_{nj}(k+1)} \\ \Delta\theta_{dj}(k+1) = \theta_{dj}(k+1) - \theta_{dj}(k) = -\eta_{dj}\frac{\partial e(k+1)}{\partial \theta_{dj}(k+1)} \end{aligned} \tag{3.4}$$

where η_{nj} and η_{dj} are the learning rates.

The integrated parameter estimation procedure can be summarised in terms of the BP computation as follows:

I. Initialisation

Set all model parameters to uniformly distributed random numbers with zero mean and small variance, or set up the initial parameters using the prior information extraction technique, which is proposed later in this study.

II. Parameter estimation

For l = 0 to L-1 (epoch number)

For k = 1 to N (training data length)

Forward computation

$$\begin{aligned} a(k+1) = \sum_{j=1}^{num} p_{nj}(k+1)\theta_{nj}(k) \quad & b(k+1) = \sum_{j=1}^{den} p_{dj}(k+1)\theta_{dj}(k) \\ \hat{y}(k+1) = \frac{a(k+1)}{b(k+1)} \quad & e(k+1) = y(k+1) - \hat{y}(k+1) \end{aligned} \tag{3.5}$$

Backward computation

$$\begin{aligned} \frac{\partial e(k+1)}{\partial \theta_{nj}(k+1)} = -e(k+1)\frac{1}{b(k+1)} p_{nj}(k+1) \\ \frac{\partial e(k+1)}{\partial \theta_{dj}(k+1)} = e(k+1)\frac{a(k+1)}{b^2(k+1)} p_{dj}(k+1) \end{aligned} \tag{3.6}$$

$$\theta_{nj}(k+1) = \theta_{nj}(k) - \eta_{nj}\frac{\partial e(k+1)}{\partial \theta_{nj}(k+1)}$$
$$\theta_{dj}(k+1) = \theta_{dj}(k) - \eta_{dj}\frac{\partial e(k+1)}{\partial \theta_{dj}(k+1)} \tag{3.7}$$

end

end

Initialisation

The first step in the above computation is to initialise the parameters. A sensible choice will provide tremendous help for the resultant parameter estimation. A wrong choice of the initial parameters can lead to two critical phenomena, premature saturation and divergence. In the case of premature saturation, the instantaneous sum of squared errors remains almost constant for some period of time during the learning process. In the case of divergence, the sum of squared errors tends to become very large. A general initialisation technique (Haykin 1994) is to uniformly randomise the initial parameters with zero mean and small variance. In this study, the initial parameters will be determined using a prior information extraction technique. Since the model structure is assumed to be known in advance, an ordinary or extended least squares algorithm can be applied to obtain initial estimates. Although there will be biased they should not be far away from the true parameters in Euclidean distance.

Learning Rate

The BPE algorithm provides an approximation to the trajectory in parameter space computed in terms of steepest descent. The smaller the learning rate, the smaller the change to the model parameters will be from one iteration to the next and the smoother the trajectory will be in parameter space. However a small learning rate will possibly cause slower convergence of the parameter estimation. If the learning rate is set too large, the estimator may become unstable. The choice of learning rates are discussed below:

i) Constant

$$\eta_{nj} = constant_1 \quad \eta_{dj} = constant_2 \tag{3.8}$$

The disadvantage is that the learning rate cannot properly adjust in the early stage (large learning rate) and in the later stage (small learning rate) of the learning process.

ii) Inversely decayed sequence

$$\eta_{nj}(k) = \eta_{dj}(k) = \frac{1}{k} \tag{3.9}$$

The disadvantage is that the learning rate decays too fast and this may not be suitable when long training data sequences are available.

iii) Generalised delta rule (Rumelhart, Hinton, and Williams 1986, Jacobs 1988)

$$\theta_{nj}(k+1) = \theta_{nj}(k) - \eta_{nj}\frac{\partial e(k+1)}{\partial \theta_{nj}(k+1)} + \alpha_{nj}\left[\theta_{nj}(k) - \theta_{nj}(k-1)\right]$$
$$\theta_{dj}(k+1) = \theta_{dj}(k) - \eta_{dj}\frac{\partial e(k+1)}{\partial \theta_{dj}(k+1)} + \alpha_{dj}\left[\theta_{dj}(k) - \theta_{dj}(k-1)\right] \qquad (3.10)$$

where α_{nj} and α_{dj} are positive numbers called momentum constants. This has been extensively studied and can increase the learning rate and avoid instability. However, this choice involves more computations and requires the know-how to set up the parameters α_{nj} and α_{dj}.

iv) Linearly decayed sequence

To overcome the above disadvantages, a new learning rate sequence is proposed in this study as follows

$$\eta_{nj}(k) = \eta_{dj}(k) = -\frac{\eta_0 - \eta_{end}}{LN}(lN + k) + \eta_0 \qquad (3.11)$$

where L is the number of iteration epoch, N is the training data length, l (= 0, …, L-1) for the lth epoch, and the $\eta_0 \succ \eta_{end}$ are positive constants for the initial and end learning rates respectively. Figure 2 shows the learning rate variation.

Stopping Criterion

Two criteria (Kramer and Sangiovanni-Vincentelli 1989) have been used to stop neural network training. These are formulated by means of local or global minimum of the error surface.

i) Gradient stopping criterion

This states that the BP converges when the Euclidean norm of the gradient vector reaches a sufficiently small gradient threshold. The drawback of this criterion is longer training and complicity in the computation of the gradient vector.

ii) Error measure stopping criterion

This states that the BP converges when the absolute rate of change in the mean squared error per epoch is sufficiently small. Typically 0.01 percent per epoch is used. The drawback is that the error may still be correlated for non-linear models, even though the mean squared error or its variation rate is very small (Billings and Woon 1983, Billings and Zhu 1994b).

To overcome the above drawbacks, a higher order correlation test (Billings and Zhu 1994b, 1995), introduced for non-linear model validation, can be used as a stopping criterion. The tests are described below:

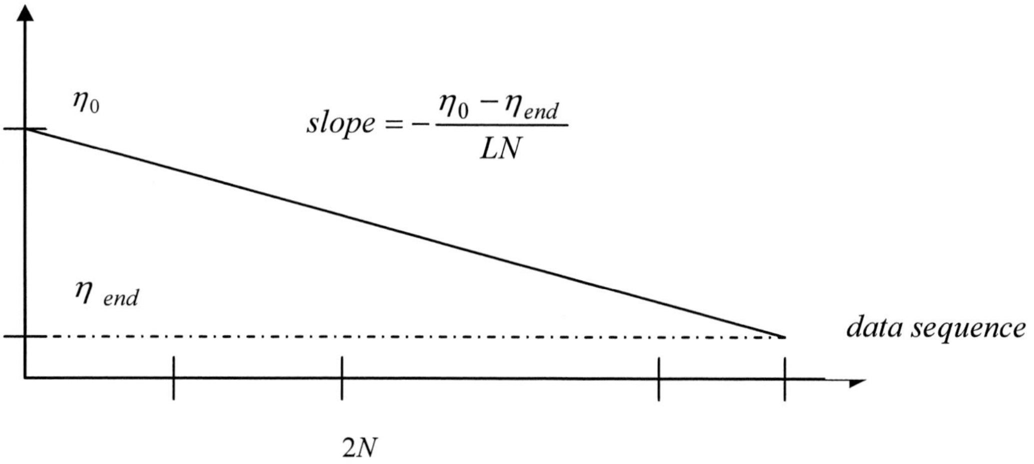

Figure 2. Linearly decayed learning rate.

$$\Phi_{\alpha e^2}(\tau) = \frac{\sum_{k=1}^{N}\left(\alpha(k)-\overline{\alpha}\right)\left(e^2(k-\tau)-\overline{e^2}\right)}{\sqrt{\left[\sum_{k=1}^{N}\left(\alpha(k)-\overline{\alpha}\right)^2\right]\left[\sum_{k=1}^{N}\left(e^2(k)-\overline{e^2}\right)^2\right]}}$$

$$\Phi_{\alpha u^2}(\tau) = \frac{\sum_{k=1}^{N}\left(\alpha(k)-\overline{\alpha}\right)\left(u^2(k-\tau)-\overline{u^2}\right)}{\sqrt{\left[\sum_{k=1}^{N}\left(\alpha(k)-\overline{\alpha}\right)^2\right]\left[\sum_{k=1}^{N}\left(u^2(k)-\overline{u^2}\right)^2\right]}} \tag{3.12}$$

where

$$\alpha(k) = y(k)e(k) \qquad \overline{\alpha} = \frac{1}{N}\sum_{k=1}^{N}\alpha(k)$$

$$\overline{u^2} = \frac{1}{N}\sum_{k=1}^{N}u^2(k) \quad \overline{e^2} = \frac{1}{N}\sum_{k=1}^{N}e^2(k) \tag{3.13}$$

When the higher order correlation functions $\Phi_{\alpha e^2}(\tau)$ and $\Phi_{\alpha u^2}(\tau)$ satisfy

$$\Phi_{\alpha e^2}(\tau) = \begin{cases} k \succ 0 & \tau = 0 \\ 0 & otherwise \end{cases}$$

$$\Phi_{\alpha u^2}(\tau) = 0 \quad \forall\tau \tag{3.14}$$

the estimated parameters are considered to be unbiased. Otherwise the training procedure will continue until the above conditions are satisfied. In practice the 95% confidence limits, about $\pm 1.96/\sqrt{N}$, are used as the stopping thresholds.

3.2. Statistics of the Estimator

For the bias

$$
\begin{aligned}
Bias\,\theta_{nj} &= \theta_{nj} - E\left[\theta_{nj}(k+1)\right] = E\left[e(k+1)\right]E\left[\frac{1}{b(k+1)}p_{nj}(k+1)\right] = 0 \\
Bias\,\theta_{dj} &= \theta_{dj} - E\left[\theta_{dj}(k+1)\right] = E\left[e(k+1)\right]E\left[\frac{a(k+1)}{b^2(k+1)}p_{nj}(k+1)\right] = 0
\end{aligned}
\tag{3.15}
$$

Therefore the estimator will produce unbiased estimates of the non-linear rational model parameters. For the variance

$$
\begin{aligned}
Var\,\theta_{nj} &= E\left[\left(\theta_{nj}(k+1) - E\left[\theta_{nj}(k+1)\right]\right)^2\right] = \sigma_e^2\sigma_{nj}^2 \\
Var\,\theta_{dj} &= E\left[\left(\theta_{dj}(k+1) - E\left[\theta_{dj}(k+1)\right]\right)^2\right] = \sigma_e^2\sigma_{dj}^2
\end{aligned}
\tag{3.16}
$$

where

$$
\sigma_{nj}^2 = E\left[\left(\frac{1}{b(k+1)}p_{nj}(k+1)\right)^2\right] \quad \sigma_{dj}^2 = E\left[\left(\frac{a(k+1)}{b^2(k+1)}p_{dj}(k+1)\right)^2\right]
\tag{3.17}
$$

Therefore the variance of each parameter estimate is associated with the noise variance and the second order moment of the regression term.

3.3. Summary

The BP based algorithm provides an alternative approach to estimate rational model parameters. Obviously there is no need to expand rational models into linear in the parameters expressions. Some techniques such as learning rate and stopping threshold playing important roles in the iterative training and learning process need to be carefully considered in the computations.

4. Orthogonal Least Squares (OLS) Algorithm for Term Selection and Parameter Estimation (Zhu and Billings 1993, Billings and Zhu 1994)

4.1. Orthogonal Parameter Estimation

Consider an orthogonal transformed expression of (2.1)

$$\begin{aligned}Y(k) &= \sum_{j=1}^{num} p_{nj}(k)\theta_{nj} - y(k)\sum_{j=2}^{den} p_{dj}(k)\theta_{dj} + \zeta(k) \\ &= \omega(k)G + \zeta(k) \\ &= \sum_{j=1}^{num} w_{nj}(k)g_{nj} - y(k)\sum_{j=2}^{den} w_{dj}(k)g_{dj} + \zeta(k)\end{aligned} \quad (4.1)$$

where

$$\begin{aligned}\omega(k) &= [\omega_n(k) \quad \omega_d(k)] = [\omega_{n1}(k) \quad \cdots \quad \omega_{nnum}(k) \quad \omega_{d2}(k)y(k) \quad \cdots \quad \omega_{dden}(k)y(k)] \\ G^T &= [G_n \quad G_d] = [g_{n1} \quad \cdots \quad g_{nnum} \quad g_{d2} \quad \cdots \quad g_{dden}]\end{aligned} \quad (4.2)$$

Define the orthogonal regression matrix W as

$$\begin{aligned}W &= \Phi T^{-1} \\ W^T &= [\omega^T(1) \quad \omega^T(N)] = \begin{bmatrix} \omega_n^T(1) & \cdots & \omega_n^T(N) \\ \omega_d^T(1) & \cdots & \omega_d^T(N) \end{bmatrix} \\ T &= \begin{bmatrix} 1 & t_{12} & . & . & . & t_{1n} \\ 0 & 1 & . & . & . & . \\ 0 & 0 & 1 & . & . & . \\ . & . & . & . & . & . \\ . & . & . & . & . & t_{n-1n} \\ 0 & 0 & . & . & . & 1 \end{bmatrix}\end{aligned} \quad (4.3)$$

where Φ is the regressor matrix defined in (2.4). There are several typical algorithms to compute the elements of T, such as Gram-Schmidt, Householder, or Givens transforms (Billings and Zhu 1994a). The orthogonality of the matrix W gives

$$W^T W = D = diag\{d_{n1} \quad \cdots \quad d_{nnum} \quad d_{d2} \quad \cdots \quad d_{dden}\} \quad (4.4)$$

The unbiased parameter vector G is estimated with

$$\hat{G} = \left[W^T W - \sigma_e^2 \Psi_{orth}\right]^{-1} \left[W^T Y - \sigma_e^2 \psi_{orth}\right] \tag{4.5}$$

where

$$\Psi_{orth} = diag\left\{\overline{e_{n1}^2} \quad \cdots \quad \overline{e_{nnum}^2} \quad \overline{e_{d2}^2} \quad \cdots \quad \overline{e_{dden}^2}\right\}$$

$$\psi_{orth} = \left[\overline{p_{d1}e_{n1}} \quad \cdots \quad \overline{p_{d1}e_{nnum}} \quad \overline{p_{d1}e_{d2}} \quad \cdots \quad \overline{p_{d1}e_{dden}}\right]^T \tag{4.6}$$

where over bar denotes time average value. The computation of the elements in (4.6) can be found in (Zhu and Billings 1993). Accordingly the original parameter vector in (2.1) can be calculated by

$$\Theta = T^{-1} G \tag{4.7}$$

4.2. Structure Detection (Term Selection)

The orthogonal property can be applied to select regression terms and estimate the associated parameters simultaneously. A criterion of Error Reduction Ratio (ERR) has been introduced to select the most important regression terms to contribute to reduce estimation errors from a large candidate term base (Billings and Zhu 1994a). In definition, individual ERR from numerator and denominator polynomials is calculated respectively by

$$err_{nj} = \frac{g_{nj}^2 \overline{w_{nj}^2}}{\sigma_Y^2 \sigma_b^2} - \frac{g_{nj}^2 \overline{e_{nj}^2} \sigma_e^2 + 2 g_{nj} \overline{e_{nj} b} \sigma_e^2}{\sigma_Y^2 \sigma_b^2}$$

$$err_{dj} = \frac{g_{dj}^2 \overline{w_{dj}^2}}{\sigma_Y^2 \sigma_b^2} - \frac{g_{dj}^2 \overline{e_{dj}^2} \sigma_e^2 + 2 g_{dj} \overline{e_{dj} b} \sigma_e^2}{\sigma_Y^2 \sigma_b^2} \tag{4.8}$$

$$\sigma_Y^2 = \frac{1}{N} \sum Y^2(k) \quad \sigma_b^2 = \frac{1}{N} \sum b^2(k)$$

It should be noticed that the larger the ERR, the more contribution of its regression tem to reduce estimation errors. Therefore the candidate term with the largest ERR in each selection sequence should be taken in as the model regression term.

An iterative learning procedure for rational structure detection has been developed as follows:

Step 1, Fit a deterministic rational model (that is no noise model) initially with 3 to 4 regression terms (choice of the terms comes from experience, actually it is not critical) in numerator and denominator polynomials respectively. Therefore the model error sequence $e(k)$ and its variance σ_e^2 can be initially obtained.

Step 2, Set up a series of ascending weights till a full scale to the credibility of the model variance σ_e^2 (such as 0.1, 0.25, 0.5 0.75, 0.95, 1, 1, ..., 1). Set up a cut-off point (COP), which can be determined by trail and error approach, as

$$COP \geq 1 - \frac{1}{\sigma_b^2} + \sum bias[err] + \frac{\sigma_e^2}{\sigma_Y^2}$$

$$bias[err] = \begin{cases} \dfrac{g_{nj}^2 \overline{e_{nj}^2} + 2g_{nj}\overline{e_{nj}}b\sigma_e^2}{\sigma_Y^2\sigma_b^2} & \text{numerator terms} \\ \dfrac{g_{dj}^2 \overline{e_{dj}^2} + 2g_{dj}\overline{e_{dj}}b\sigma_e^2}{\sigma_Y^2\sigma_b^2} & \text{denonimator terms} \end{cases} \quad (4.9)$$

Step 3, Do a search through a pre-set-up full rational model candidate term base to select significant terms according to the ERR values and estimate their associated parameters.

Step 4, Repeat steps 2 and 3 until the computations converge, the preset maximum number of the iterations exceeded, or a specified number of terms are selected in the model.

4.3. Summary

This algorithm delivers a more realistic procedure to fit models to data in case of which terms should be included in the models before estimating their associated parameters.

5 Model Validation (Zhang, Zhu, and Ashley 2007, Zhu, Zhang, and Ashley 2007)

5.1. Introduction to Correlation base Model Validation

Consider a generalised single input and single output (SISO) nonlinear parametric model.

$$y(k) = \hat{y}(k) + \varepsilon(k) = \hat{f}(\mathbf{y}^{k-1}, \mathbf{u}^{k-1}, \boldsymbol{\varepsilon}^{k-1}) + \varepsilon(k) \quad (5.1)$$

where $\hat{f}(\cdot)$ is the identified nonlinear model. $\mathbf{y}^{k-1}$, $\mathbf{u}^{k-1}$, and $\boldsymbol{\varepsilon}^{k-1}$ are measured output, input and residual (the difference between the measured output and the model predictive output) vectors with delayed elements from 1 to r respectively. It should be noticed that when above model is properly identified, residual $\varepsilon(k)$ should be reduced to a random noise sequence denoted by $e(k)$ in this section (without being confused with those used in other sections) with zero mean and finite variance (Billings and Zhu 1994). In other words, a residual sequence which is obtained from using a valid identified model should be uncorrelated to the delayed residuals, inputs and outputs (Ljung, 1999).

It should be noticed that all validation methods developed based on nonlinear models have included all linear model validation as their simplified cases. However the linear model based validation test methods are failed to be applied to nonlinear model validation in most of the cases.

5.2. Combined ODACF and ODCCF for Model Validation

Combined omni-directional auto-correlation functions (ODACF) is defined as follows,

$$\begin{cases} if \ |\max(R_{\varepsilon\varepsilon})| > |\min(R_{\varepsilon\varepsilon})| \\ \quad \rho_{\varepsilon\varepsilon}(\tau) = \max(R_{\varepsilon\varepsilon}) \\ else \\ \quad \rho_{\varepsilon\varepsilon}(\tau) = \min(R_{\varepsilon\varepsilon}) \end{cases} \tag{5.2}$$

Combined omni-directional cross-correlation functions (ODACF) is defined as follows,

$$\begin{cases} if \ |\max(R_{u\varepsilon})| > |\min(R_{u\varepsilon})| \\ \quad \rho_{u\varepsilon}(\tau) = \max(R_{u\varepsilon}) \\ else \\ \quad \rho_{u\varepsilon}(\tau) = \min(R_{u\varepsilon}) \end{cases} \tag{5.3}$$

In above definition

$$\begin{cases} R_{\varepsilon\varepsilon} = [r_{\alpha'\alpha'}(\tau), r_{\alpha'\varepsilon'}(\tau), r_{\varepsilon'\varepsilon'}(\tau), r_{\varepsilon'\alpha'}(\tau)] \\ R_{u\varepsilon} = [r_{\beta'\alpha'}(\tau), r_{\beta'\varepsilon'}(\tau), r_{u'\varepsilon'}(\tau), r_{u'\alpha'}(\tau)] \end{cases} \tag{5.4}$$

$$\begin{cases} \alpha'(k) = |\varepsilon'(k)| - \frac{1}{N}\sum_{k=1}^{N} |\varepsilon'(k)| \\ \beta'(k) = |u'(k)| - \frac{1}{N}\sum_{k=1}^{N} |u'(k)| \end{cases} \tag{5.5}$$

where $r_{**}(\tau)$ denotes correlation function and the prime ' in (5.4) to (5.5) denotes that the mean level has been removed from the corresponding data sequence,

Then the validity tests for a properly identified model are derived as

$$\begin{cases} \rho_{\varepsilon\varepsilon}(\tau) = 1, \tau = 0 \\ \rho_{\varepsilon\varepsilon}(\tau) = 0, otherwise \end{cases} \tag{5.6}$$

$$\rho_{u\varepsilon}(\tau) = 0, \forall \tau \tag{5.7}$$

Compared with the other correlation tests based methodologies, this approach enhances the power of nonlinear model validity tests and significantly reduces the number of correlation plots. For large N the correlation function estimates given in (5.2) and (5.3) are still asymptotically normal with zero mean and finite variance in accordance with the central limit theorem (Bowker and Lieberman 1972). The standard deviations are $1/\sqrt{N}$ and the 95% confidence limits are therefore approximately at $\pm 1.95/\sqrt{N}$.

5.3. Summary

It should be mentioned that the method is not only applicable to rational models, but also for validation of identified neural networks and fuzzy logic models.

6. CONCLUSION

With regarding to the presented algorithms, the ELS algorithm sets up a solid and concise basis for expanding this classical approach into nonlinear rational model term selection and parameter estimation. The insight presented in the study could be a useful indication to stimulate other complex model identification with proper extension of classical least squares algorithms. The BP algorithm provides a different angle to study the model parameter estimation issues. However the model structure detection via PB computation (i.e. regression term selection) is still remained unresolved. The OLS algorithm provides a step by step procedure to select model regression terms and estimate their associated parameters simultaneously. Finally the new model validation method provides a generic routine to effectively examine the model validity with significantly reduced correlation plots. It should be noted again that the structure of rational model brings various advantages, which cannot be easily implemented through polynomial model framework. On the other hand rational model presents a number of new challenges for research and demonstration.

REFERENCES

Billings, S.A. and Chen, S., Identification of non-linear rational systems using a prediction error estimation algorithm, *Int. J. Sys., Science,* Vol. 20, pp. 467-494. 1989.

Billings, S.A. and Mao, K.Z., Structure detection for nonlinear rational models using genetic algorithms, *Int. J. Sys. Science, Vol.* 29, pp. 223-231, 1998.

Billings and Woon, W.S.F., Structure detection and model validity tests in the identification of nonlinear systems, *IEE proceedings, PtD, Vol.* 130, pp. 190-199, 1983.

Billings, S.A. and Zhu, Q.M., Rational model identification using extended least squares algorithm, *Int. J. Control,* Vol. 54, pp. 529-546, 1991; A structure detection algorithm for nonlinear dynamic rational models, *ibid., Vol. 59,* pp. 1439-1463, 1994a; Nonlinear model validation using correlation tests, *ibid., Vol.* 60, pp. 1107-1120, 1994b; Model validation tests for multivariable nonlinear models including neural networks, *ibid., Vol. 62,* pp. 749-766, 1995.

Bowker, A. H., and Lieberman, G. J., *Engineering Statistics,* Prentice Hall, 1972.

Chen, S. and Billings, S.A., Representations of nonlinear system: the NARMAX model, *Int. J. Control,* Vol. 48, pp.1013-1032, 1989.

Dimitrov, S.D. and Kamenski, D.I., A parameter estimation method for rational functions, Computers and Chemical. *Engineering,* Vol. 15, pp. 657-662, 1991.

Ford, I., Titterington, D.M., and Kitsos, C.P., Recent advances in non-linear experimental design, *Technometrics,* Vol. 31, pp. 49-60, 1989.

Haber, R. and Unbehauen, H., Structure identification of nonlinear dynamic system --- a survey on input/output approaches, *Automatica,* Vol. 26, pp.651-677, 1990.

Haykin, S., *Neural networks,* Macmillan College Publishing Company, Oxford, 1994.

Jacobs, R.A., Increased rates of convergence through learning rate adaptation, *Neural Networks,* Vol. 1, pp.295-307, 1988.

Kambhampati, C., Mason, J.D., and Warwick, K., A stable one-step-ahead predictive control of nonlinear systems, *Automatica,* Vol. 36, pp. 485-495, 2000.

Kramer, A.H. and Sangiovanni-Vincentelli, A., Efficient parallel learning algorithms for neural networks, In *Advances in Neural Information Processing Systems* (Touretzky, D.S., ed.), pp. 40-48, San Mateo, CA: Morgan Kaufmann, 1989.

Leung, H. and Haykin, S., Rational function neural network, *Neural Computation,* Vol. 5, pp. 928-938, 1993.

Ljung, L, Analysis of recursive stochastic algorithms, IEEE, *Transactions on Automatic Control,* Vol. 22, pp.551-575, 1977.

Ljung, L., *System Identification Theory for the User* 2nd Edition, Prentice Hall, Upper Saddle River, NJ, 1999.

Narendra, K.S. and Parthasapathy, K., Identification and control of dynamical systems using neural networks, *IEEE Transactions on Neural Networks,* Vol. 1, pp. 4-27, 1990.

Ponton, J.W., The use of multivariable rational functions for non-linear data presentation and classification, *Computers and Chemical Engineering,* Vol. 17, pp.1047-1052, 1993.

Proll, T. and Karim, M.N., Model predictive PH control using real time NARX approach, *AIChe J.,* Vol. 40, pp. 269-282, 1994.

Rumelhart, D.E., Hinton, G.E., and Williams, R.J., Learning representations by back propagating errors, *Nature (London),* Vol. 323, pp.533-536, 1986.

Sontag, E.D., Realization theory of discrete time nonlinear systems: Part I --- the bounded case, IEEE transaction on Circuits and Systems, Vol. 26, pp. 342-356, 1978; *Polynomial response maps --- lecture notes in control and information sciences,* Vol. 13, Springer-Verlag, Berlin, 1979.

Wang, L.X, Adaptive fuzzy systems and control, Prentice Hall, Englewood Cliffs, NJ, 1994.

Zhang, L.F., Zhu, Q.M., and Longden, A., A set of novel correlation tests for nonlinear system variables, *Int. J. Systems Science,* Vol. 38, pp. 47-60, 2007.

Zhu, Q.M. and Billings, S.A, Recursive parameter estimation for nonlinear rational models, J. Sys. Engineering, Vol. 1, pp. 63-67, 1991; Parameter estimation for stochastic nonlinear models, *Int. J. Control,* Vol. 57, pp. 309-333, 1993.

Zhu, Q.M., Ma, Z., and Warwick, K., Neural network enhanced generalised minimum variance self-tuning controller for nonlinear discrete time systems, *IEE Proc. – Control Theory Appl.,* Vol. 146, pp. 319-326, 1999.

Zhu, Q.M., A back propagation algorithm to estimate the parameters of nonlinear dynamic rational models, *Applied Mathematical Modelling,* Vol. 27, pp. 169-187, 2003.

Zhu, Q.M., An implicit least squares algorithm for nonlinear rational model parameter estimation, *Applied Mathematical Modelling,* Vol. 29, pp. 673-689, 2005.

Zhu, Q.M., Zhang, L.F., and Longden, A., Development of omni-directional correlation functions for nonlinear model validation, accepted , *Automatica,* Vol. 43, pp. 1519-1531, 2007.

INDEX

A

B

C

D

E

F

P

Q

R

S

T

U

V

W

Y

Z